BUILDING EVALUATION

BUILDING EVALUATION

Edited by
Wolfgang F. E. Preiser
University of New Mexico
Albuquerque, New Mexico

PLENUM PRESS • NEW YORK AND LONDON

Library of Congress Cataloging in Publication Data

Building evaluation / edited by Wolfgang F. E. Preiser.
p. cm.
"Based on the Symposium on Advances in Building Evaluation: Knowledge, Methods, and Applications, held as part of the Tenth Biannual Conference of the International Association for the Study of People and Their Physical Surroundings, July 5-8, 1988, Delft, Netherlands"—Verso of t.p.
Includes bibliographical references.
ISBN 0-306-43337-0
1. Buildings—Environmental engineering—Congresses. I. Preiser, Wolfgang F. E. II. Symposium on Advances in Building Evaluation: Knowledge, Methods, and Applications (1988: Delft, Netherlands)
TH6025.B86 1989 89-16365
690'.22—dc20 CIP

Based on the Symposium on Advances in Building Evaluation: Knowledge, Methods, and Applications, held as part of the Tenth Biannual Conference of the International Association for the Study of People and Their Physical Surroundings, July 5-8, 1988, in Delft, The Netherlands

A Division of Plenum Publishing Corporation
233 Spring Street, New York, N.Y. 10013

Printed in the United States of America

PREFACE

This book is about building evaluation in the broadest sense and it transcends the meaning and conventional boundaries of the evolving field of "post-occupancy evaluation" by focusing on evaluation throughout the building delivery process. This process is seen not just as being linear with a product in mind, i.e., the completed and occupied building, but rather, it is seen as a cyclic evolution which has as its goal the continuous improvement of the quality of buildings. This goal can only be accomplished if evaluation occurs throughout the building delivery process, and if:

1. the evaluation that does occur is systematic and rigorous,
2. the data that is obtained can be fed into data bases and clearinghouses for use in future generations of buildings, and;
3. there is continuity in information flow.

The idea for this book originated with a symposium that was part of a conference held at the Technical University in Delft, Netherlands, in July of 1988, i.e., IAPS 10, the tenth biannual conference of the "International Association for the Study of People and their Physical Surroundings." Authors presented papers based on their book chapters, and discussions ensued about the expanded boundaries of the field, about theoretical, methodological, and practical issues, as well as applications in building evaluation. Other relevant topics were identified and several additional authors were invited to participate in order to round out the contents of this book.

The purpose of the symposium "Advances in Building Evaluation: Knowledge, Methods, and Applications" was to describe recent developments in building evaluation, including post-occupancy evaluation (POE). The chapters which follow range from new uses and boundaries in the topic area, and a performance-based conceptual framework for rigorous and systematic evaluations, to advances in methods as well as applications. The developments that may represent innovation in the field are identified. These include an apparent increase in the volume and acceptance of building evaluations, and shifts in the sponsorship and in the types of evaluation programs that are run by public and private sponsors. Possible advances included the integration of behavioral and technical assessments, moving toward the application of "total building performance," and the development of greater sophistication in dealing with organizational issues and the clearer discrimination of multiple levels of evaluation.

The epilogue of this book raises a number of questions regarding issues in building evaluation, including but not limited to the following concerns, which are in no particular order of priority:

- The building delivery processes in differently-sized client organizations need to be better understood and addressed accordingly.
- Evaluations are bound and influenced by the current societal, economic, and cultural context, e.g., concern for handicapped accessibility in less-developed countries is minimal.
- The relevance of building evaluation to facilities management, organizational behavior, and decision-making needs to be emphasized.

- Values are implicit in evaluations; however, conflicts arise and mediation methods may be appropriate to help resolve them.
- Standards, criteria of performance, "benchmarks," and evaluation methods are only now being developed, e.g., by ASTM.
- The present lack of theory to be addressed.
- Methodological shortcomings, e.g., semantic differential applications, need to be dealt with, including replication studies for validation purposes, and proper representation of various user groups.
- Control of information and proprietary data resulting from evaluation studies is necessary to protect clients.
- Liability concerns in evaluations need to be addressed, e.g., unions are potentially interested in and affected by the data generated in evaluations.

In an edited book like this some overlap in topics and repetition among the individual contributors is unfortunately unavoidable. This compendium of papers on building evaluation is intended to complement an earlier book on *Post-Occupancy Evaluation* (Preiser, Rabinowitz, and White, 1988), which described the evaluation process in a technical and practical "how to" manner, by showing the many facets of methodology, theory, and model building, as well as innovative uses and their applications.

Some of the ideas expressed in this book are new and unproven, whereas others point to unrecognized potential for improvements in the building delivery process. It is hoped that this volume will inspire further debate about the necessity and the benefits of building evaluation. By process we mean the planning, programming, design, construction, activation, and occupancy of new or remodeled facilities.

Wolfgang F. E. Preiser
April 1989

ACKNOWLEDGMENTS

Thanks are owed to Professor Niels Prak from the School of Architecture, Technical University of Delft, Netherlands, for having organized the IAPS 10 conferences in July 1988. He invited us to hold two half-day sessions for the symposium on building evaluation which laid the foundation for this book project.

I thank the chapter authors and the various clients, both in the public and private sectors, who have encouraged innovation, especially the Veterans' Administration, Office of Facilities and the VAMC Albuquerque Chief of Engineering, Ron Richter, for the activation evaluation project in Albuquerque, reported in chapter 8.

Based on multiple grants from the National Science Foundation and a grant from the National Endowment for the Arts, we were permitted to carry out a number of case studies in building evaluation. These, in turn, led to a companion volume to this book, entitled *Post-Occupancy Evaluation*, a practical "how-to" text published in 1988 by Van Nostrand Reinhold.

I thank the School of Architecture and Planning at the University of New Mexico for encouraging building evaluation to be taught on a required basis prior to graduation. Over the years we have been able to train close to 750 students in the methodologies and practical know-how of building evaluation.

Melanie Yelity, Editor at Plenum Publishing Corporation, facilitated an expeditious publishing effort.

Arthur Blume, Editorial Assistant, compiled more than thirty different word-processor disks from around the world to create the book you are now reading. He also edited many of the chapters.

Finally, I wish to thank my wife, Cecilia, for her support throughout this project.

CONTENTS

PART IV: ADVANCES IN APPLICATIONS

CHAPTER 1

TOWARDS A PERFORMANCE-BASED CONCEPTUAL FRAMEWORK FOR SYSTEMATIC POES

Wolfgang F. E. Preiser

School of Architecture and Planning
University of New Mexico
Albuquerque, New Mexico

INTRODUCTION

This chapter suggests a more systematic and rigorous approach to POEs through the adoption of the performance concept in building evaluation. Reconceptualizing basic evaluation approaches can improve POE in fundamental ways, resulting in changes to current practice that will integrate previous improvements and proposed new ones. Increased methodological rigor and improved utility of POE results will benefit both the public and private sectors by enhancing the quality of their buildings.

This chapter presents the two basic parts of a systematic framework for POEs: the concept of building performance, and the performance evaluation process. These aspects of a POE assume high-quality measures on a performance basis, a formal and rigorous methodology for evaluating measures against appropriate criteria, and a system of dissemination that is useful and accessible to practitioners and researchers alike.

For purposes of clarification, a definition of POE is offered: POE is subsumed by the higher order type of evaluation called "building diagnostics" which has both diagnostic and prognostic capabilities. POE evaluates buildings in use and has short, medium, and long-term implications, the latter being evolutionary, based upon feedforward of POE generated information. Furthermore, POE focuses on the requirements and performance of building occupants' needs, and therefore, technical performance is only considered in so far as it affects the occupants of buildings.

THE PERFORMANCE CONCEPT IN THE BUILDING PROCESS

The "Performance Concept" proposes that POEs be built into design and construction programs of agencies from the beginning as an integral part of the building delivery process. Planning for a POE should begin in the programming phase for a new facility.

Systematic and rigorous POEs are predicated upon the use of the performance concept in the building process. The performance concept facilitates an objective

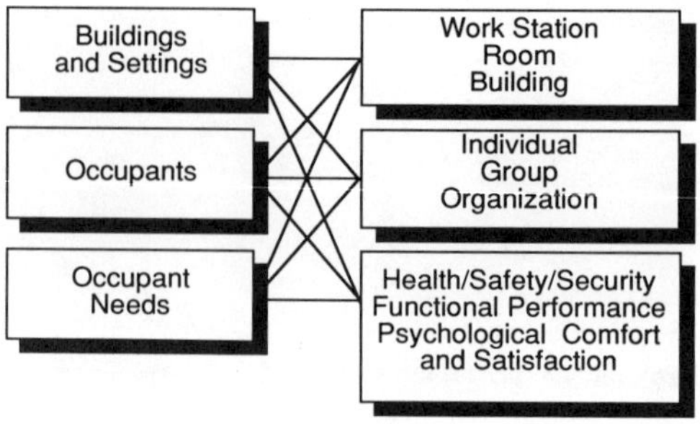

Figure 1. Elements of building performance.

evaluation method by comparing explicitly stated performance criteria for buildings with the actual performance as measured or perceived by building occupants and evaluators.

The performance concept is based on the assumption that a building is designed and built to support, and enhance, the activities and goals of its occupants. Early work on building performance was conducted by Ezra Ehrenkrantz and his associates on the School Construction Systems Development Project in California (Educational Facilities Laboratories, 1967). This work had been inspired by concepts developed at the Institute of Advanced Technology of the National Bureau of Standards (Eberhard, 1965). Subsequent projects executed by the National Bureau of Standards for the Department of Housing and Urban Development and the General Services Administration built upon these initial efforts (Wright, 1971).

Performance Evaluation

Performance evaluation and feedback relates client goals and performance criteria to the actual, objectively and subjectively measurable building performance. The performance concept can help improve the evaluation process by increasing objectivity and clarity of measurement, enhancing communication, providing incentives for innovation, development of alternatives, aiding decision making, and advancing professionalism.

Performance criteria used in evaluation are developed from goals and objectives which in themselves are derived from values held by individuals, groups, and organizations. Frequently, there are differences in values among various groups or units of the same organization.

Building Performance and Evaluation

Because the performance concept in the building process views buildings as dynamic entities, it requires a comprehensive attitude in evaluation. Performance measures are compared to performance criteria, and the differences are used as feedforward into improved planning, programming, design, and construction of future buildings, as well as the creation of data bases or information clearinghouses on building types, attributes, and occupant groups.

The elements of performance that are measured, evaluated, and used in POEs to improve buildings include three major categories: technical, functional, and behavioral.

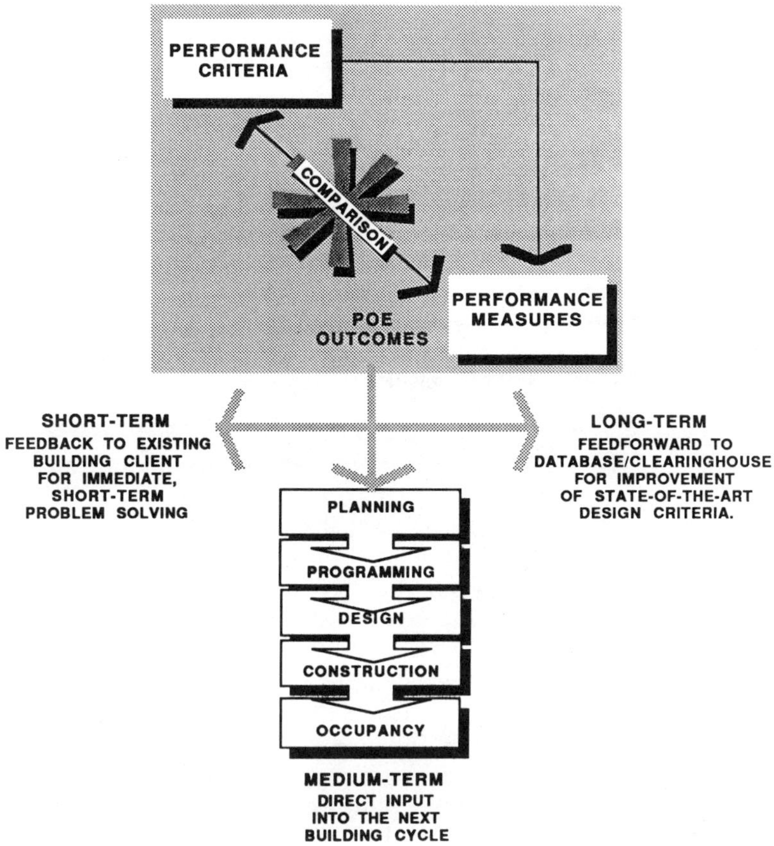

Figure 2. The performance concept in the building process.

Although there are other building performance elements such as location and economics that influence physical performance and affect owners, organizations, and building occupants, these three elements are the most important.

THE PERFORMANCE EVALUATION RESEARCH FRAMEWORK

The performance evaluation research framework (figure 1) connects the evaluation of buildings with:

1) measurement technology,
2) data bases and information systems (including clearinghouses), and:
3) the development of performance criteria for buildings.

Measurement Technology

Measurement technology employs all those techniques and technological aids that are used in data collection and analysis of POEs. They include interviews, questionnaire surveys, direct observation, mechanical recording of human behavior,

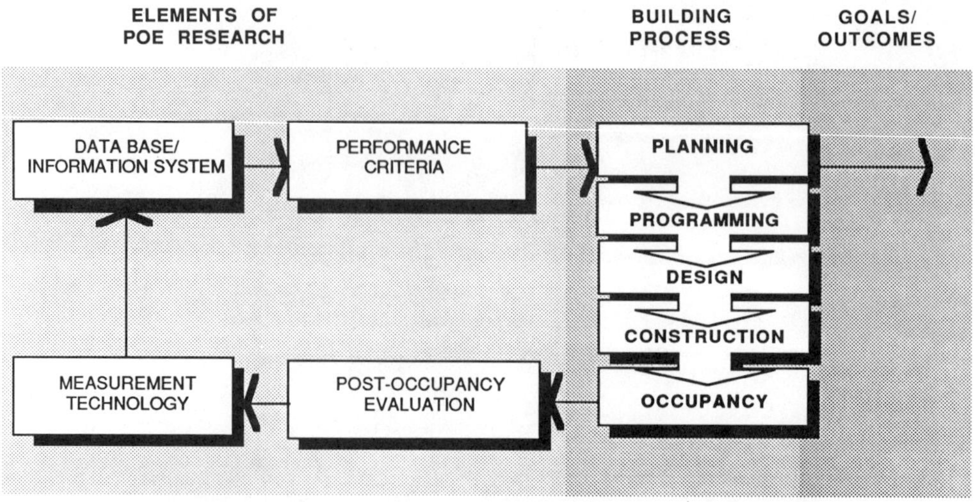

Figure 3. The performance evaluation research framework.

measurement of light and acoustic levels, video recording, mapping of behavior, and still photography.

Programmatic Statements and Performance Criteria

It is essential that performance measures collected by POEs be compared with specific performance criteria in the form of programmatic statements which are contained in the program for a given facility. While these criteria can be of a general nature describing the design intent as expressed in the program (e.g., provide visual and acoustical privacy), a POE should document how the design was expected to meet these criteria. In this way, the findings from each POE can be compared to other POE findings which address similar issues.

Performance criteria and guidelines are frequently developed from data bases and information systems for a given agency and/or building type and from the programmatic criteria for a given facility. These criteria and guidelines are usually documented in technical manuals, design guides, or in specialized data bases. The criteria are building-specific and address particular sets of occupants and building functions. As such they are an evolving and improving set of performance "benchmarks" for a given building type. Performance criteria and guidelines feed the entire building process, and thereby the cyclic evolution of improved building performance can be accomplished.

Performance Evaluation Criteria

Explicit performance criteria need to be developed for purposes of evaluation and use in POEs. One needs to differentiate among the following:

- Criteria concerning the current use of a building.
- Criteria pertaining to the original, intended use of a building, as documented in the program.
- Criteria that pertain to the state-of-the-art in a given building type.

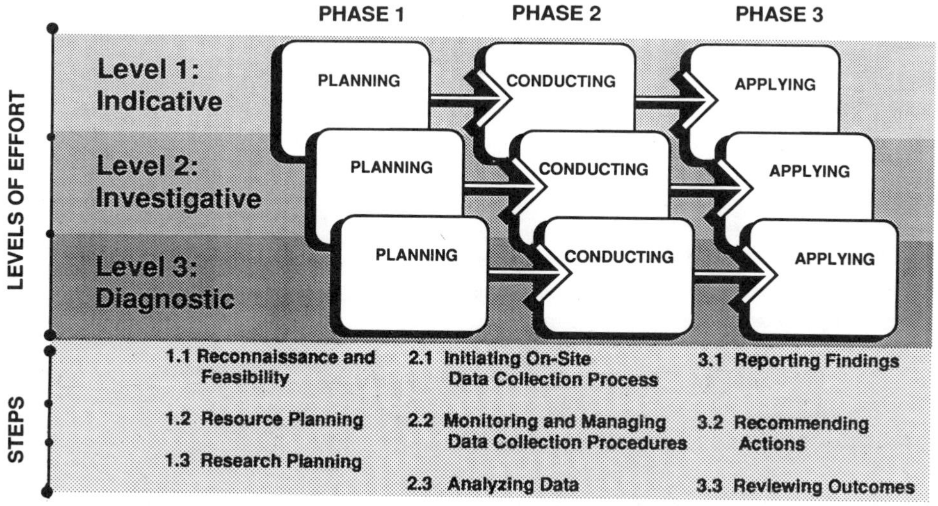

Figure 4. Post-occupancy evaluation process model.

- Criteria which relate to management of the client organization versus those which pertain to the end users/occupants, and:
- Criteria as internalized knowledge and experience which the evaluators/experts may apply regarding certain building types.

A POE Process Model

General models of the POE process have been described by several authors in their writings (e.g., Daish et al., 1980; Marans and Spreckelmeyer, 1981). While there are variations in the process, depending on the nature and objectives of the respective POEs, three levels of effort can be generally distinguished in POE work. Preiser and Pugh (1986) described this as the "POE Process Model" and used it to outline the levels of effort involved in a typical POE. Thus, the model presented in figure 2 is a further development of that POE Process Model (National Research Council, 1987, and Preiser, Rabinowitz and White, 1988).

Levels of effort refer to the amount of time, resources, and personnel, the depth and breadth of investigation, and the implicit cost involved in conducting a POE. The three levels are: 1) indicative, 2) investigative, and 3) diagnostic. Each higher level requires more data gathering and is more comprehensive than the previous level, as depicted in figure 2. The three levels of effect in POEs can be characterized as follows:

1) Indicative POEs give an indication of major strengths and weaknesses of a particular building's performance. They usually consist of a walk-through and selected interviews with knowledgeable informants.
2) Investigative POEs go into more depth whereby objective evaluation criteria are explicitly stated.
3) Diagnostic POEs require considerable effort and expense, are time consuming, and utilize sophisticated measurement techniques. They correlate physical environmental measures with subjective occupant

response measures, thus providing a higher degree of credibility for the results.

In carrying out a POE, there are three basic phases with three steps in each, as shown in figure 2 above.

Benefits and Limitations of Current POE Practice

Each of these POEs can result in several benefits and uses. Recommendations can be tied back to the client to correct problems. Lessons learned can influence design criteria for future buildings, as well as provide information about buildings in use to the building industry. This is especially relevant to the public sector which designs buildings for its own use on a repetitive bases.

The many benefits which result from conducting POEs provide the motivation and rationale for developing POE programs for the following reasons:

1) *Short-Term Benefits*

- Identification and solutions to problems in facilities.
- Pro-active facility management responsive to building user values.
- Improved space utilization and feedback on building performance.
- Improved attitude of building occupants through active involvement in the evaluation process.
- Understanding of the performance implications of changes dictated by budget cuts.
- Informed decision making and better understanding of consequences of design.

2) *Medium-Term Benefits*

- Built-in capability for facility adaptation to organizational change and growth over time, including recycling of facilities into new uses.
- Significant cost savings in the building process and throughout the building life-cycle.
- Accountability for building performance by design professionals and owners.

3) *Long-Term Benefits*

- Long-term improvements in building performance.
- Improvement of design databases, standards, criteria and guidance literature.
- Improved measurement of building performance through quantification.

CONCLUSION

The performance concept and framework for systematic evaluation of the built environment as outlined in this chapter is a much needed and timely methodological approach toward achieving higher quality in buildings, accountability in the building process, and ultimately, better building utilization and user satisfaction. Making explicit the performance requirements that are expected from a building, designing a building accordingly, and eventually comparing the actual performance of the building with that which was initially stated in the building program is the basis of the performance concept advocated for use in POEs.

NOTES

Figures 1 and 2 were produced by Architectural Research Consultants, Inc. of Albuquerque, NM and first appeared in *Post-Occupancy Evaluation* by Preiser, W. F. E., Rabinowitz, H. Z., and White, E. T., New York: Van Nostrand Reinhold, 1988. Permission of the Building Research Board and Van Nostrand Reinhold for the use of these figures and excerpts in this chapter is gratefully acknowledged.

REFERENCES

Eberhard, J. P., 1965, Horizons for the Performance Concept in Building, in: *Proceedings of the Symposium on the Performance Concept in Building. Building Research Advisory Board,* Washington, D.C., National Academy of Sciences, pp. 93-98.

Ehrenkrantz, Ezra., 1967, *SCSD: The Project and the Schools,* New York, Educational Facilities Laboratories, Inc.

Marans, R. and Spreckelmeyer, K., 1981, *Evaluating Built Environments: A Behavioral Approach,* Ann Arbor, MI, The University of Michigan: Institute for Social Research and College of Architecture and Urban Planning.

National Research Council, Building Research Board, 1987, *Post-Occupancy Evaluation Practices in the Building Process: Opportunities for Improvement,* Washington, D. C., National Academy Press.

Post-Occupancy Evaluation of Government Buildings, Wellington, New Zealand, Victoria University of Wellington, School of Architecture.

Preiser, W. F. E. and Pugh, R. R., 1986, Senior Centers: A Process Description of Literature Evaluation, Walkthrough Post-occupancy Evaluations, a Generic Program and Design for the City of Albuquerque, in: Wineman, J., Barnes, R., and Zimring, C., (Eds.) *The Costs of Not Knowing: Proceedings of the 17th Annual Environmental Design Research Association Conference,* Washington, D. C., EDRA, Inc.

Preiser, W. F. E., Rabinowitz, H. Z., and White, E. T., 1988, *Post-Occupancy Evaluation,* New York, Van Nostrand Reinhold.

Wright, J. R., 1971, Performance Criteria in Building, *Scientific American,* 224:17-25.

CHAPTER 2

THE USES AND BOUNDARIES OF POST-OCCUPANCY EVALUATION: AN OVERVIEW

Harvey Z. Rabinowitz

School of Architecture and Urban Planning
University of Wisconsin – Milwaukee
Milwaukee, Wisconsin

INTRODUCTION

Post-Occupancy Evaluation is recognized and valued as a process that can improve, and help explain, the performance of the built environment. During the past twenty-five years it has emerged as a distinct area of research, scholarly activity, and application, and work in the field has rapidly expanded – it now encompasses significant activity beyond that found in the standard environment-behavior literature. This chapter examines the evolution of, and contemporary activities in, Post-Occupancy Evaluation (POE), to distinguish patterns in the use and boundaries of POE. Specific attributes of POE activities that are examined include: (1) the intentions of the POE; (2) the type, size and complexity of the buildings investigated; (3) the types of variables that were included in the evaluations; and (4) the relationships among the variables studied.

The Evolution of Post-Occupancy Evaluation

Post-occupancy evaluation has progressed through three distinct periods over the past quarter-century. These periods could be classified as those in which POE became successively "useful," "usable," and "used." Utilized by Goldberg (1984) as three criteria for judging simulation models, these three characteristics also are appropriately used in describing the evolution of POE as well as other fields.

- Post-occupancy evaluation was first found to be a useful area of research. The results of POEs – that is, the formal and comprehensive examinations of buildings – were found to have benefits for users and owners of buildings as well as for architects designing similar buildings.
- Systematic processes and research methods to conduct building evaluations were then developed, which were reliable, replicable, and could be used by other evaluators. POE evolved qualitatively

from basic to applied research and thus was accessible to and usable by a larger audience.
- The benefits of POE became valued, the POE process became widely known, and POE became frequently used. Post-occupancy evaluation has become a commercially accepted and valued enterprise, though basic and applied research continues.

Post-occupancy evaluation is only twenty-five years old; it is still a young field. A review of the three periods of its development will address some of its attributes, directions and limitations.

The 1960's – Pioneering Studies: Useful Results From Early POEs

The term post-occupancy evaluation was not used to describe such research in the 1960's: these studies were part of the nascent field of environmental design research and were called "environmental analyses," or, in Britain, "building appraisals." Typical environmental design research was quite focused, and, in contrast, these early POE studies examined buildings comprehensively.

The "Dorms at Berkeley" Study (Van der Ryn, 1967), the School and Office studies conducted at the Building Performance Research Unit (Markus, 1972) and the Pilkington Research Unit (Manning, 1965), respectively, were the seminal projects which created the area of post-occupancy evaluation. The projects demonstrated that this type of analysis or appraisal was eminently "useful."

The intentions of these projects were experimental – the researchers were charting new territory. It was a time when architects could seriously ask the question, "If you're designers, what business do you have dabbling in sociology, psychology and all of that?" (Van der Ryn, 1967, p. 11) And researchers could credibly report that they found POE "provides organized information for the designer and reduces the realm of uncertainty in which he works." (Van der Ryn, 1967, p. 7) This early work paralleled the ongoing research and development of theory in environmental design. Concepts such as sociometrics, proximity, territoriality, and privacy, first developed in environmental design research, were applied in POEs.

A critical factor in the success of these early POE projects was that the nature of the buildings and variables studied was limited. For instance, little funding was available, but dormitories *were* available, and contained willing and cooperative students. The "Dorms at Berkeley" study (Van der Ryn,1967) was followed by other dormitory studies. Typically these early evaluations investigated institutional buildings characterized by boundaries and formal rules, such as mental health centers, children's environments, and hospitals. In larger buildings, only selected areas were evaluated. Evaluations also limited the numbers of users studied; in turn, they had limited choices. The physical environment was relatively small in scale, fixed, and provided few options. Though in most cases a number of variables were examined, each one was typically a focused subset of the whole evaluation. These variables were limited to such areas as sociometrics in dormitories, perceived levels of daylighting, and the adequacy of desk workspace and storage.

These first evaluation studies demonstrated the value of POE, and their limited scope and complexity was appropriate for pioneering efforts. Problems in buildings were identified, significant improvements demonstrated, and results disseminated. Post-occupancy evaluation was thus found to be "useful" and the stage was set for a second generation of evaluations.

The 1970's – Systematic POEs: Developing Usable Methods

The scope, number, rigor, and size of POEs increased significantly in the 1970's. By the mid-1970's John Zeisel, in his monograph *Sociology and Architectural Design*, could say "There is a growing interest in visiting and evaluating buildings after they are occupied" (1975).

The intentions of projects during this decade were still largely research-oriented. Evaluation was conducted; however, these projects were equally concerned with the process and methods of evaluation, as well as exploring the relationships between the design of the physical environment, behavior, and building performance.

Several qualitative factors changed during this decade. The number of evaluations increased substantially. The scope and magnitude of evaluations also changed – a number of sophisticated, well-funded evaluation projects were conducted. Housing, especially housing for the elderly and public housing, were often the subjects of these evaluations, many of which were funded by the federal government.

A major change also occurred in the methodologies and research techniques employed. The first multi-building studies were conducted. Technical and functional factors were added to the scope of POE activities after the earlier emphasis on strictly behavioral research (except in Britain) and sophisticated data gathering and analysis methods were used as part of POEs. The performance concept, with its corollary criteria and measurement techniques, became a widely accepted model for evaluation research: this is described in detail in this volume's introductory chapter, "Towards A Performance-Based Conceptual Framework for Systematic POEs," by W. F. E. Preiser.

Of the dozen milestones in POE research in the 1970's, the work by Newman (1973) particularly stands out in terms of scope – it examined data from 100 housing projects – and in influence. Newman's work linking the incidence of crime to housing form and disposition, site design, and circulation, was provocative and well-publicized – it even appeared in *Time* magazine. This study effectively changed housing policy on the national level as well as stimulating the renovation of existing public housing projects – renovation that still continues today and is largely based on the principles resulting from the evaluation research conducted by Newman.

Although the "defensible space" project may have been the most well publicized POE, others were also influential. Researchers at the University of Illinois included project management as part of a housing evaluation study and demonstrated its importance to the residents' satisfaction with their environment (Francescato et al., 1979). This evaluation also included a number of housing projects and utilized sophisticated statistical analysis. McLaughlin's 1972 evaluation of hospitals was also a multi-facility study. This innovative POE focused on the frequency of change and renovation in a hospital's functional areas and related these findings to the life-cycle costs of providing built-in flexibility to accommodate change – a concept widely used today. This still remains one of the few POEs that ties building performance to economic costs and benefits.

The projects mentioned above are representative of many advances that occurred in post-occupancy evaluation during that decade. The studies used multiple buildings for data gathering and comparative analysis – a major change of direction for POE research. The buildings were larger, though they were still for the most part institutional settings – a result of much increased funding from government agencies. Studies in the 1970's also began to find strong links between the design of the environment and behavior, such as the incidence of crime. Multi-method approaches to POE were also initiated during this decade, as were projects which investigated a comprehensive set of environmental factors, including non-physical factors such as management, not as isolated variables but to assess their relative importance to the users of the facilities.

The post-occupancy evaluation projects of the 1970's were major events for the growing community of scholars interested in this topic area. Research was quickly assimilated, improved upon, and applied by researchers on subsequent projects. POE was found to be a useful and valued concept in the 1960's; in the 1970's a second generation of POEs established a body of usable techniques and precedents which became the foundation of this discipline.

The 1980's – Applied POEs: Becoming Routinely Used

Post-occupancy evaluation, based on the two earlier periods of development, has become a widely used activity in the 1980's. POE is now routinely used by government agencies, and many private sector companies involved in facilities are also using this process as a standard operating procedure; it has become a standard activity in the functional responsibilities of the growing group of facilities managers; and the number of successful consulting practices in POE is growing.

Post-occupancy evaluation is now commercially accepted – and used. During the 1970's, government agencies funded POE projects and in many cases used the results. Wener, Frazier, and Farbstein (1985), for example, document three generations of POE utilized by the Federal Bureau of Prisons to continuously evaluate, construct, modify and refine a series of prototype jails from the early 1970's through the 1980's. A number of large U.S. government agencies with extensive construction responsibilities have used POEs as routine procedures for a number of years, emphasizing technical and functional evaluations (Building Research Board, 1987). As discussed later in this chapter, the private sector is quickly catching up and will probably be the major user of POE in the long term.

The variety of POE methods and processes has seen convergence in terms of the basic descriptions employed in the field. For instance, the performance concept and its applications are in wide use as terminology and method. The levels of effort involved in a POE are now often categorized at three levels and described as "walkthrough" or "indicative" POEs, "investigative" POEs, and "diagnostic" POEs (Preiser, Rabinowitz, and White, 1988). Among consulting organizations doing POEs, an emphasis has become the successful implementation of the results of such studies, rather than the actual POE processes used (see chapters by Joiner and Ellis, as well as Farbstein, in this volume)

The emergence and growth of the field of facilities management in the 1980's is also an indicator of increased POE activity. Facilities management is the integration of real estate, design, and building management functions with a goal of coordinating and rationalizing these traditionally disparate fields. POE is a standard part of the facilities management function and responsibility. The rapid growth of IFMA, the International Facility Management Association (it was founded in 1980 and has grown to over 6000 members in just 6 years) is an indicator of the interest in this topic area and has direct implications for the utilization of POE. In contrast, the Environmental Design Research Association (EDRA), the traditional research organization which has been in existence for some 20 years and whose members did most of the POE research during the 1960's and 1970's, has approximately 750 members.

Notwithstanding the success of the utilization of POE in the 1980's, most of its use are low-tech applications based on normative criteria. However, a stream of sophisticated research-based POEs still continues: the BOSTI office study (Brill, 1984) is a significant example. This 5-year study examined over 70 private and public sector office environments and included responses from over 5000 workers. The evaluation explored the relationships among specific physical factors, job satisfaction, productivity, and performance, as well as ease of communication in the workplace. Ongoing research includes the Marans study of office lighting, described in this volume, which relates responses from over 1000 employees as well as technical evaluations to the standard lighting systems used in offices.

After some twenty-five years, POE can now be considered a success in that it is routinely used as the part of the environmental design process. While its use is still not universal, it is frequently employed by many leading organizations to provide a competitive edge that will be emulated by other design professionals. POE has demonstrably become "useful," "utilized," and "used."

PATTERNS IN THE USE AND BOUNDARIES OF POST-OCCUPANCY EVALUATION

There are new directions and constraints in POE that have important implications for the field. These include the 1) the continued growth of POE as an applied tool; 2) the emphasis on its use in the private sector; 3) the role of technology in POEs; and 4) the emergence of research on the relative effects of the physical and the non-physical environment on users – an important boundary limit to present POE work.

The Use of POE as an Applied Tool

The 1970's saw POEs funded as research whose results had varying levels of of influence in improving building performance. POE in the 1980's has been accepted by a number of major building organizations as a standard part of the building construction and management process. The Naval Facilities Engineering Command, for example, has been using POEs for over 10 years and conducts about a dozen POEs each year. The U.S. Postal Service does the same as standard operating procedure (Building Research Board, 1987).

The size and number of buildings affected by POEs has increased greatly – ongoing POE projects will have significant impacts. This includes the Canadian Hospital Evaluation done at the national scale, the California Department of Corrections project, which will influence over a billion dollars' worth of construction, and the U.S. Postal Service Project, which will potentially affect post offices across the U.S. POEs are routinely used in a number of other building types: hotels, retail stores, shopping centers, and office buildings, particularly by owners who manage large numbers of facilities and who have ongoing development and renovation programs.

The Use of Post-Occupancy Evaluation in the Private Sector

In the 1960's and 70's, most funding for POEs was provided by the government, and this is reflected in the proceedings of earlier conferences of the Environmental Design Research Association. However, over 90% of new buildings in the U.S. are developed and constructed by the private sector. In the 1980's, the private sector is increasingly using POE as a standard activity. Sophisticated developers are finding that the results of POE can be reflected directly at the bottom line.

The emergence of the very large private sector developer, the builder, owner and lessor of real estate properties on a national scale, began in the mid-1970's. These developers have focused on the retail, office, and lodging sectors, where most of the POE efforts have occurred. Efforts by this group are applied research POEs due, in part, to competitive factors. For instance, the Marriott Corporation is now well into the process of building 300 "Courtyard" hotels across the country – a plan that calls for opening one hotel each week for the next several years. Post-occupancy evaluation was used in the design of this new "product" aimed at the mid-priced market. Full size rooms were built and modified as hundreds of customers were surveyed on room attributes. This process included the use of full-scale final prototype hotels that were used to "tinker" with the final product (Wall Street Journal, 1985). It was found, for example, that guests objected to a narrower room but hardly noticed a lessening of the room's length. A reduction of 18 inches in each room of a 150 room hotel is the equivalent of an $80,000 savings (in 1985 dollars) to the developer.

In the private sector, large organizations of facility owners and managers can provide data sources, based on existing practice, that provide criteria for evaluating facilities. The International Facility Management Association, for example, has published the first 'benchmarks' for office buildings (IFMA, 1987), based on national surveys. This not only includes square footage for employees, by industry, but also rental and office location information. These benchmarks probably will become de facto normative criteria which facilities managers will use to evaluate and manage their buildings. Other groups (the Building Owners and Managers Association (BOMA), for example) issue data, on microcomputer disks as well as hard copy, which includes details on office building income and expenses, by region, city, size of building, and age of building. Organizations in the lodging and retail fields already disseminate similar information.

Some of the most intense POE-related activities occur in retailing. Large retail chains use demographic statistics which are directly linked to market survey data, to produce local lifestyle and purchasing profiles. Information is now available for individual sites which can document, for example, what the residents in a relatively small area watch on television, what brands they purchase of almost every type of product, and what financial products they own. This information is used to locate and target individual stores. Retailers regularly survey their customers and sales to monitor shopping habits and modify their product selection. While many stores in a national chain may share the same name, each store is customized to respond to its specific user population.

Early work in POE was primarily focused on environments and users that were constrained or regulated, and often in the public sector – such as hospitals, schools, elderly housing, dormitories, and public housing. The success of these studies and their usefulness was due in part to the constrained nature of these environments and/or their users. The private sector is characterized by much more elective behavior. Far fewer evaluation studies have been done in larger, unregulated, and more complex environments such as retail centers, recreational, and entertainment facilities, and mixed-use complexes, though there has been important and continuing work done on public plazas (Whyte, 1980). Environments with such choices are much more complex; however, the current work in retail environments and urban spaces shows important directions in this area.

Expanded Boundaries of POE Due To Technological Advances

During the next 10 years, technological considerations should play an important role in the POE process. This will include advances in three areas: 1) monitoring the environment; 2) the development of databases and clearinghouses; and 3) simulations of the built environment. On the other hand, technological considerations will be making the environment itself more complex and dynamic. What will happen to existing settings as computers and communications technology change the way people work, shop, and recreate?

POEs have developed considerably over the past 25 years, but the environments they have evaluated have also changed and changed considerably in some cases. For instance, current office planning practice utilizes open office concepts which were introduced 25 years ago. Trends in technology also point to significant shifts in activity patterns over the next decade. The use of independent workplaces, many at home, will increase due to the use of communications and computer technology. Some sectors in the corporate world have already made progress in this direction. For instance, one major insurance company has about one-sixth of its claims adjustors working at home on a piece-rate basis and connected to the company's mainframe computer. "Although they are about one-sixth of the claims work force," says Baran (1985), "the 'homeworkers' adjustors produce about one-half of the company's total claims. Overhead costs are low and the company is freed from the expense of paying employee benefits." This shift of the environmental boundary may be an significant opportunity for POE research.

The same technology provides opportunities for the use of clearinghouses to share data and for the networking of practitioners in this field. POE has had a poor record of coordination among researchers in terms of the development of measures, methods, criteria and analysis techniques. Up to now this may have, in fact, been advantageous for the field. The variety of approaches to POE provided a large "test bed" for the emergence of the more successful strategies. At this time coordination of such techniques and data may be appropriate through central organizations, probably entities associated with specific building types, such as IFMA or other trade associations.

Bechtel, in chapter 15, advocates the separation of Post-Occupancy Evaluation from Pre-Design Research. Pre-design research (PDR) uses the results of POEs but focuses on the specifics of the particular building and its users. Computer simulation becomes a tool that has critical significance in this PDR activity. Advances in the computer modeling of environments in both three-dimensional and environmental characteristics provides a means to evaluate proposed design solutions and their implications for such factors as cognition, energy, and economics.

Expanded Boundaries in Understanding the Links Between Environmental and Non-Environmental Factors

What is the role of environmental and non-environmental factors in affecting behavior? Is the environment less of a factor in user satisfaction and productivity than has been assumed by a field that emerged from the physical design disciplines? Are there specific venues in which the environment has a stronger influence? In Peters and Waterman's *In Search of Excellence* (Peters, 1982), which analyzes factors in the success of well-run corporations, physical facilities are rarely mentioned as contributing to excellence. In fact, the culture of the "skunk works," located in a "dingy loft" some miles away from the main corporation, is attributed with considerable success. Communication is mentioned, as well as its relationship to proximity and opportunity for discussion, but by and large the physical environment is not considered significant. Is the field of Post-Occupancy Evaluation (of buildings) mature enough to recognize the effectiveness of such evaluations?

Marans and Spreckelmeyer (1981) developed a model of the work environment which emphasizes the relationship between environmental and non-environmental factors. *The Negotiable Environment* (Williams, 1985) focuses on the importance of such non-environmental factors as corporate culture, individual cognitive style, and the type of work being performed, as well as the physical environment. This is an aspect of POE research which needs considerable effort and provides a dilemma in the utilization of existing POE methodologies.

Can a scale or measurement technique be developed for gauging the relative influence of physical variables in different situations? Research work is now proceeding in this direction, but the difficulty of this work cannot be overstated. As more and softer factors are added to the stew of variables, there is a real danger of the outcome being nothing more than comprehensive incomprehensibility.

The Future of POE

Post-Occupancy Evaluation has emerged from its roots in environmental research as a successful commercial enterprise. The hopes for this type of "environmental analysis" described 25 years ago have largely been achieved. POEs do provide "organized information" to designers and building owners and they considerably "reduce the uncertainty in making design decisions." Similar to the acceptance of programming as a standard activity in design a decade before, POE is developing a consistent set of vocabulary and techniques and a critical mass of practitioners. The leadership in the use of post-occupancy evaluation consists of facilities managers, owners, and independent consultants specializing in this field. While a few architects do conduct POEs, in general the profession is not known for providing this service.

Today's POEs are a third-generation product. A fourth generation of POE is headed in two directions – one concerned with reinforcing its value in the commercial sphere and one concerned with research.

In the commercial world there is an emphasis on "action" research and the implementation of POE findings within organizations. POE efforts in the past have traditionally stopped short of implementation, perhaps due to the research roots of many of its practitioners. The success of a POE in the marketplace depends upon the correct implementation of its results, and there are efforts to integrate the evaluation and implementation processes, as elaborated by Farbstein in Chapter 21. Economic justification of POE efforts, as described by McLaughlin (1972) and Brill (1984), can also provide leverage to the bottom line of private sector POE projects. A number of chapters in this volume are directly concerned with the implementation aspects of the evaluation process. Peter Jockusch writes in Chapter 4 about highly participatory POEs used in the programming process for competitions in Germany and public commentary used in helping to select competition winners. Jockusch would prefer to call this process "prebriefing evaluation." "Preoccupancy evaluation" of another sort is the subject of Laurance Molloy's Chapter 5. Here the concern is implementing the move to new quarters of larger and complex organizations. Evaluation not only concerns the evaluation and requirements of the corporation, but also the search for, and evaluation of, new spaces for occupancy, as well as implementing the construction and finish of the new space and the move itself.

As the POE process has proven itself to be "useful," "usable," and "used," it may have come to its most difficult juncture: many of the 'easy' answers have already been found. The next breakthroughs in POE will be in understanding larger, more dynamic environments with more choice and with the physical environment being only one of many factors influencing the performance of the people and the facilities they function in.

REFERENCES

Baran, B., Ross, J. and Cohen S., 1985, Technological Innovation and Deregulation: The Transformation of the labor process in the Insurance Industry, *Berkeley Roundtable on the International Economy*, University of California at Berkeley, Berkeley, Calif.

Brill, M. et al., 1984, *Using Office Design to Increase Productivity, Vol. 1*, Workplace Design and Productivity Inc., Buffalo, N.Y.

Building Research Board, 1987, *Post-Occupancy Evaluation Practices in the Building Process: Opportunities for Improvement*, National Academy Press, Washington D.C.

Francescato, G. et al., 1979, *Residents' Satisfaction in HUD-Assisted Housing: Design and Management Factors*, U.S. Department of Housing and Urban Development, U.S. Government Printing Office, Washington D.C.

Goldberg, Michael, 1984, Assessing Land Use Impacts of Transportation Improvements, in: *Land-Use Impacts of Highway Projects: Proceedings of the Wisconsin Symposium on the Land-Use Impacts of Highway Projects*, Center for Urban Transportation Studies, University of Wisconsin, Milwaukee, April, 1984.

Harvey, J. and Hennings, H. (Eds.), 1987, Environmental Design Research Association, *EDRA 18 Conference Proceedings*, Environmental Design Research Association, Washington D.C.

International Facility Management Association, 1987, *IFMA Facilities Benchmarks*, Houston, Texas.

Manning P. (Ed.), 1965, *Office Design: A Study of Environment, Pilkington Research Unit, Department of Building Science*, University of Liverpool, Liverpool, England.

Markus, T. et al., 1972, *Building Performance*, Halstead Press, New York.

Marans, R. and Spreckelmeyer, K., 1981, *Evaluating Built Environments: A Behavioral Approach*, The University of Michigan, Institute for Social Research and the College of Architecture and Urban Planning, Ann Arbor, Michigan.

McLaughlin, H., Kibre, J. and Mort, R., 1972, Patterns of Physical Change in Six Existing Hospitals, in: W. Mitchell (Ed.), *Environmental Design: Research and Practice*, University of California, Los Angeles.

Newman, O., 1973, *Defensible Space: Crime Prevention Through Urban Design*, Collier Books, New York.

Preiser, W. F. E., Rabinowitz, H. Z. and White, T., 1988, *Post-Occupancy Evaluation*, Van Nostrand Reinhold, New York.

Peters, T., and Waterman, R. Jr., 1982, *In Search Of Excellence*, Harper and Row, New York.

Van der Ryn S. and Silverstein M., 1967, *Dorms at Berkeley*, University of California, Center for Planning and Research, Berkeley, Calif.

Wall Street Journal, September 18, 1985, Basic Bedrooms: How Marriott Changes Hotel Design to Tap Mid-Priced Market, Wall Street Journal, New York.

Wener, R., Frazier, W., and Farbstein J., 1985, Three Generations of Evaluation and Design of Correctional Facilities, in: *Environment and Behavior*, 17:1, M. Kantrowitz, and A. Seidel (Eds.), Sage Publications, Beverly Hills, Calif., Jan. 1985.

Whyte, William H., 1980, *The Social Life of Small Urban Spaces*, The Conservation Foundation, Washington D.C.
Williams, C., Armstrong, D., and Clark, M., 1985, *The Negotiable Environment*, Facility Management Institute, Ann Arbor, Michigan.
Zeisel, J., 1975, *Sociology and Architectural Design*, Russell Sage Foundation, New York.

CHAPTER 3

POST-OCCUPANCY EVALUATION FROM THE CLIENT'S PERSPECTIVE

Edward T. White

Florida A&M University
Tallahassee, Florida

OVERVIEW

This chapter discusses concerns, questions, problems, and opportunities of POE (post-occupancy evaluation) from the client's perspective. Its thesis is that the planning of POE must not only satisfy professional criteria such as reliability and integrity, but that the fundamental strategy and planning of the POE should also be sensitive to the client's personal, organizational and political context. Specifically, the chapter addresses the importance of making planning factors such as purpose, building sample, content, rigor, tools and techniques, participants, scheduling, cost, and reporting method responsive to client concerns such as avoiding embarrassment, defending results, satisfying administrative superiors, boosting morale, minimizing disruption, resolving conflicts, and obtaining maximum public relations benefit from the POE study. Neglect of these contextual factors can lead to POE results which satisfy professional criteria, but which are useless and even counterproductive for the client's organization.

GREATER POE FLEXIBILITY HAS INVITED EVEN FURTHER EXPANSION OF POE USES AND METHODS

The increased capacity of the building evaluation process to accommodate a greater variety of projects and project situations has significantly expanded the number of ways in which POE can be made valuable and useful. The evolution of the field has positioned post-occupancy evaluation where it can invite even further development in terms of expanding the list of possible goals, extending the potential uses by various professions, increasing the number of success criteria and refining the sophistication of POE preparation and implementation. These dimensions are now considered open-ended in the sense that all can be further extended, and through this extension, POE may evolve to even higher levels of maturity, relevance, and value.

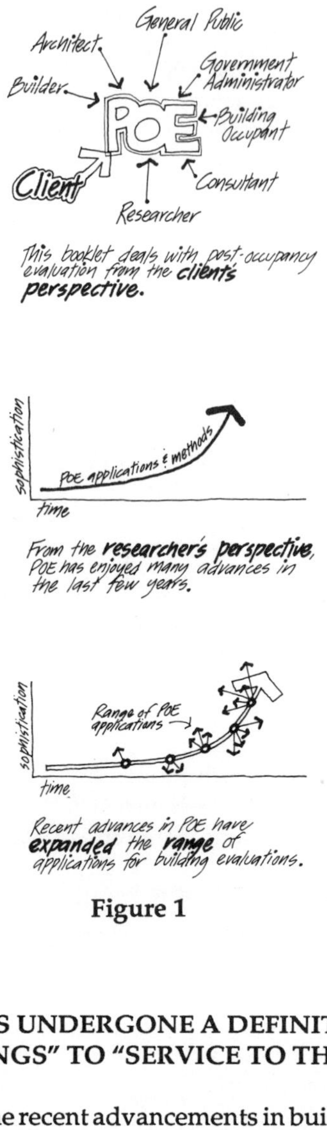

Figure 1

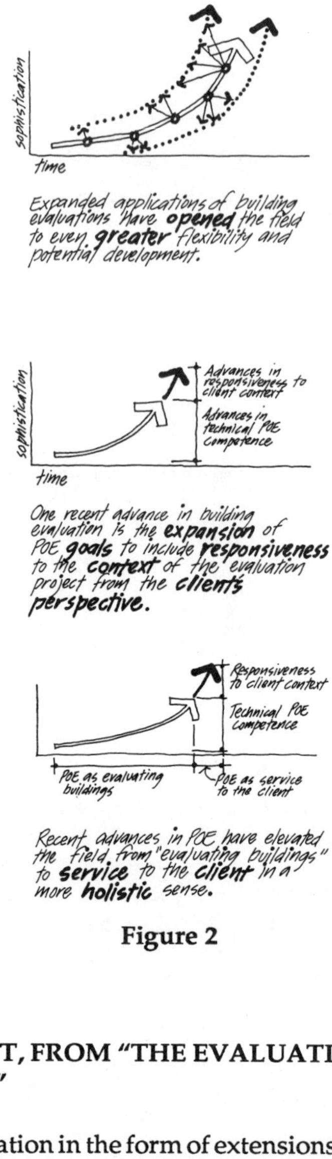

Figure 2

POE HAS UNDERGONE A DEFINITION SHIFT, FROM "THE EVALUATION OF BUILDINGS" TO "SERVICE TO THE CLIENT"

The recent advancements in building evaluation in the form of extensions to POE goal potentials, to success criteria, and to the influences on planning and implementation of evaluation studies has led to a gradual shift in the definition of POE from the "evaluation of buildings" to "service to the client" in a more holistic sense. The same definition shift is also occurring in other professions such as medicine where the meaning of "responsible medical practice" now extends beyond technical competence to include concern for the health of the whole patient. Attention to the whole patient in medicine or to the whole client situation in POE both involve an expansion of the boundaries of the professional evaluator's responsibility. Both indicate a broadening of concern which extends beyond the official, formal, explicit products and effects of professional services to include the deliberate management of factors which have, in the past, been considered outside the limits of professional responsibility. Many professions have come to understand that the by-products of their work are often more

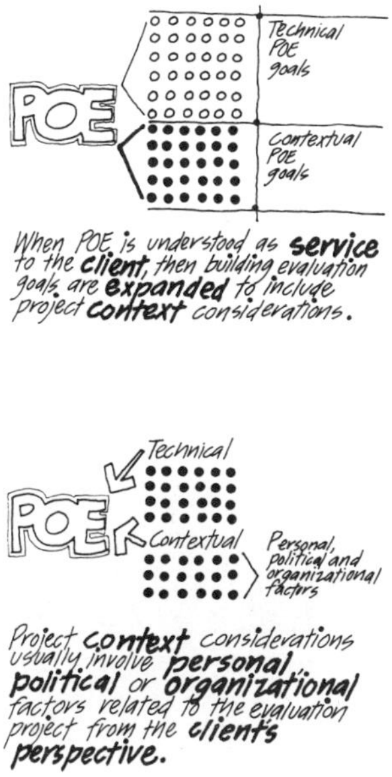

Figure 3

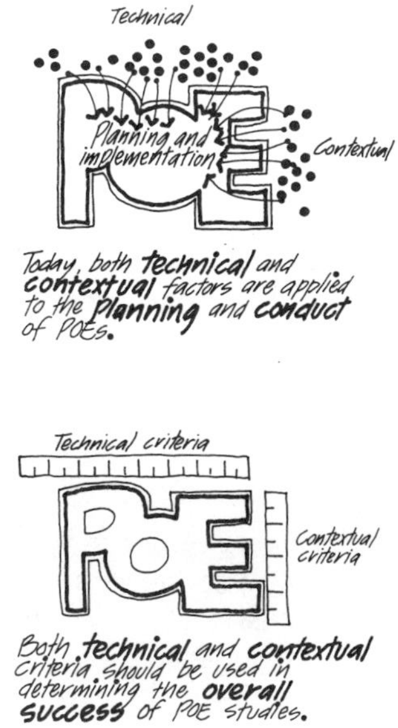

Figure 4

important than the products and that unmanaged and inadvertent outcomes within the halo of contextual issues surrounding their work can be damaging for the client, even though, from a technical standpoint, the work meets professional standards.

POE AS SERVICE TO THE CLIENT HAS MEANT EXPANDING THE LIST OF POSSIBLE POE GOALS TO ACCOMMODATE CLIENT CONTEXT AND CONCERNS

From the professional researcher's viewpoint, the goals of a POE relate mainly to the direct uses of the evaluation itself as a building performance measurement tool. A POE may provide a means for identifying and solving immediate problems in a facility or may be helpful in fine-tuning a building before or after move-in. Evaluation methods can enable continuous feedback in particular aspects of long-term performance such as energy utilization, space adequacy, or circulation efficiency. POEs have been used to document building deficiencies as part of the justification of new construction or remodeling projects. Within the building evaluation field, a particularly important goal of POE is the continual testing and updating of planning standards and the long-term upgrading of resource literature and decision guidelines for the design professions.

In addition to the POE goals just mentioned, clients may have other motivations for initiating evaluation projects. These motivations may be singular or multiple within a given project and can relate to desired POE effects which are both internal and external to the client's organization. Internal POE goals may involve the desire to have the study serve as one component of a broader quality and productivity improvement program. The building evaluation can be a means of boosting client employee morale and an approach to strengthening employee relations by expressing management's concern for

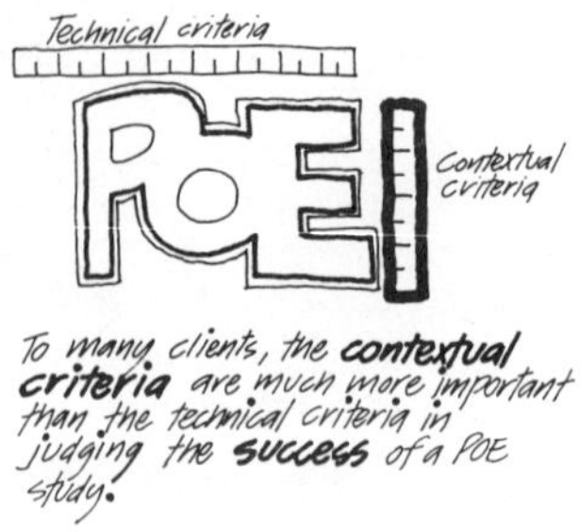

Figure 5

the quality of the work environment. Clients may initiate POEs to increase their personal standing and value or their department's standing and value in the larger organization or to expand and solidify their domain of responsibility within their institution. POE projects can be undertaken to signal a renewal or redirection within the client's department or to simply provide client staff the chance to observe and learn from POE experts.

Clients may also pursue external POE goals such as enhancing their organization's competitive standing in the marketplace, improving their company's public image and reputation, or announcing to colleagues in similar institutions that they are now in the POE business. A client may implement a building evaluation in order to fend off threats of environmental regulation by outside agencies by showing that the client's organization has the will and the means to police itself. POEs can facilitate long term client contacts and relationships with evaluation experts which produce continuing benefits beyond the projects themselves. Example benefits include the accumulation of POE reference material, staying abreast of current practices and trends in the field and participating in national conferences which raise the profile of the client's company. Post-occupancy evaluation may be used to put design professionals on notice that building performance is taken seriously by the client's institution and that design consultants hired by the client will be held accountable for the consequences of their planning decisions.

These kinds of POE goals used to be described by building evaluation professionals as hidden agendas or ulterior motives and were often ignored in the planning and conduct of POEs. They were often seen as unworthy and undignified project goals, irrelevant to the essential purposes of the study or even a threat to the integrity of the evaluation process. Today, the sensitive POE consultant probes for these client motives as part of a thorough preparation for the POE project and, insofar as possible, respects these client concerns in the way that the project is carried out and reported.

POE AS SERVICE TO THE CLIENT HAS MEANT ADDITIONAL CRITERIA FOR JUDGING POE SUCCESS

Traditionally, the criteria by which the value, quality, and overall success of POEs have been determined have primarily been derived from the cannons of research. These guidelines largely pertain to technical competence and concentrate upon considerations such as the clarity of the problem definition and boundaries of the study, the integrity,

Figure 6

logic, and internal consistency of the research plan, the appropriateness of all samples, and the reliability of the data collection instruments. It is important to the POE professional that the evaluation methods be current, that data be accurate and thorough, and that the most effective and telling data analysis techniques be employed. POE success depends upon the discovery of significant correlations and the isolation of cause-effect relationships among building and behavioral factors. In a well-done POE, the conclusions drawn and recommendations made are logical, reasonable and rooted in the evidence of the data and its analysis. The building evaluation field values POEs which result in clear, well-organized, action-oriented reports and in positive change to the environment. Professionals expect that POEs will be replicable and verifiable and, ideally, that evaluation studies will contribute to the advancement of the state-of-the-art.

While they may appreciate the importance of these professional guidelines for POEs, clients will often bring their own criteria to the project which, from their viewpoint, are as important or even more important than the research principles just mentioned. To many clients, the technical competence of the post-occupancy evaluation

Figure 7

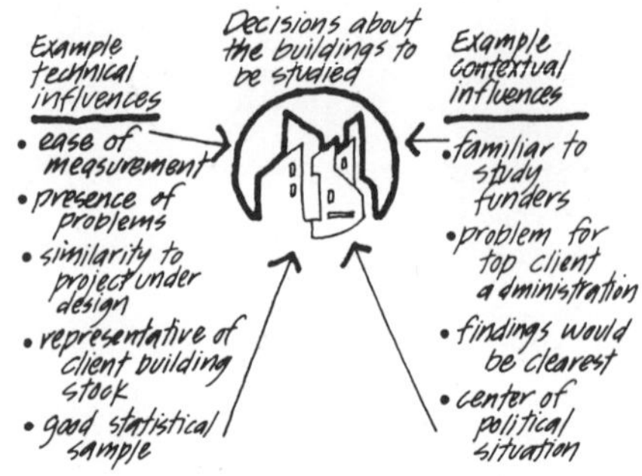

Figure 8

is a given. It is an expected minimum which any competent professional should be able to achieve and is only one set of a larger number of criteria which the client hopes will be satisfied by the study.

In the client's eyes, POE success may depend upon the status and reputation of the consultant who is employed, the amount of public relations value that can be obtained from all phases of the POE project, or the scope of service that is negotiated for a given consultant fee. POE success may be measured by the avoidance of disruption in the client organization during the project or the extent to which the study sensitizes client employees and top administrators to the importance of the environment. Technical competence must never be accomplished at the expense of embarrassment or in ways that lay blame or expose the client to legal action. In some projects, the overriding concerns are confidentiality of POE results, gentleness and sensitivity in the implementation of the study and balance in the reporting of building successes and shortcomings. The political situation surrounding the POE project may require extraordinary skill in couching findings in especially careful language or may demand unusual attention to the management of the POE so that it impresses the client's administrative superiors. The social dynamics of a POE can make it crucial that the findings are defendable and obvious on face value and that the benefits of the study be clearly worth the cost. A client may initiate a POE project because a building evaluation is seen as a way of resolving long-standing conflicts between employees and management. Sometimes conflicts arise between the client and the other building occupants over what the POE emphasis should be or over how the study should be conducted. These conflicts can often be resolved if the professional looks closely at the POE success criteria for each group and searches for areas of agreement which, before, were not apparent. In some POEs, clients are fearful that the study will not capture and report building problems that everyone in the client organization already knows about, thus calling to question the thoroughness of the POE and the reliability of its methods to discover the real and practical issues of the building's performance.

As in the case of the POE goals discussed earlier, the sensitive evaluation consultant today will explore these types of criteria with the client at the beginning of the project and will respond to these types of concerns in the formulation of the POE work plan.

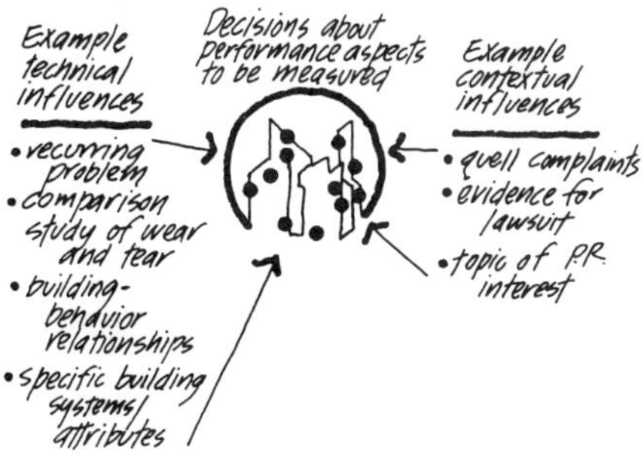

Figure 9

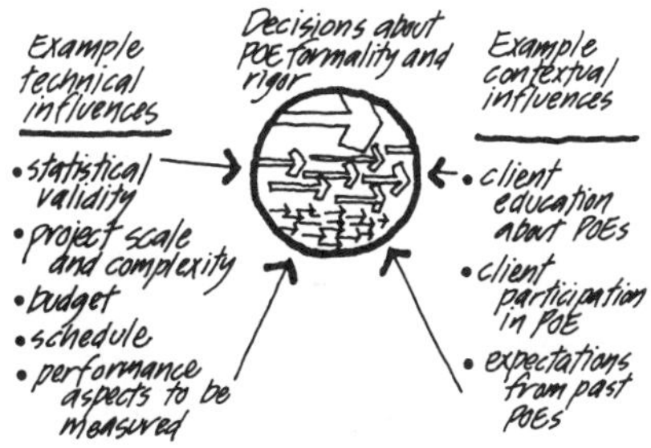

Figure 10

PROFESSIONALS NOW PERMIT THE CLIENT'S CONTEXTUAL GOALS AND CRITERIA TO INFLUENCE POE PLANNING DECISIONS

The recent attention to the client's contextual goals and criteria in the evaluation field has meant a change in the way that professionals understand, prepare for, carry out and conclude POEs. To some extent, researchers have responded to these contextual influences at the level of day-to-day project operations such as meeting with the client more frequently, routing conference notes to additional client staff, taking special care in how certain project issues are written about, or adapting a certain working style and personality when interacting with particular project participants. Beyond these tactical responses to the client's contextual influences, of greater significance to recent advances in POE has been the impact of the client's contextual goals and criteria upon the strategic planning of POEs and upon the fundamental structure and organization of the project work plan.

Formulation of a POE work plan requires decisions in a number of areas such as what the purpose of the POE will be, which buildings will be evaluated, what aspects of building performance will be measured, how rigorous and formal the POE will be, which evaluation and analysis tools are most appropriate, who should participate in the study, when the POE should occur and for how long the study should last, what the study should cost, and how to most effectively report the findings.

Planning for a technically competent POE from the professional's perspective involves making decisions within each of these aspects of preparation, and very importantly, understanding how decisions about each of these planning factors affect and are affected by decisions about all the other factors. For example, decisions about the best evaluation techniques to use can be influenced by earlier decisions about POE purpose, building sample, performance measures, level of rigor, timing, participants, and cost. In turn, a commitment to certain POE tools and techniques can influence all these other factors in the evaluation work plan.

The reflection of client contextual concerns in the POE work plan means that the kinds of client goals and criteria discussed earlier are allowed to influence POE planning decisions and that the building researcher deliberately organizes the evaluation with attention to both professional POE standards and non-technical client issues. Some instances of POE planning from both the technical perspective and the client's perspective are outlined below.

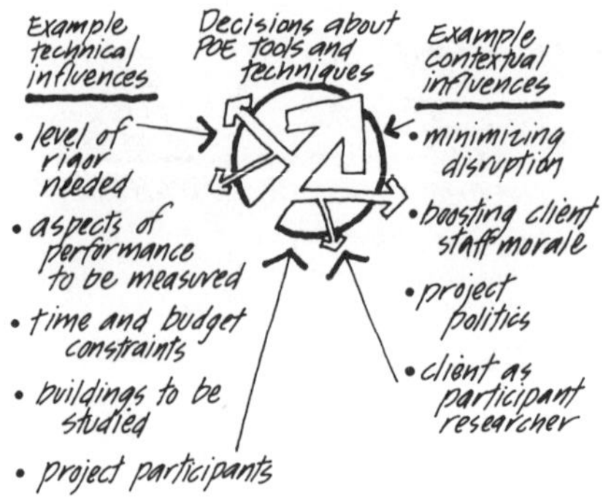

Figure 11

EXAMPLES OF TECHNICAL AND CONTEXTUAL INFLUENCES UPON POE PLANNING DECISIONS

The POE planning factors mentioned above will be discussed here, first in traditional terms that relate to technical competence, and then in terms that relate to the project context from the client's perspective.

Purpose of the POE Study

In most building evaluation projects, the determination of the purpose of the study is the first step in planning a POE. What is the primary reason for the evaluation? What, in the broadest terms, should be accomplished? What is the fundamental motivation for undertaking the POE? Who will use the results and how will they use them? Traditional technical reasons for implementing a POE may include learning the level of client and user satisfaction with the building, understanding how and where the building is wearing due to use over time, determining the true cost of building ownership, establishing the patterns and causes of vandalism, learning the actual ways that spaces are used, discovering the correlations between productivity and environment, defining the impact of building upon the surrounding context or validating design standards and guidelines.

The determination of the purpose of the POE has a direct impact upon the other POE planning decisions. The basic intent behind the POE can affect the choice of buildings to be evaluated, the selection of building performance aspects to be measured, the establishment of the rigor with which the study is conducted, the choice of evaluation tools and techniques to be employed, the selection of POE participants, the calculation of the cost of the study, the scheduling of the job and the determination of the method for reporting the study results. In some projects, the example technical purposes of the POE mentioned above may be secondary and may actually be determined by broader and more governing purposes which derive from the context of the job. Example context-driven purposes are generating publicity for the client's organization, gaining a competitive edge with a rival department in level of funding, convincing governmental regulatory agencies that the client organization is self-policing, or bringing the attention of the client's top administration to the fact that the quality of the work

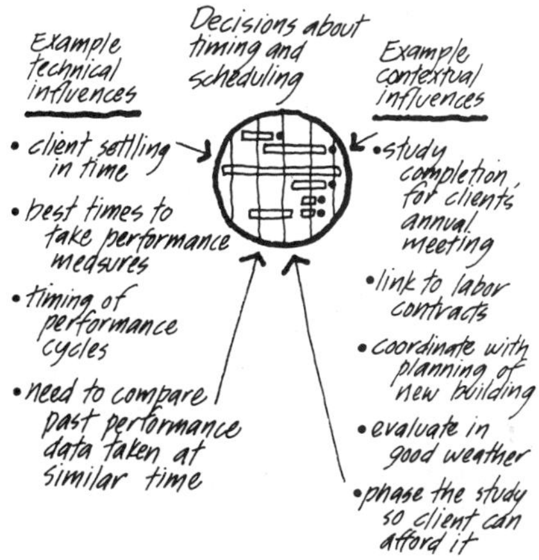

Figure 12

environment is important. It is not difficult to imagine how purposes such as these might affect other POE planning decisions about the buildings to evaluate, aspects to measure, study rigor, evaluation tools, participants, cost, scheduling and reporting method. In this instance, it is clear how a context-related purpose can influence the fundamental structure and organization of the POE work plan.

Buildings to be Evaluated

A vital component of any POE work plan is the selection of the building or buildings to be evaluated. Often this decision is an obvious one, but in some POEs purpose, aspects to be measured, participants, scheduling and cost may be pre-established and the selection of the building to be studied may follow from these earlier decisions. Given the basic reason for the POE, which buildings should be evaluated? Which building would yield the best data on the performance aspects to be measured? Does the POE budget and schedule allow several buildings to be studied or just one or two? What do the participants suggest in terms of choice of building to be studied? Technical reasons why certain buildings might be selected for study may involve the ease of measurement of the aspects of performance to be evaluated, the presence of a particular problem needing study in a building, the similarity between the building about to be designed and the buildings which are candidates for evaluation where lessons learned from the POE will be applied to the new design or the satisfaction of the requirements of statistical sampling within the study.

The selection of the building sample for the POE study may also be affected by contextual factors within the project. For example, a particular building may be chosen for evaluation because it is most familiar to the funders of the study, it is a particularly problematic building for an influential officer in the client organization, it promises to reveal results that are most clear to the audience who must understand the findings or it is at the center of a political situation whose resolution is tied to the study results. The decision to evaluate a particular building on the basis of contextual issues may then influence subsequent decisions about project participants, evaluation techniques, cost, schedule and reporting method.

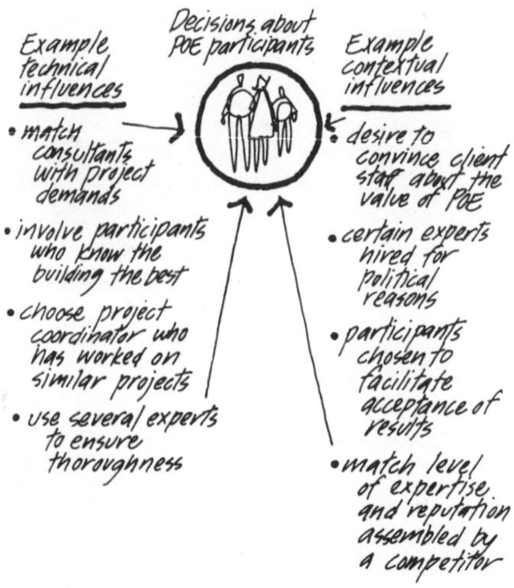

Figure 13

Building Performance Aspects to be Measured

Any POE work plan must include decisions about the aspects of building performance to be measured. Which dimensions of user satisfaction should be evaluated? What elements of the building's technology ought to be studied? Should the POE evaluate aspects of the building delivery process as well as or instead of evaluating the building itself. Should energy conservation be studied? Thermal conductivity? Lighting? Temperature mediation? Air quality? Should the POE evaluate behavioral factors such as orientation and way finding? Privacy? Territoriality? Vandalism? If determined strictly on the basis of technical competence, reasons for choosing particular aspects of building performance to be measured might include the desire to study and resolve a recurring problem in the client's building stock, comparison of alternative installation strategies for a particular building component, determination of which building elements or systems experience the earliest and most severe wear and breakage through normal use, study of the behavioral and performance interrelationships between building components or between building attributes and occupant behavior or evaluation of occupant satisfaction with lighting quality. Particular aspects of the building delivery process may be chosen for study because the client wants to resolve a recurring problem in his/her planning process, a government agency has decided to reorganize the way it delivers buildings or a corporation is interested in learning how its management of the building process affects the performance of the resulting facilities in the area of image and symbolism.

Contextual issues within the POE project may also affect the selection of the aspects of building performance to be measured. The subjects for the evaluation may be based upon the client's desire to show employees that complaints about the work environment are being acted upon, the need to gather evidence for a lawsuit or the public relations value of conducting a study in an area of building performance which is of national interest.

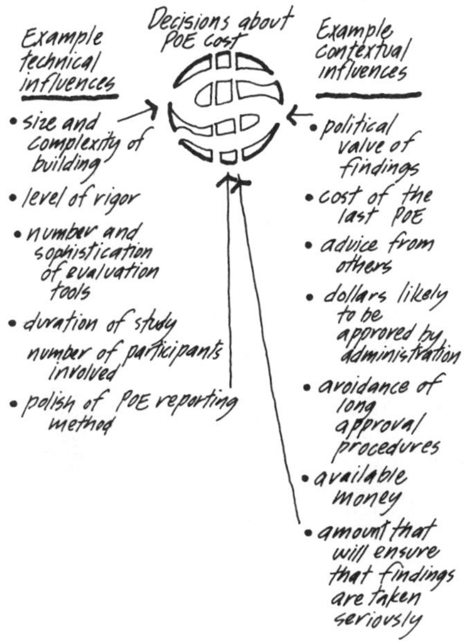

Figure 14

Level of Rigor

A POE work plan usually includes a deliberate decision about the level of effort, formality and rigor with which the study will be implemented. Generally, the more formal the POE, the more careful the planning process, the more systematic the evaluation method, the more care in analysis and documentation, and the greater the effort to share and apply the findings. Level of effort may range from a simple phone call to the client to see if the building is performing satisfactorily in general terms to a full scale study taking several months and involving research protocols and computerized statistical analysis. Typical reasons why a certain level of rigor might be adopted for a POE include the requirement that the study produce statistically defendable findings, the need for a level of care and organization due to the scale and complexity of the project, the impact of the client's POE budget on the allowable level of effort, the effect of the window of time opportunity upon the level of POE effort, or the influence of the aspects of performance to be measured upon the selected degree of formality of the evaluation.

In addition to technical considerations, the context of the POE can have an influence upon the rigor and formality of the job. For example, the adopted level of effort may be responsive to the desire of the client to learn how to do POEs by observing the consultant at work, to the need for the client to take over and operate the long-term POE study when the consultant has completed the initial phase, or to the expectations of the client's board of directors regarding level of rigor which were gained from past POE studies done for the organization.

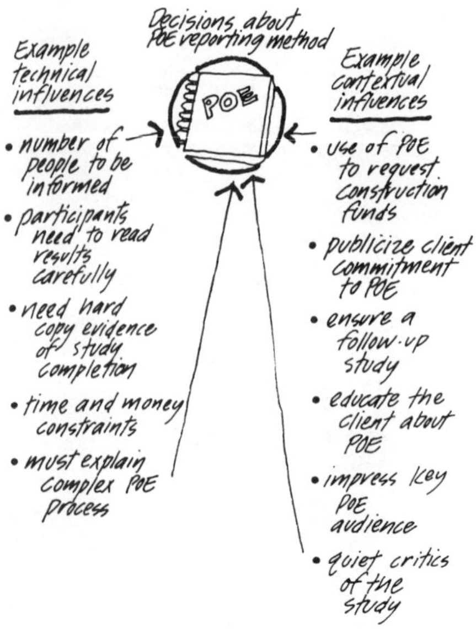

Figure 15

Evaluation Tools and Techniques

Central to any POE work plan is the selection of the evaluation tools and techniques to be employed in measuring the chosen aspects of building performance. Given the POE purpose, buildings, performance aspects, rigor, cost, schedule and participants, which POE tools are most appropriate, effective and telling for the evaluation? Which techniques will uncover the key issues? What tools best match the job to be done? Should multiple tools be applied to the study for reliability and quality control or will one technique suffice? The building evaluation consultant has a wide variety of tools available which may be used singularly or in combination in a POE. These include interviews, questionnaires, walk throughs, user diaries, decision tracking, simulation, behavioral traces, computer modeling, design logs, behavioral mapping, physical measures, photography, videos, observation, audits, remote sensing, movement detectors, infrared and chemical analysis. Each of these techniques may be applied at several levels of rigor ranging from very informal to very formal and systematic. Traditional reasons for choosing evaluation tools based upon technical factors include matching the tools with the required level of POE rigor, using tools that are especially suited to measuring the aspects of building performance to be studied, applying techniques that fit within the time and budget constraints or choosing tools that will be most effective in relation to the buildings and building occupants under study.

Contextual project considerations may also be influential in selecting POE tools and techniques. For example, particular tools might be less disruptive for the client operation, certain techniques may be more effective in boosting client employee morale, specific tools might be most effective in involving a large number of participants for political purposes or particular evaluation methods may work more successfully in allowing the client to participate in the collection of the evaluation data.

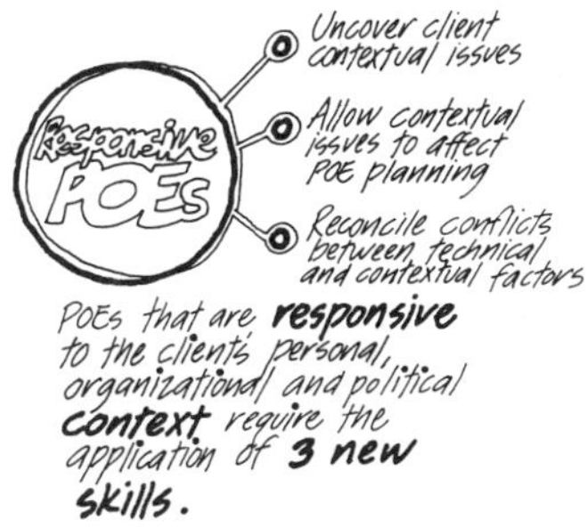

Figure 16

POE Timing and Scheduling

The timing and scheduling of the POE may be an early constraint in the planning of the study or may be determined from earlier decisions about the other planning factors. For example, timing could be determined by the allowable project budget, by the limited availability of certain key POE participants or by a narrow window of opportunity for measuring the desired aspects of building performance. On the other hand, timing could be established early in the POE planning process and then influence later planning decisions such as the building sample, the aspects of performance to be measured, the evaluation tools to be used, the cost of the study, the participants, and the reporting method. When should the evaluation be conducted? Should the POE be one continuous effort or broken into several efforts over a longer time span? Over what time span should the evaluation take place? Planning decisions about the scheduling, longevity, and phased versus single effort character of the POE are normally based upon factors such as the length of time it takes the client to settle into a routine within the new facility, the appropriateness of taking certain kinds of performance measurements at certain times of year, month, week or day, the occurrence of events or performance aspects at particular points in time, the need to measure performance aspects over several cycles or patterns, or the desire to compare data from this study with other studies done under special scheduling circumstances.

Contextual factors within the project situation can sometimes have an influence upon timing and scheduling along with the traditional issues related to the integrity of the research. Examples include the need to complete the POE in time for the client's annual meeting of the board of directors, the desire to initiate the study before renegotiation of labor contracts begin, scheduling the POE so that the results may be applied in the planning of a new facility, planning the POE for the good weather months or breaking the study into several phased efforts so that the project fits within the client's funding cycle.

Project Participants

Determining the participants who will be involved in the POE study usually requires a number of decisions about several potential team members such as the POE consultant, sub-consultants, advisors, client decision makers, committees, building occupant groups, various building-related constituencies, and governmental/regulatory agencies. Decisions about whether to involve certain participant groups, whom specifically in these groups to involve, how many from each group to involve and how to assign roles and responsibilities to them are all part of the formulation of the POE work plan. These decisions may be affected by earlier choices made about the purpose of the study, the buildings to be evaluated, the aspects of performance to be measured, the level of POE rigor to be applied, the evaluation tools to be employed, or the selected POE timing, cost and reporting method. In those instances where the participants have been determined early in the POE planning process the participants chosen can influence later decisions about POE purpose, building sample, aspects to be measured, rigor, evaluation tools, timing, cost and reporting method. Typical research-related reasons why certain participants might be selected for involvement in the POE include the desire to have a particular POE expert on the job to match their expertise with the job to be done, the appropriateness of involving certain building occupants because of their knowledge of certain building performance attributes, the logic of naming an individual as project coordinator because of his/her experience in past POE studies in the client's organization or the inclusion of other POE consultants on the evaluation team to ensure the most thorough possible coverage of a certain aspect of building performance. The determination of participant roles and responsibilities within the evaluation study is normally based upon who can do what best, upon who can speak most articulately and knowledgeably about the aspects of performance to be measured, upon creating the clearest possible lines of communication and upon defining efficient and appropriate lines of authority and decision-making.

Project context may influence decisions about POE participants in a number of ways such as involving particular people from the client's staff in order to change their minds about the value of POE, inviting certain experts to participate for political reasons, involving specific individuals to facilitate acceptance of the POE findings, or tailoring the evaluation team to compare favorably with a POE team assembled by the client's key competitor. Contextual factors can also affect the assignment of participant roles and responsibilities in the study. For instance, a team member may be put in charge of an aspect of building performance that their department is notorious for complaining about, a team chairperson may be named to placate a vocal group of POE critics within the client's organization, or a political appointee to the POE team might be assigned duties that will not jeopardize the research integrity of the project.

POE Cost

The determination of the cost of the POE study may be a matter of learning the allowable budget from the client and then backing into the affordable work plan or working out to the required cost from a careful formulation of the POE work plan. Typical kinds of influences upon POE labor cost include the number, size and location of buildings to be evaluated, the number and complexity of the building performance aspects to be measured, the rigor with which the study will be implemented, the types and number of evaluation tools to be used, the number of participants, the POE scheduling and the reporting method to be employed. Overall POE cost is also affected by other costs such as travel, consultants, computer time, materials, photocopying, overhead, and profit. Where overall POE cost is pre-established by the client in the form of an allowable total budget for the complete study, the costing process is reversed. The total dollars available are allocated to each of the factors listed above and then decisions are made about each work plan component based upon its individual budget. Dollars allocated to direct labor affect decisions about the building sample, the performance aspects to be measured, the tools to be used, the POE rigor, the participants, the timing and scheduling and the reporting method. Budgets are also established for the non-labor POE components such as consultants, travel, and photocopying. In calculating the true cost of the POE to the client, the lost productivity within the client's organization due to the involvement of his/her employees in the study should be estimated. In addition to direct loss of productivity, the POE may reduce productivity further due to indirect affects upon the client's staff such as distraction, disruption of work routines, loss of morale, increased expectations regarding potential changes to the work environment, or heightened awareness of building problems with resulting increases in complaints.

Where contextual factors have influenced the determination of any of the work plan components (purpose, buildings, participants, etc.), then project context will have also necessarily affected POE cost. Contextual considerations may also directly affect cost in situations where the client establishes the project budget before the POE work plan is formulated. For instance, the client may base the allocated dollars for the study upon the political value of the anticipated findings, upon the cost of the last POE study, upon the advise of a friend who is knowledgeable about POEs, upon the amount that is likely to be approved by the client's home office, upon an amount that does not need to endure a lengthy bureaucratic approval process, upon the amount of money left over at the end of the client's fiscal year, or upon the client's desire to have the POE results taken seriously.

Method for Reporting POE Findings

A thorough POE work plan should contain decisions about the method or methods for reporting the results of the study. Should the findings be delivered to the client in the form of a written report? A slide presentation? A video? Who should learn the results of the study and why? How should the report of the findings be organized and formatted? What should the content of the report be? What are all the ways that the POE findings may have value? Decisions about the POE reporting method usually are responsive to earlier decisions about the POE purpose, building sample, performance aspects to be measured, study rigor, evaluation tools, project participants, scheduling and cost. For example, the reporting method may be influenced by the need to inform a large number of people of the findings, the desire to have the participants read over the results at their leisure, the requirement that a report be produced so that there is hard copy evidence that the study was completed, the time and money constraints on the report length and polish and number of copies printed or the need to explain and justify a particularly complex and involved evaluation method.

The influence of the project context upon the reporting method may be based upon factors such as the client's intent to use the POE as a basis for requesting construction and remodeling funds, the announcement of the findings as a way to publicize the client's commitment to POE, or delivery of the results in a manner that will ensure the funding of a follow-up study. The content and format of the report may be affected by project context in that the evaluation method must be explained in a way that educates the client about POE so that he/she may implement them in the future, the graphics may be designed to illustrate the key ideas to a particular political constituency in the client's organization, a special chapter may be included to prove to antagonists of the study that a certain aspect of building performance was studied carefully, or a certain POE expert's work may be referenced often to facilitate his/her hiring for the next POE project.

NEED FOR NEW POE SKILLS

In order, then, to effectively deal with POE goals and criteria from the client's perspective, the professional consultant now appears to need three new skills. The mastery of these skills by many building evaluation experts has led to the recent advances in POE discussed in this chapter. The skills are:

- The ability to uncover client contextual issues in the early stages of POE project planning.
- The skill and commitment to allow these contextual issues to affect POE planning decisions and to integrate client concerns with technical competence considerations in the POE work plan.
- The capacity to reconcile conflicts where a response to client contextual factors in the POE work plan threatens the technical integrity of the study or where the satisfaction of technical POE criteria reduces the value of the POE from the client's perspective.

BIBLIOGRAPHY

Building Research Board, 1987, *Post-Occupancy Evaluation Practices in the Building Process: Opportunities for Improvement*, Washington, D.C., National Academy Press.

Marans, R. and Spreckelmeyer, K., 1981, *Evaluating Built Environments: A Behavioral Approach*, The University of Michigan, Ann Arbor.

Preiser, W., Rabinowitz, H., and White, E., 1988, *Post-Occupancy Evaluation*, New York, Van Nostrand Reinhold.

White, E., 1983, *The Value of Post-Occupancy Evaluation to the Architect in Government*, Tucson, Arizona, Architectural Media Ltd.

White, E., 1985, *Building Evaluation in Professional Practice*, Tucson, Arizona, Architectural Media Ltd.

White, E., 1988, *Programming, Post-Occupancy Evaluation and the Financial Success of the Architect*, Tucson, Arizona, Architectural Media Ltd.

White, E., 1988, *An Automated Facility Management System for the Florida Department of Transportation*, Tallahassee, Florida, Institute for Building Sciences.

Zimring, C., 1986, *Final Report: Post-Occupancy Evaluation of the School of Architecture at Florida A&M*, Atlanta, Georgia Institute of Technology.

CHAPTER 4

POST-OCCUPANCY EVALUATION AS A TOOL FOR THE PREPARATION OF ARCHITECTURAL COMPETITIONS

Peter Jockusch

Department of Architecture
University of Kassel
West Germany

INTRODUCTION

In West Germany, the homeland of architectural competitions, all major public building projects are subject to competitions in order to give equal opportunity to all architects, especially to not-yet-well-known architects. Furthermore, in big town planning and urban redevelopment projects, it is mandatory to involve the general public in the preparation and critique of planning proposals.

In this chapter I am presenting a case study in which the involvement of the general public and the preparation of a competition were combined, and POE was used as a tool to investigate present shortcomings as well as future objectives. I will draw some conclusions about the appropriateness of POE in the preparation of architectural competitions.

The case study deals with an inner-city location in the City of Kassel, West Germany, called the Königsplatz (Kings Plaza). It is a round plaza of 130 meters diameter and it is situated in the axis of Königsstrasse, the main shopping and pedestrian street of the city. This became the subject of an urban redevelopment competition for architects and town planners. In order to involve the general public in the preparation of the competition brief, a public contest was arranged in which people could express what they disliked, and what they would like to have changed in the Königsplatz.

Our research group specializes in programming and in the preparation of architectural competitions. In the case of the Königsplatz competition, we applied elements of POE, among other investigation techniques, just like we had done previously in many other West German architectural competitions and feasibility studies for institutions, universities, and public social service organizations.

Typically, we start the process by asking the users of existing facilities about the shortcomings and problem areas of these facilities. At the same time, we conduct a survey of the existing building stock and its potential for reusability. Users' individual complaints are contrasted with the quasi objective data from the building stock survey, and they are complemented by an institutional diagnosis both in terms of organizational patterns and operational criteria.

Image studies done directly by interview or indirectly by semantic differential scaling methods are also carried out. We look into the history, development, and present state of the area around the institution or site which we investigate. We want to detect the deficiencies and failures of the present situation, and we ask people on the street about their criticism, their likes and dislikes, as well as their opinions about the future. To stimulate the latter, we may show to people those design options that we develop ourselves or distill from discussion with experts and the actual users concerned.

We assessed the results of the professional competition, and they were subsequently published after the jury concluded its work. The general public responded to the results of the professionals' competition, carried on hot debates in front of the design competition entries, and inquired how their special requirements were met by the competing architects. The winning scheme is well known to the general public and at present, first phases of the realization are starting on the site.

The activities outlined above are in fact the inception of a longer social process of participatory planning, decision making, and realization of large urban redevelopment schemes.

In the case of the Königsplatz contest, we used the local newspaper to stimulate peoples' pride and awareness about their city, but also, to focus the public debate on criticism of the status of the inner city. We then published in the local newspaper the results of a basic data analysis, which we did ourselves to show alternative future scenarios in an argumentative but not promotional way. Then we conducted a public contest in which every citizen was able to participate, either by ticking "Yes" or "No" boxes in a questionnaire, by verbal comments, or by making drawings.

These attempts were based on the theoretical axiom that lay persons cannot produce, create, and conceive of alternative solutions for the design and use of their physical surroundings, but that they are quite capable of deciding and opting for or against those design alternatives which are explained to them.

The public contest results were evaluated and assessed by a joint group of people from the municipal town planning department, a journalist, and representatives of local store owners. Then we published the jury's results in the local paper and invited the public to see the exhibition of contest results. One evening we discussed in a public forum the issues and alternatives, and the public commented on the process as such.

Parallel to this process we had been commissioned to prepare the program for a professional competition for town planners, architects, and landscape designers for exactly the same area, the Königsplatz. We managed to have the results of the public contest for citizens to be included in the professional competition.

RESULTS RELEVANT TO POE IN ARCHITECTURAL COMPETITIONS

As stated above, we found that lay persons cannot produce, create, and conceive alternative solutions for the design and use of their physical surroundings. As a result of our work another axiom is that the method of presenting alternatives to citizens must be imaginative in such a way that people can recognize elements of their everyday experience, something they would not do in the usual, abstract architectural drawings.

Normally lay persons have no particular understanding of planners' and architects' tools. However, they can express themselves pretty well by using verbal and graphic codes which we analyzed in some detail. Communication techniques are predominantly words contained in our printed materials. Also, symbolic representation of certain facts can be found. Furthermore, on typical plans design elements drawn as elevations are turned 90 degrees, and mostly out of scale, so that important things are drawn larger, thereby contributing to confusion.

Another type of representation of the physical environment is called mental mapping. Building users who are given a graphic code frequently copy elements of graphic representation and use them for the expression of their own concerns and priorities, thus pretending at professional presentation.

Concerning the social process that POE stimulates, we had the pleasure of seeing that one of the prize-winning entries of the contest was produced by a class of secondary school children. The pupils used our basic materials to run a special workshop for the production of their entry. Now they are even more excited after winning the prize. By going more into details of urban ecology, traffic planning, architectural history, and preservation of monuments, etc., they use many of our methods and keep asking us about available unpublished data and materials.

Regarding POE as means for stimulating the social process, the entire contest provided a prime opportunity for those political powers which were already at odds about the project issues and objectives. The local transport authority had conflicts with the town planners; the store owners around the square fought with other store owner groups of an adjacent area, or with the municipality. Investors would use the whole process to explore new investment opportunities. Those who had already invested in the Königsplatz area were in fear of losing their market share, and so on. Furthermore, our preparatory activities unveiled hidden powers such as the hot dog vendor. When we published the general idea of removing the ugly hot dog stand from the center of the plaza, influential people behind the scene insisted on selling hot dogs in the center of the plaza because long-term contracts should not be violated.

Despite many ideas that resulted from the contest, there is a dilemma in that not many uses can be identified which can take place in open air, particularly since investment for beautification purposes only was not permitted. On the other hand, wanting to revitalize a public plaza means the need to define new uses for it.

Some people who thought they were going to lose if the plaza were improved complained: According to the vote of the majority of people a public market should be relocated to the Königsplatz from another site, and consequently, the people who now run the market in that location there were in fear of losing it.

The contest also reminded the general public of the fact that many "introvert" design solutions had been realized lately, including covered streets, arcades through inner-city courtyards, etc. Thus, public streets and plazas became more and more empty so that their ugliness and under use resulted from the complete loss of their public functions.

Lay persons, if asked about their concerns mostly argued about minor details, and overall we found a lack of systematic design solutions. Therefore, one can assume that many people conceive of their city as a good one if only their prime concerns are being met. Also, they sometimes want to get rid of things which concern them, but they cannot suggest as to where they should be put instead.

This occurred when the track of the streetcars and the buses were discussed. Issues changed as the debate went on. While we published our findings before the public contest, some people said public transport (the streetcars) had to be taken out of the plaza. When competing investors said "Oh well, then let them stop in front of our new shopping precinct," the store owners on the Königsplatz said "No, the streetcars must remain." We introduced statistical data and made people aware of the 27,000 people

(continued on page 55)

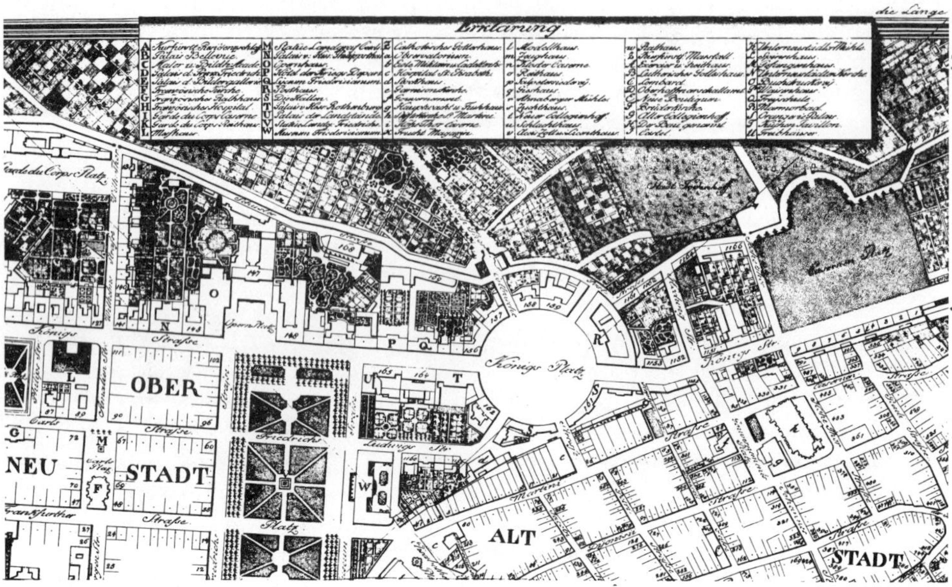

Figure 1. Map of central Kassel, 1803, by H. J. Martens.

Figure 2. The Königsplatz, Kassel.
Sketch taken after a watercolor painting by Ph. L. Feidel, 1820.

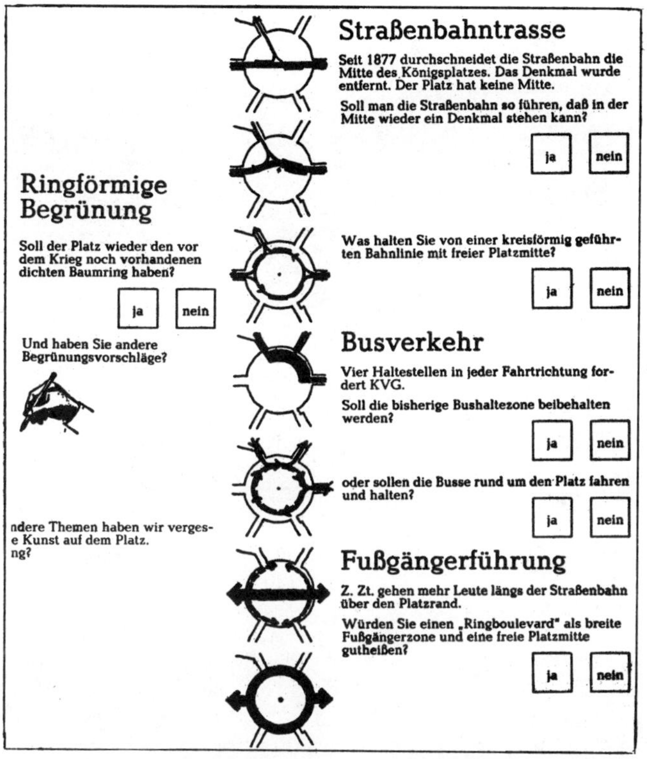

Figure 3. Extract from the contest document, showing pictographic and verbal statements to which people could react by "yes/no" statements, or by writing and drawing their own comments or concepts.

Nr. 26 **Stadt Kassel** Freitag, 31.

Königsplatz
Rathausumfeld

Innenstadt-Preisausschreiben (I)

Das Thema 1 ist der Königsplatz

Kassel. Heute geht's um den Königsplatz. Hier ...

Königsplatz
Rathausumfeld

Innenstadt-Preisausschreiben (II)

Kassels „Kö" macht Probleme

Kassel (red). Nichts kann der Oberen Königsstraße den Rang als Haupteinkaufsstraße Kassels ablaufen. Die rund 200 Jahre alte steinerne Achse der Ober... mancherlei Hinsicht zu wünschen übrig. Wie schon auf dem Königsplatz ...

Stadt Kassel

Königsplatz
Rathausumfeld

Innenstadt-Preisausschreiben III

Rund um unser altes Rathaus

Kassel (red). Kurz vor dem Endpunkt der Oberen Königsstraße, dem Brüder-Grimm-Platz, ist das Kasseler Rathaus d a s markante Bauwerk. In der Umgebung vieler Nachkriegsbauten wirkt der mächtige Altbau fast fremdartig. In Zeiten der bürgernahen Verwaltung macht das von Bürgerstolz geprägte Gebäude einen eher unnahbaren Eindruck, der durch die wenig einladende Gestaltung des Vorplatzes noch verstärkt wird.

Fünf Türen führen hinein

Das Kasseler Rathaus besteht aus insgesamt drei Einzelbauten,
- dem Altbau aus dem Jahre 1909,
- dem Anbau im Hof aus den Fünfziger Jahren,
- dem Neubau an der Oberen Karlsstraße aus den Siebziger ...

... ses an der Oberen Königsstraße, sind allerdings mühevoll, die Zugangstüren wehrhaft und winzig. Die gewaltige Treppenanlage ist für Behinderte nicht überwindbar.

Will man die 30 Treppenstufen der Haupttreppe umgehen, ist man auf die vier Nebeneingänge angewiesen. An der Fünfensterstraße befinden sich zwei ...

Die Umgebung

Von Westen her ist bislang die wenig attraktive Fußgängerunterführung Hauptverbindung zum Rathaus. Überwege schülen bessere Verbindungen zum gegenüberliegenden Block „Dache/Hotel Hessenland", der Teil des Sanierungsgebietes ist.

Mit dem Rathaus-Neubau wurde die Obere Karlsstraße in ihrer Achse verschoben. Ihre alte Achse läuft durch den Innenhof hinter dem Rathaus und ist an der Wilhelmsstraße noch ...

Figure 4. Extracts from three one-page articles from the local newspaper, which we prepared to inform the public about the key concerns and issues and to attract public interest to participating in the contest.

Figure 5. The city planning department head, Christiane Thalgott, among the winners of the citizen contest.

Kassel (W.R.). Drei Stunden rauchten die Köpfe, sichtete und debattierte die achtköpfige Jury die vielen Einsendungen, dann war es soweit, standen die Gewinner des Bürgerpreisausschreibens zur Neugestaltung der Kasseler Innenstadt fest. 389 Bürgerinnen und Bürger hatten sich Gedanken zu den Themenbereichen Königsplatz, Rat- hau Obe mei Sog

Bürgerpreisausschreiben zur Neugestaltung der Innenstadt

Jury von Ideenreichtum und Niveau beeindruckt

Einstimmig schloß sich das Gremium der positiven Einschätzung der Stadtbaurätin Christiane Thalgott an, lobte das Niveau und den Ideenreichtum und beschloß, die Möglichkeit zu nutzen, durch eine andere als vorgesehene Preisverteilung dem vielfältigen Angebot gerechter zu werden.

Statt, wie vorgesehen, fünf gestaffelte Geldreise und fünf „Ankäufe" im Gesamtwert von 10 000 Mark (ausgesetzt vom Magistrat und der HNA) zu vergeben, einigte sich Jury darauf, in zwei Preisgruppen jeweils drei Arbeiten mit je 2000 Mark und drei weitere mit je 1000 Mark zu prämiieren, die restlichen 1000 Mark auf fünf „Ankäufe" zu verteilen.

Die Preisträger und die fünf Gewinner der Verlosung (fünf dreitägige Busreisen nach Berlin für jeweils zwei Personen) können ihre „Gewinne" beim Bürgerforum am Freitag, 23. Oktober, 18 Uhr, im Bürgersaal des Rathauses aus der Hand von Stadtbaurätin Christiane Thalgott entgegennehmen.

Und noch in einem anderen Punkt wich die Jury vom Geplanten ab. Nicht einer der ersten Preisträger wird in die Jury des geplanten Profi-Wettbewerb eingeladen, sondern ein(e) Sprecher(in) der Geo-Arbeitsgemeinschaft der Jacob-Grimm-Schule, deren eindrucksvoller Arbeit auch ein zweiter Preis zuerkannt wurde. Er/sie soll dort gewissermaßen einen Kontrapunkt zu den berufsmäßigen Profis bilden.

Der Jury gehörten neben der Stadtbaurätin Dipl.-Ing. Christiane Thalgott an: Dipl.-Ing. Klaus Humpert, Architekturlehrer, ehemals Stadtbaurat in Freiburg, Dr.-Ing. Werner Durth, Professor für Baugeschichte, der gegenwärtig eine Stadtbeschreibung über Kassel erarbeitet, Gerhard Viereck, Ortsvorsteher Kassel-Mitte, Karl Schumann, Vorsitzender des Einzelhandelsverbandes, Thorsten Evers, Sprecher der Interessengemeinschaft Obere Königsstraße, Guntram Klocke, Sprecher der Interessengemeinschaft Königsplatz, und Wolfgang Rossbach, stellvertretender Chefredakteur der HNA.

Die Fragen und Tendenzen der Bürgervoten

Vorgeprüft und gesichtet wor- Friedrichsplatz „anbinden" und bündelt aufnehmen sollen.

Figure 6. Extracts from the article published in the daily local newspaper, informing the public about the results of the contest, summarizing tendencies of citizens' critical remarks, and naming the winners of the contest.

Figure 7. The first-prize-winning proposal from the professionals' competition contains many items called for by citizens.

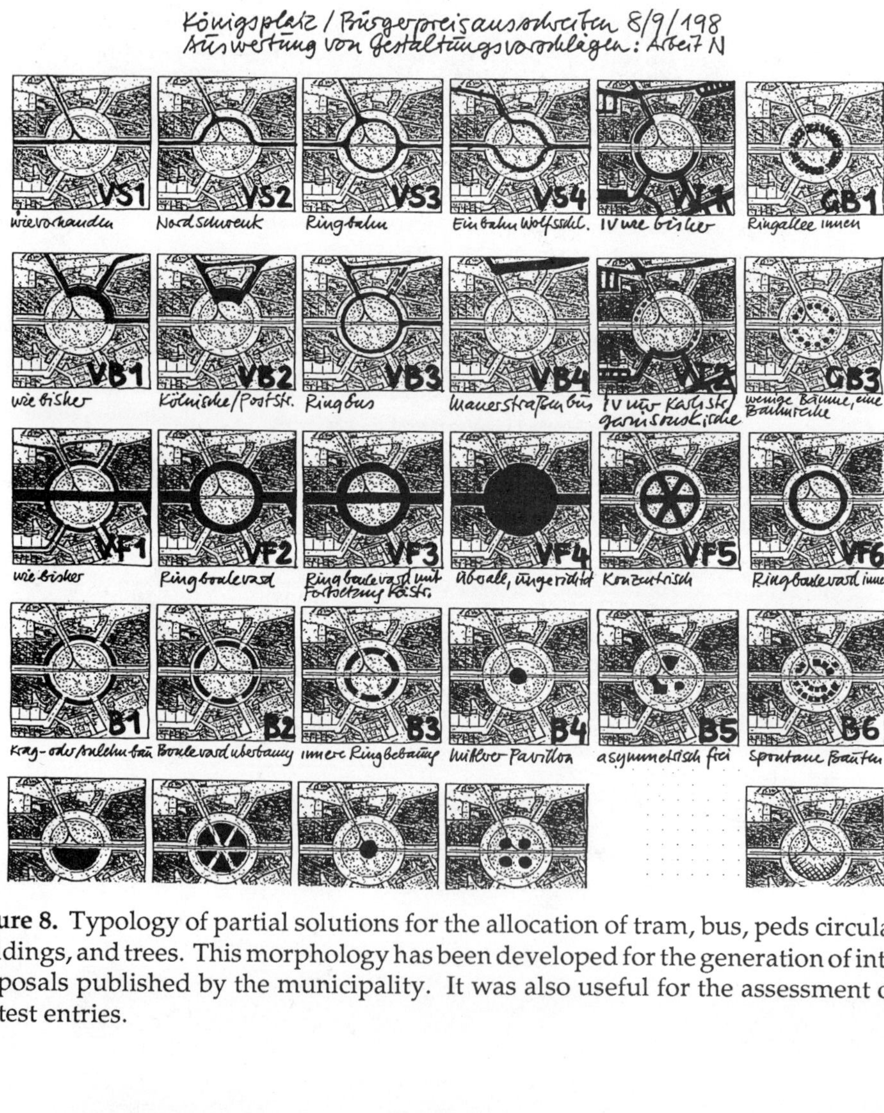

Figure 8. Typology of partial solutions for the allocation of tram, bus, peds circulation, buildings, and trees. This morphology has been developed for the generation of integral proposals published by the municipality. It was also useful for the assessment of the contest entries.

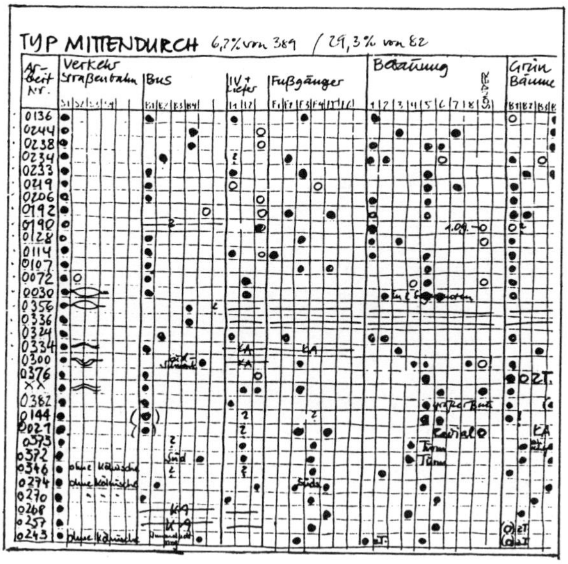

Figure 9. Contest evaluation chart, allocating participant's solution to elements of the typology.

Figure 10. Extremely restricted code of presentation only by writing future functions on places where they ought to be, plus some linear drawing of the rail track and of the boundary of a central coffee-house.

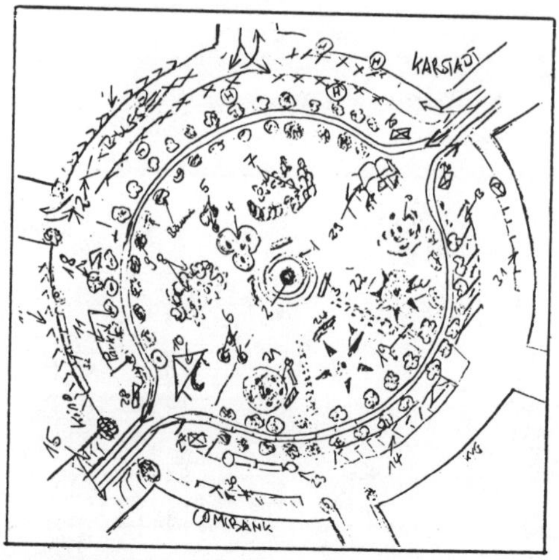

Figure 11. A happy mix of perspective views, graphical symbols, decorative patterns, and index figures related to longish verbal comments (not shown here). Items shown are out of scale and do not really territorialize the central space; they represent mostly activities and their necessary settings.

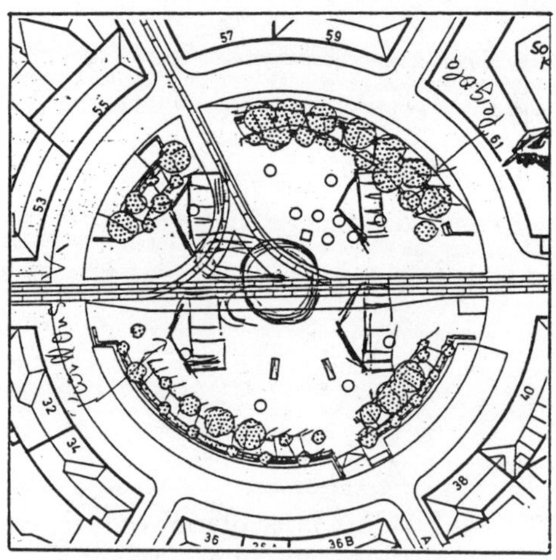

Figure 12. Four pavilions and a central fountain, drawn as turned-down elevations and totally out of scale – a very restricted and naïve example.

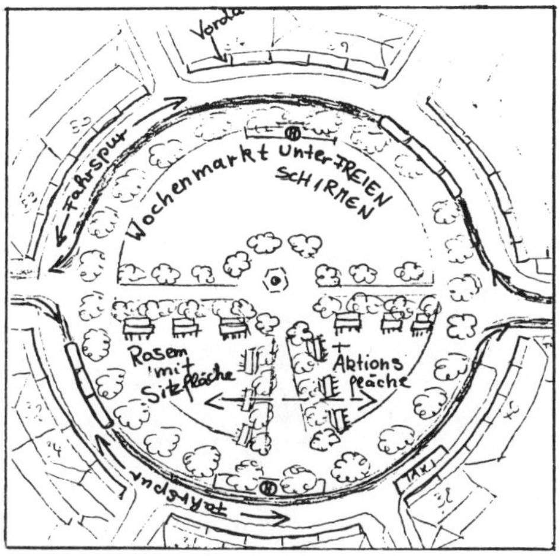

Figure 13. More advanced drawing with verbal and symbolic codes for tram, bus lane, and market area, but graphical symbols for trees in right scale and iconic presentation of totally oversized benches.

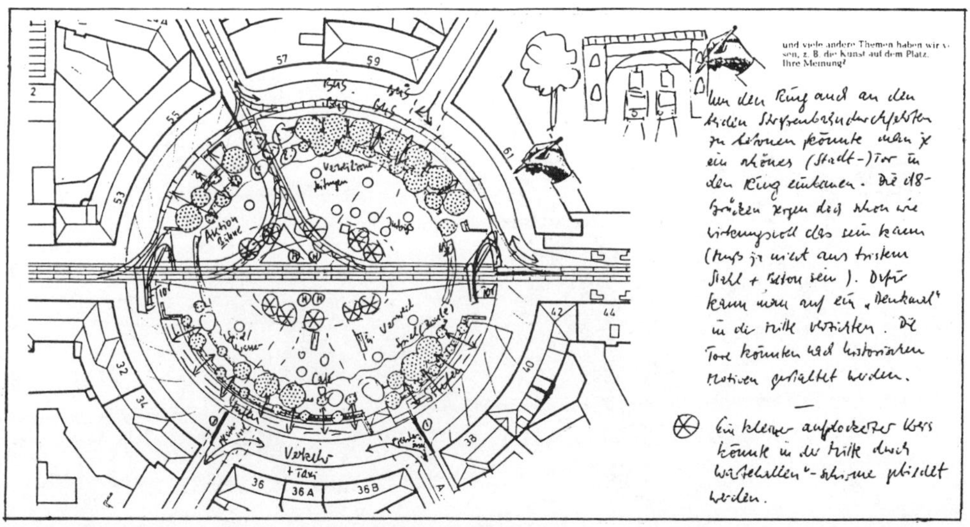

Figure 14. Mixing words, arrows, graphical symbols with iconic perspective elements; the latter being used for the visually most important non-functional elements of metaphoric load: two gates, one of which is in addition being shown in an elevation.

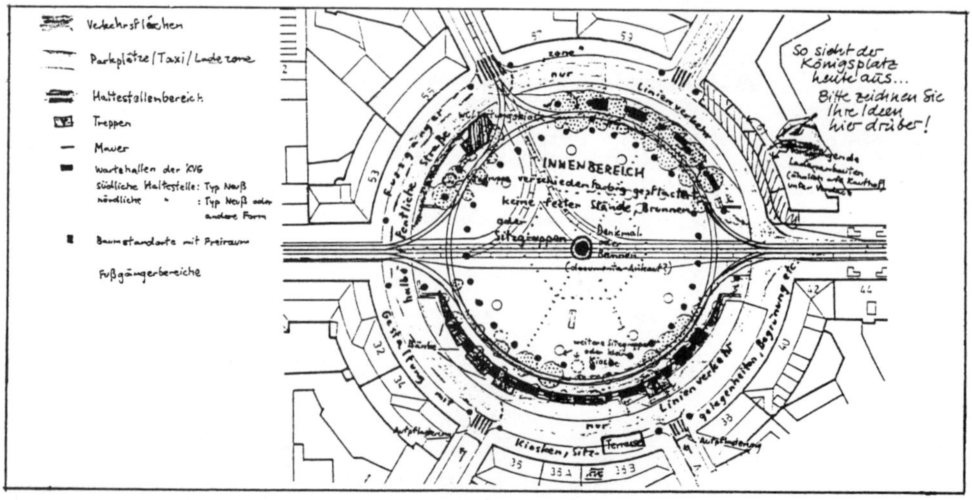

Figure 15. Lays learn to adopt planners' language; they are aware that their drawings visualize only in a symbolic form of presentation, partly out of scale. Therefore, they explain their inventory of symbols.

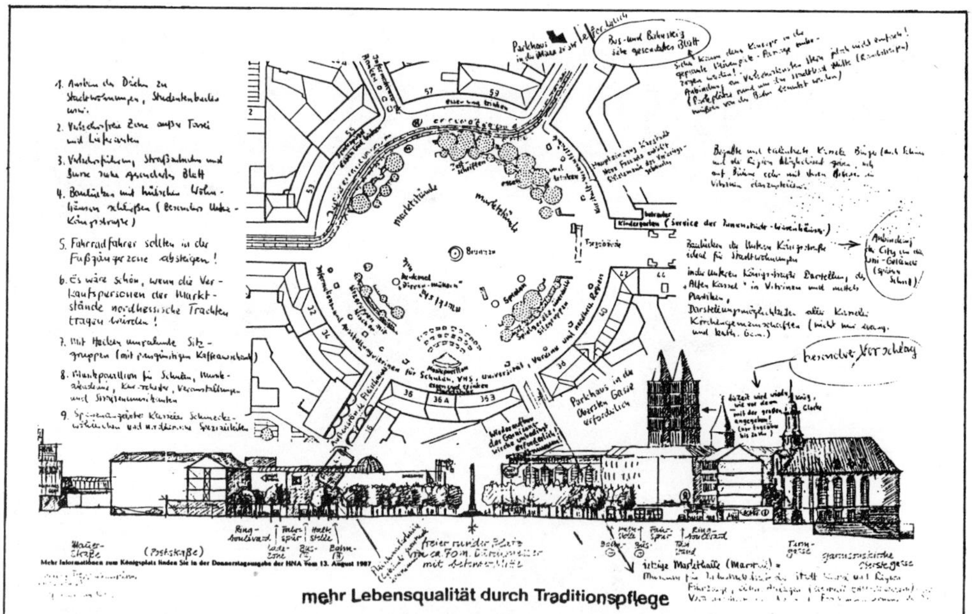

Figure 16. A very "narrative" example which allocates nicely verbal criticism and proposals for improvements to those areas and items where they apply. Again, the most important proposals are drawn (music pavilion and seat rows, as well as tram rails).

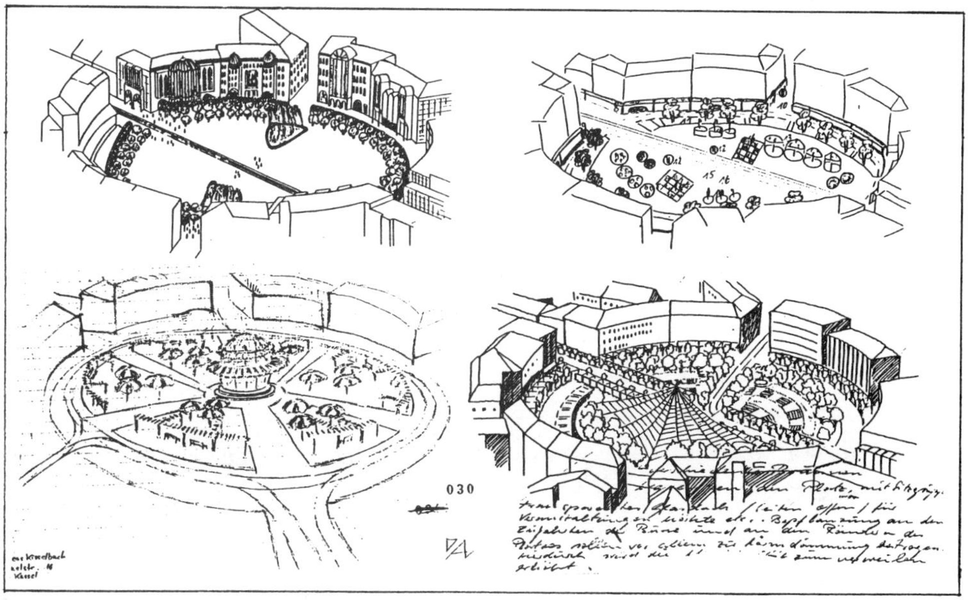

Figure 17. Iconic three-dimensional drawings handed in by participants who have drawn on tracing paper over photographs which we published in our introductory newspaper articles.

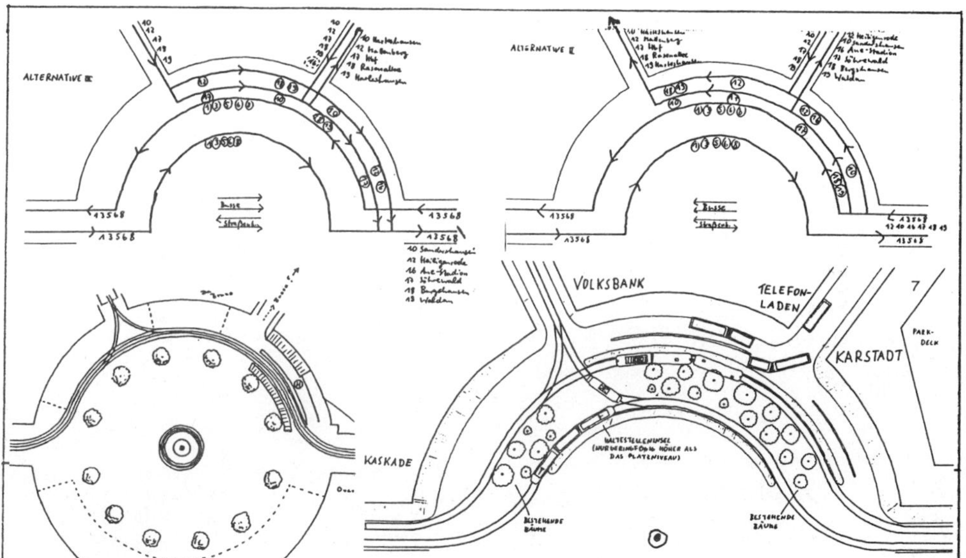

Figure 18. The quasi-professionals among the lay persons, especially in the domain of traffic routing, gave precise assessments of obvious and hidden flaws in the present system of time tables, frequency of connecting lines, waiting queues, exchange rates, etc., and made very correct and outstandingly detailed proposals as if the whole complex planning problem could be solved by reducing it to public transport problems.

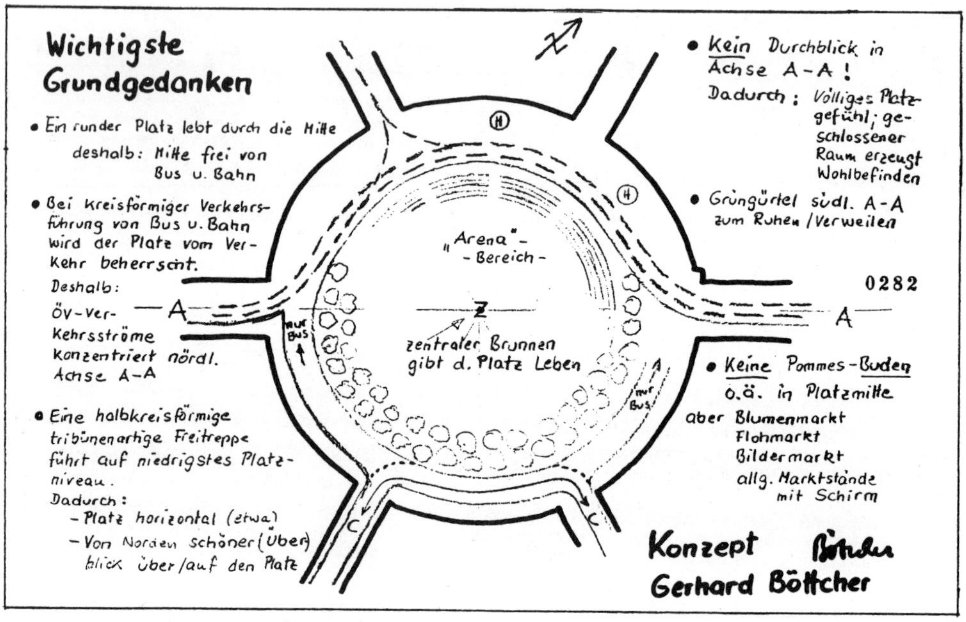

Figure 19. Someone telling first what he does not want and why, then showing what he proposes as a replacement. The proposal only contains very general lines of development, but its graphical and verbal code is clear.

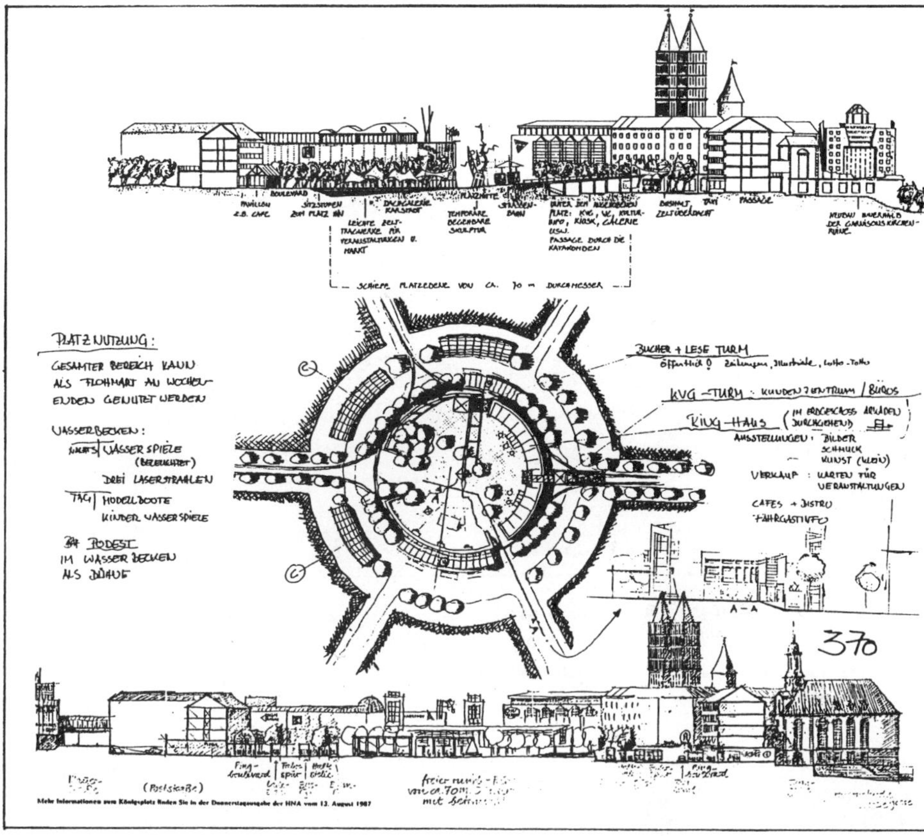

Figure 20. A very imaginative, witty drawing of a proposal that inverts the slope of the place leaving the tram in the center and creating space under the place at the lower part for servicing functions, and another lot of professionally drawn proposals which adopt our published code of iconic visualization.

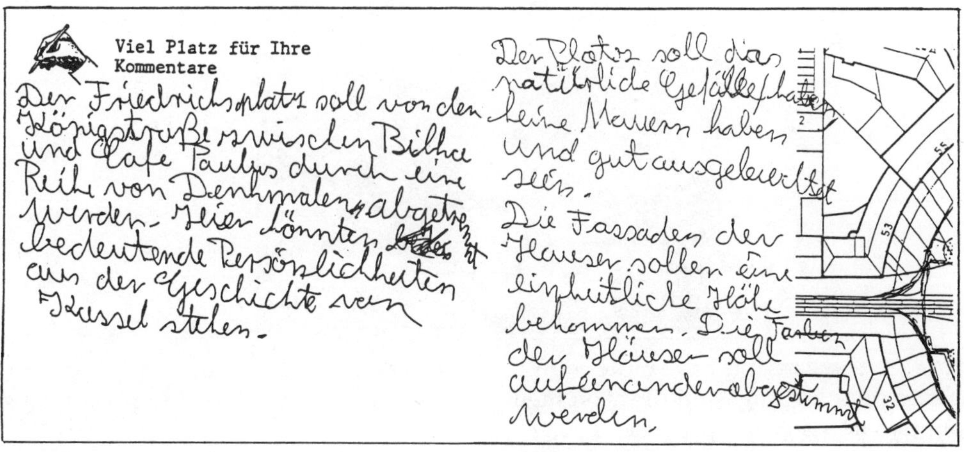

Figure 21. Proposal, from a ten-year-old schoolboy, for placing a series of statues of personalities who have had important impacts on the history of Kassel, and calling for beautification of facades around the place.

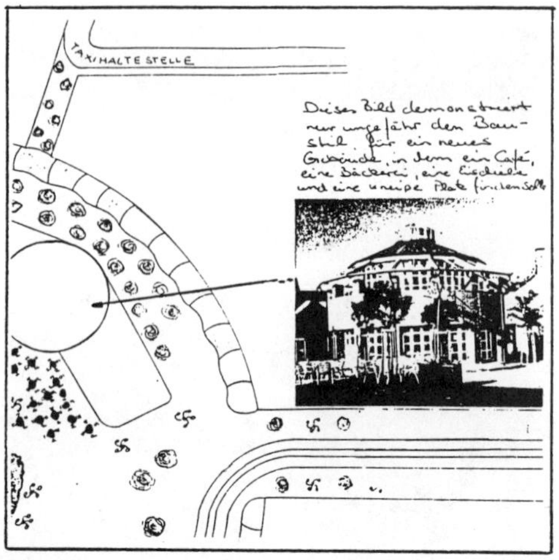

Figure 22. Overcoming drawing inability: taking well-known building examples as photographs to show details of a concept. Written text in drawing: this picture demonstrates only vaguely the building style for a new building intended to house a café, a bakery, an ice cream shop, and a pub.

Figure 23. Some people have more comprehensive ideas than they can draw; therefore, they use press cuttings and draw on them. This is an example: the creator complains about rain and dirt and proposes a solar dome and arcades covering adjacent streets.

Figure 24. The contest results stimulated prize-winning pupils to investigate more comprehensively the worries of people about their urban environment.

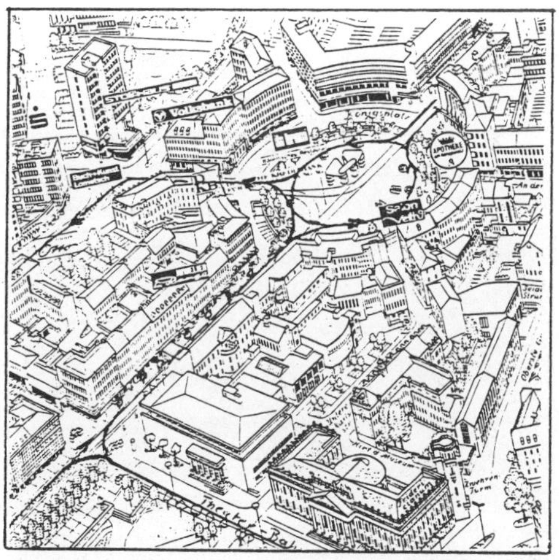

Figure 25. Some people use further visual materials for a better expression of their proposals, especially if they want to put the tram somewhere else.

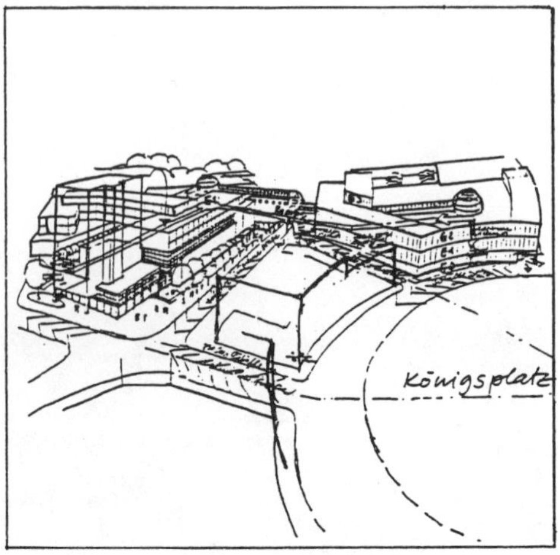

Figure 26. This is the 80-million-DM redevelopment project which came up while improvements for the Königsplatz were being discussed. Managers proposed to translocate the bus traffic from Königsplatz to their project. The project also shows that public urban investment stimulates private investment of ten times the value of the former.

Figure 27. Many participants want to get rid of the central sausage stand, which is thought to be visually too dominant.

Viel Platz für Ihre Kommentare

Bettler und Penner sind unerwünscht!
"Es ist kein schönes Stadtbild!

Bitte mehr Abfall-behälter auf allen Gebieten!!

Andern Straßenbelag

Immer wieder: schafft Parkplätze!!

ein sauberes wo man hin gehen kann!!

Platznutzung

Figure 28. Some proposals are concerned with one important worry: "clean public toilets," "keep beggars and vagrants away because it's unpleasant to look at them," "abolish bikes in pedestrian areas," etc. And the perennial favorite, "create parking space!" As if life would be much easier if just one worry could be taken care of!

Viel Platz für Ihre Kommentare

Verschandelt mir meine, unsere Stadt nicht so.
Sie ist doch so gerade gut!

„Kassel ist schön"
Keine großen Veränderungen!
Die Stadt muß sparen!

Figure 29. An outcry of protest against any change (such as our concept proposal for the Königsplatz): "Don't damage my, our city this way – it is just fine as it is." Or another: "Kassel is beautiful. No big changes! The city must cut down expenses."

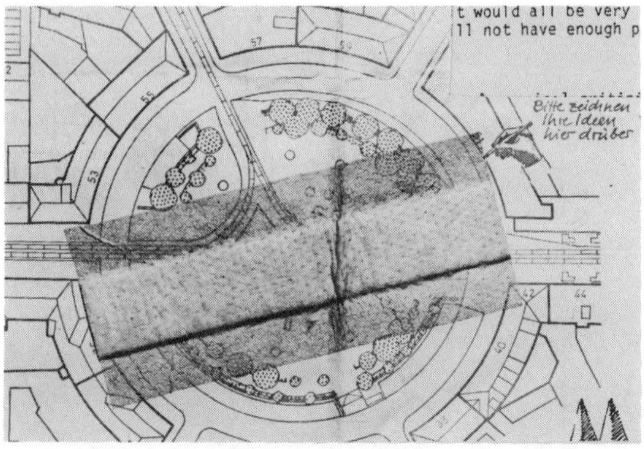

Figure 30. A cynical critic with an extremely pessimistic and depressed mood, complaining about planning professionals, and then making a metaphoric picture, playing with the double meaning of the German word "Pflaster," a homonym meaning both "patch" and "paving." "Patch" is his metaphor for a necessary healing process of a derelict city, "paving" represents for lack of naturality, ecological indifference, etc. (The stamp at above right means "reanimation trial: positive/negative.")

Figure 31. Symbolic criticism of the city's being dominated by business: at the center, the coffeehouse and visitor platform in the form of a huge shopping bag.

Figure 32. "Dreadful, sterile! Send your staff to Hamburg, Berlin, Barcelona – or best, send them home without pay!" reads one angry comment on our published proposal. Symbols at the center of the place: the writer, as an outlet for anger against planning authorities, shows Lord Mayor Hans Eichel, as Mickey Mouse, on top of the central monument.

who get on and off the streetcars and busses on the plaza everyday, and further, that these people are all potential shoppers. Finally, solutions were sought that created a compromise between the need for public transport and the beauty of the historic, round plaza, the Königsplatz.

Defining design issues in a public, participatory process also means that there is an opportunity for changes in issues and also, for assuming more power. Some people argued for abolishing all commercial uses and for making the plaza a non-commercial, restored historic plaza. Others argued that the established commercial uses should be moved and that only flea market type commercial uses should be permitted. Others again said that there is not enough incentive to go downtown at night, e.g. for the performances of non-established artists, or, for ad hoc festivals and fun, etc.

Some said that there was no need for action at all, blaming us (the project facilitators) for raising the wrong issues by arguing that the problems of the inner city cannot be solved by town planning and architectural means alone because they are primarily social ones. Other issues were argued to be even more crucial, such as pollution and environmental hazards, or that in residential districts of Kassel, problems were much more critical than in the inner city. At the Königsplatz, some argued store owners would stand to make more profit by public investments in civic infrastructure, and there would not be much benefit for the citizens at large.

We also had a strange feeling based on many of the statements raised by contest participants, such as the general mistrust in the whole approach. For example, they implied that the municipality would never be able to raise the money needed for the development, or, that it would be very nice to change the city but that our political system will not have enough power to implement those changes.

Much pride was evident in the statements made by many people with arguments like "hands off our beautiful city," or "the professor (the author) seems not to be from this city, otherwise he would not have made such proposals," etc. We also see a certain tendency towards reducing conflict in arguments like "our city is not all that bad off," "we can survive with what we have."

We would like to comment on the role of the local press as a medium in the post-occupancy evaluation process. The press is the organ of the local population; it is a tool for propaganda by the municipality; and it is also an organ of the political influence of journalists and editors. Fortunately, after hard debates, the editors refrained from too much propaganda in support of the town planners' issues and the newspapers became more a sounding board of the arguments of the general public.

Strangely enough, some design issues were agreed on by very different interest groups for different reasons. The straight boulevard and streetcar tract brought life into the center of the plaza. Our proposal for a boulevard with a circular streetcar track was applauded by most people who said that it keeps the plaza open for different activities and that it brings life to shops around the plaza. Store owners objected to more shops, but they eventually agreed once they recognized the opportunity to invest there themselves.

We would like to conclude this discussion with some symbolic issues. Before streetcars were introduced in 1877, the round plaza had a monument in the center. We raised the argument that the center of the plaza should by all means be able to accommodate something in its center again. Where a solution provokes non-functional issues, people think about symbolic values, and they discussed in detail what symbols there should be. Firstly, people remember the tradition and want to re-establish what has been there once. Secondly, people want to use the opportunity to commemorate those important people or ideas for which the city has never had a monument. Thirdly, in a more cynical approach, people tried to find symbols *ex negativo* by proposing a huge shopping bag as an ugly symbol for commercialization of the city. Fourthly, people became more aggressive by proposing that the Lord Mayor is accused of having organized the entire competition as part of a selfish power-play and, therefore, that his sculpture should stand as a monument in the center of the plaza.

ADEQUATE APPROPRIATIONS OF POE METHOD AND APPROACH

The process which I described above is certainly not used in all competitions in Germany, but in more complex cases where promoters feel obliged to involve the general public our approach is generally being used.

We use POE as starter for any programming efforts (not only for competition programs) because we believe in the necessity to start careful definitions of future projects by identifying the concerns that presently exist and by analyzing what are the causes of those concerns. This implies an empirical approach, oriented towards not only the issues, but also to find an approach to deal with the people who are concerned. We apply social science techniques, mixed with investigation techniques used for facilities assessment, and we dig deeper if and when we notice great contradictions in the data.

Investigations of peoples' opinions about their existing environments, their needs and opportunities for change are by no means the only ways to develop architectural competition programs. There have been other rational and more precise ways to shape the future of the built environment. That implies institutional diagnostics, facilities evaluation, precise calculation of space needs, as well as a rigorous definition of future environmental performance criteria.

POE is by no means the only tool used in our work, but it proves to be an indispensable one in certain phases of a project. The aim, of course, is to have both —

the rational and the scientific approach, as well as the emotional and affective aspects as part of every project.

Rational approaches must be agreed to by the people concerned. On the other hand, peoples' emotions must be confronted with hard facts. Viable solutions depend largely on a fair balance between both sides – experts and lay persons, i.e. the people concerned and the people who are in power.

The evaluation aspect in POE may be too analytical – we must go beyond this in order to contribute to problem solving. I propose participatory programming in order to define what might be called pre-programming evaluation, or *Bedarfsanalyse*, which means "needs analysis."

In my country we do not have POE as a generally established application tool for planners and architects. However, there are well-known applications, i.e. the so-called preparatory investigation, assessments of basic planning information, as well as pre-design studies. Within these tasks, POE is only one part of the theories and methods that are used. These investigations have also quantitative empirical results, but this is by no means the most important aspect. In getting people involved in a participatory process, in a public debate, it is important to keep the process of planning open and argumentative. Furthermore, POE should be supportive of making expert decision makers and their actions legitimate, combining affective statements with hard facts, giving a chance to non-quantifiable issues and arguments and creating a culture of "being" rather than "having."

REFERENCES

Bürger-Preisausschreiben Königsplatz-Rathausumfeld, 1987, 3 introductory articles in the local daily paper, and a separate inlay folder with the contest brief and questionnaire.

Competition results, June 1988, published in *Wettbewerbe Aktuell*, a special magazine for architectural competitions, Munich, vol. 6/88, pages 369-382.

Jockusch, P., Hegger, M., and Ettinger-Brinckmann, B., 1984, German Lesson 1: Just how successful in reality are the organization and results of the much-praised German competition system?, *The Architect's Journal*, London, pp. 39-46.

Städtebaulicher Ideenwettbewerb Königsplatz-Rathausumfeld 1987-88, Kassel – Wettbewerbsunterlagen, October 1987.

CHAPTER 5

PRE-OCCUPANCY EVALUATION IN FACILITIES MANAGEMENT

Laurence B. Molloy

Molloy Corporation
New York City

INTRODUCTION

Molloy Corporation, an environmental and facilities consulting firm in New York City, advises corporations and professionals involved with a move to new quarters. Molloy evaluates locations and spaces, investigates environmental conditions, advises on architectural and real estate issues, and often manages construction for its clients. In short, Molloy Corporation manages a move from inception to installation.

Moving a business in New York involves hundreds of professionals: facility consultants, space planners, architects, real estate brokers, engineers, interior designers, contractors, building owners, environmental specialists, city inspectors, movers, electricians, plumbers, carpenters, painters, etc. Unfortunately, most clients, particularly the Chief Executive Officer, hereinafter "CEO," have no concept of the complexities in seeking, selecting, constructing, and moving into new quarters. In fact, many CEOs think that they would be able to solve the space problem in a day if only there were a free afternoon to run around the city on an organized tour.

It is therefore, important to educate clients about the building and moving process and to sort out who does what in the cataclysm to come. Several rules apply:

- Moving a corporation takes time! Although time can be saved by completing tasks simultaneously, the search, design, and construction phases are sequential. Corporate executives should plan for a moving process that normally takes two years. Always anticipate extra delays with telephones and furniture, which should be ordered during the design/draw phase. A corporation without chairs and telephones cannot function.

Approximate Time Allocations
For a Corporate Move

• Define program needs	2 months
• Define space needs	1 month
• Develop budget parameters	2 weeks
• Search for space	4-8 months
• Select target spaces	1 month

• Negotiate bids for spaces	3 months
• Design, working drawings, permits	8 months
• Construct and finish space	9 months
• Move	2 weeks

• Moving a corporation is expensive! Preliminary discussions often consider only the construction costs, usually no more than 40% of the total cost for a corporate move. Professionals often minimize the overall costs at the beginning in order not to alarm the client. A corporation, particularly the CEO, is better served by a frank discussion of costs at the very beginning of planning.

Emphasize at the beginning that bargains do not exist! Costs are controlled by careful planning and good management. Remind the client that the building industry is standardized. Consequently anything unusual, original, or custom made is expensive. It is always ten times less expensive to plan care-fully at the outset, than to correct later; the $200 spent to redraw details today can save $2,000 after construction begins.

Remind the client that real estate brokers, despite the limousines and lunches, most often work for the seller, not the buyer. Also remember that the broker who first brings in a space may be entitled to a commission. Brokers, therefore, often encourage their clients to undertake a dizzy survey of spaces in order to lock in their commission and their hegemony on buildings and neighborhoods.

Photograph everything! Keep files on every space discussed with your client. Within weeks of beginning a search, the client will have forgotten brokers, addresses and spaces. Buildings also have names as well as addresses. Often one broker will schedule a tour of spaces that the client has already seen with another broker.

Determine who makes decisions in the corporation; must every detail be approved by the CEO, the board, or, worse, a committee? Strongly advise your client to select someone from senior management with authority to approve designs and make decisions in the field.

One visit to a space is not enough. Ordinarily, the first tour is with the client but without scale drawings. A second opinion from another person in the consulting firm is essential. Armed with drawings and the photographs taken during the first tour, another person should tour the facility again with a camera and a tape measure.

THE LOCATION

The three most important criteria for judging real estate are location, location, and location. Clients are often unfamiliar with the characteristics of commercial neighborhoods. A list of essential needs such as transportation access, parking, safety after hours, service amenities, ambiance, and proximity to customers should be prepared before the client is toured through spaces. Be sure to ask the CEO for personal preferences. The facilities consultant should also educate the client about major real estate trends, such as which areas are improving, which are deteriorating, which are pioneer neighborhoods offering lower rents on long term leases anticipating substantial gentrification within five to ten years, and so on.

Locations also determine which important amenities are available: post offices, restaurants, food services, parking, public transit, taxi, and limousine service. Remember also to inquire about telephone service within neighborhoods. Access lines and vanity telephone numbers are often restricted in emerging commercial areas. It is also

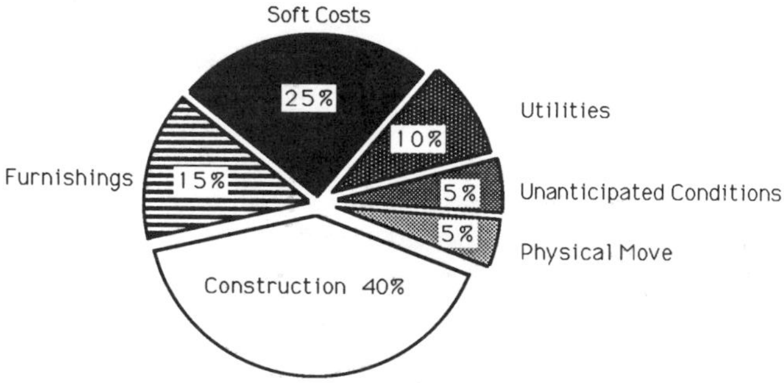

Utilities - (electrical, telephone, computer).
Soft Costs - (fees, expenses, architects, engineers, contractors, designers).

Figure 1

a good idea to survey prospective neighborhoods during weekends and at night. Public transportation and safety can vary greatly during off hours.

THE BUILDING

Winston Churchill said that we shape our buildings and thereafter they shape us. Every building both serves and deserves its tenants. The facility consultant should evaluate these differences for the client. First check the building roster for light manufacturing, non-profit groups, laboratories, rehabilitation clinics and schools. Check also that the building and your space is accessible to disabled people. Also inquire about freight and delivery access (separate guarded entrances are recommended), building security, night and weekend access, rooftop rights, and hours of operation, particularly air conditioning. One tenant in New York completed a $1 million installation in a curtain wall building only to discover that the management provided no cooling, heat or air circulation during nights and weekends. Make sure also that the tenant controls the amount and the temperature of the air conditioner. A visit to the basement can also reveal potential mechanical and environmental problems.

Building orientation is important for light and air and also because new or potential construction can suddenly alter views and sunlight. Some neighborhoods prevent or limit new construction. A long term lease on light and views in a historic district may be more secure than a short term rental in bustling midtown.

Quality of the public environment and circulation is important. The building lobby can inspire or depress staff and visitors. The number of elevators can speed or delay internal traffic, particularly if school students use lobby elevators. Make sure that there is a dedicated freight elevator separate from the passenger elevators. Also, torturous, poorly lit, or staggered public hallways are depressing and unsafe. Toilets on another floor frustrate visitors and staff, especially when they must wait for an elevator.

THE SPACE

The most significant impediment to pre-occupancy evaluation may be existing partitioned or occupied space. If the space is occupied, then testing and evaluation are limited. If partition walls exist, the facility consultant should construct a plan showing only column spacing, bearing walls, window, toilet, elevator, stair, and freight and delivery locations. Frequently, clients need to understand a space in relation to their present quarters in order to evaluate space. Overlays showing present quarters adjusted to new spaces are helpful, particularly to the CEO who may use them in presenting to the board of directors.

Evaluating the space itself depends on the client's particular needs. The facility consultant should ascertain beforehand unique needs. Design, advertising, and publishing companies, for example, need lots of natural light, whereas telephone and mail order companies working primarily on computer monitors are less particular. Similarly, the former companies have many messenger deliveries and thus expect messenger control whereas the latter are relatively self contained. In New York, delivery and messenger theft is rampant and must be carefully controlled. Often a separate supervised freight and messenger entrance is essential.

Clients also vary within an industry. Publishers, for example, range from open plan layout to small cubical arrangements. In either case, column spacing is critical and the wrong choice can waste as much as 20% of the available space. Column spacing is also critical according to the need for conference space. In general, the greater the space between columns, the more flexible and less wasteful is layout.

Besides bays, the shape of the space determines relationships and circulation. Doughnut arrangements using interior hallways with double-loaded interior and exterior offices are commonplace but often isolate departments and generate much traffic. Shotgun space with entrances in front and freight in the rear often demands a layered arrangement with executive and publicity offices in front, accounting, distribution, and mail room in the rear.

Selecting space on various floors or even different buildings is a consideration. Contiguous space between floors is often available but stairwells between floors are recommended. Companies should not be at the mercy of capricious elevator service. Space split between noncontiguous floors is not recommended because of traffic and time difficulties unless the company has neat divisions that segregate operations which can operate in different but nearby buildings effectively.

Of course, ceiling heights, general conditions, floor loads, bearing walls, access, and egress are important, depending on the client's and mechanical needs. The top floor is always an advantage (providing the roof is water tight) because the roof can house air conditioning and mechanical equipment which otherwise consumes valuable floor space. A roof also allows skylights and access to outdoor space. The roof, however, is the first line of defense against the elements. Roof leaks deteriorate roof structural members, interior plaster and sheet rock. Soggy insulation does not insulate and can cause odors and toxic growths. Given today's costs for repairs and decorating, a tight roof is absolutely necessary.

Be certain while evaluating the space to test all amenities. Some spaces are shown with fake kitchen and toilets in impossible locations. Turn on every faucet and flush every john. Check also that elevators, freight entrances, alternate lobbies, passageways, stairs, and basements are customarily open. Also interview neighboring tenants. They often have insights about the space, conditions, and the management system. Remember to ask (and later get promises in writing) about window penetration in walls, lot-line glass replacement, preservation of existing skylights, and removal of potential immovables such as penetrating bearing walls and old gas lines. Do not trust the plans and elevations provided by building management. The measurements, column spacing, and square footage should be checked assiduously.

THE ENVIRONMENT

If a deal sounds too good to be true, it probably is. Tighter buildings, curtain walls, cramped quarters, and sloppy management are helping to contaminate spaces in unexpected numbers. Building materials, cleaning solvents, industrial wastes, air, water, energy, gas, pests and sound are all culprits. Molloy Corporation has discovered nearly every conceivable toxic and environmental problem in potential office space.

Asbestos in New York is the tail that wags the dog. The Department of Environmental Protection estimates that more than 85% of the available space has asbestos or asbestos containing materials. New York requires a certified inspection before issuing building permits for renovations, but savvy facility consultants are conducting their own, more extensive search for asbestos before negotiating a lease on new space. If asbestos is discovered, get at least three independent bids on its removal. Prices vary by as much as 300%.

Even asbestos-free spaces are not exempt from problems. In one case, a corporation on a tight construction schedule leaving expensive short term quarters in Midtown New York was delayed months because a demolition company illegally removed asbestos from another floor in the new building. Environmental Protection sealed the building until cleaned. The corporation was forced to negotiate an expensive month-to-month lease arrangement with their old management. The moral is that a good pre-occupancy evaluation includes some assessment of the asbestos conditions in the building as a whole as well as the space itself. Check recent renovations and demand an asbestos assessment from the building management.

In any case, where asbestos exists, expect extra delays and expense. Besides pipe insulation, asbestos is found in walls, floors, tile, filters, fabric, electrical insulation, fire-proofing, and ventilating ducts. Many states forbid its casual removal and in some cases air testing is required. Given the sensational press coverage and the number of lawsuits pending, a cautious and conservative asbestos investigation is essential.

Water testing is also prudent. Besides asbestos, water may contain radon, lead, bacteria, mold, pesticides, fuel oil, or more than 200 other contaminants in amounts that exceed the minimum drinking levels according to the Clean Water Act – and still look clear and taste fresh. Lead and bacteria are more common problems because old industrial buildings often have lead pipes and because rooftop water tanks, often ignored and poorly maintained, allow pigeons to contaminate the water.

Office workers may be especially allergic to pollen or spores, and molds and mildew are frequent contaminants – especially in buildings with leaks or water seepage. Controlled air in office buildings tends to concentrate particulate matter and dry winter conditions in office spaces tends to disperse spores and pollen collected during the summer and fall. Remember also that many split-system HVAC installations do not, in themselves, provide fresh air. Check the ventilation ducts especially for concentrations of mold or bacteria growth – especially near HVAC compressors where water pools. Molloy Corporation investigated a New York law firm where lawyers and secretaries in one corner were always sneezing and seemed to have colds. The telephone switching closet built from unpainted plywood contaminated the space. Various molds and fungi grew on the plywood, uninhibited by light. A ceiling vent in the closet blew fresh air over the four walls of live growth and the spores spewed out the door over the nearby office workers.

A reasonable approach to air contamination is also advised. Aggressive air sampling is warranted if asbestos is present, radon testing if in the cellar or on the lower floors. Pesticides such as chlordane or insulation such as urea-formaldehyde usually can be detected by visual inspection, and air sampling may be necessary if it reveals improper or overzealous application.

A watchful eye for pests is another consideration. A double bind for office managers is that office workers complain both about pests and allergies to pesticides. The irony is that high rates of asthma in inner cities may be caused by severe levels of roach infestation. Roaches also carry viruses, bacteria, fungi, protozoans, and other contaminates. During pre-occupancy inspection, keep an eye open for the detritus of pesticides (boxes, traps, pellets, etc.) and for bait and traps for mice and rats. Check the window sills for pigeons. Mites and lice are associated with pigeons flocking on or near windows, not to mention unsightly and unhealthy deposits on masonry. Pneumocystocosis, a serious lung infection, is often caused by dry pigeon droppings borne through the air intake system or blown in through windows. A basement visit will also reveal evidence of pests. Cats and rats create particular odors in spaces.

The possibilities of toxic contamination are bewildering, but taking a building history is a good litmus test. Ask about previous tenants and the building's prior use. Check for garage facilities underneath that may cause gasoline evaporation rising in air tight buildings. Gasoline seepage from the water table is also a problem in some buildings near industrial areas. Watch out also for contamination by muriatic acid, which is often used to clean masonry, especially to remove unsuccessful sealants from basement areas. Ask also about storage use in the past. Laboratory, food packaging, or light industrial activities may have left toxic materials behind. One building in New York housed the Radium Dial Company in the 1930's and 1940's. A recent inspection revealed gamma-ray radiation levels three times higher than the highest allowable industrial exposure.

THE DEAL

Take nothing for granted! If it is not stated in the lease or on the drawings, it probably does not exist, is unsatisfactory, or will go wrong. Basic needs should be spelled out from the beginning. Anything that looks strange or is missing should be written into the lease as the owner's responsibility. One tenant in New York assumed a lease for a space with experimental lighting and air conditioning controls. Although everything worked in the beginning, the system could not be replaced when it failed. The entire installation had to be systematically rewired over the next two years.

Make sure that the building is open 24 hours a day, weekends and holidays, and that air conditioning, power, water, elevators, service entrances, and messenger deliveries are also reasonably available. A facility is not a facility when the toilets do not flush or workers have to walk to the twentieth floor.

Check also for hidden surcharges. The CEO may not expect a hidden operating charge over the length of the lease. Such charges may be handling charges on electrical surveys in unmetered spaces, union shop requirements for workers, movers, and repairmen, or usage fees for weekend deliveries or security. Also determine how future rent increases will be calculated. All conceivable surcharges and rent escalations should be wormed out of the building management and reported to the CEO.

Whatever the lease and the drawings state, conditions in the field always vary – particularly differences in square footage. Rents are often determined by an annual charge per square foot. Molloy Corporation has never evaluated space in which the actual square footage was more than the landlord claimed. In fact, the difference between the number of square feet charged and the net useable square footage can be as high as 50%. Careful measurements and precise take-offs are essential to any good pre-occupancy evaluation and particularly useful during lease negotiations.

Leases also often include a "work letter," a generally loose specification with an arbitrary dollar value assigned to the improvements to be provided by the landlord. The amount and quality of this work varies greatly. A facilities manager should make a hard-nosed assessment of this work by interviewing the building management and visiting other "work letter" installations. "Work letters" ordinarily include walls, partitions, doors, hardware, ceilings, heating, ventilating and air conditioning, electric work, telephone outlets, lighting, plumbing, structural work, floors and walls, their coverings, finishes and painting.

In the end, good communication is the most important criterion. If the CEO does not understand the report or the information in context, then the evaluation is next to useless. The length of the lease is important in context. Every building has problems. Any problem can be solved. Remind the client that the costs for good pre-occupancy evaluation and solid construction amortized over the length of the lease is negligible compared to the rental. Many leases in New York are now written for twenty years or more. Good pre-occupancy evaluation can make the difference between twenty years of compromise and misery or twenty years of good business.

CHAPTER 6

THE ROLE OF BUILDING EVALUATION IN EARTHQUAKE HAZARD REDUCTION

Michael E. Durkin

School of Architecture and Michael E. Durkin
University of Southern California and Associates

INTRODUCTION

This chapter will describe the current and potential roles of building evaluation in earthquake hazard reduction. It will discuss studies which focus on the relationship between environment and behavior before, during, and after earthquakes. It will describe why earthquakes are important, where the benefits lie, and where the field is going.

What are the basic questions that earthquake building evaluations address? They are: 1) What are the risk factors? 2) How can I avoid being injured? 3) How can I counter disruption? 4) How can I speed recovery? and 5) What kind of training is most effective?

Building evaluations have been effectively made for all four major earthquake disaster time frames – before, during, and immediately after the event, and during the recovery process. In the case of earthquakes, pre-event evaluations help identify hazards, and can help determine effective hazard reduction strategies. Evaluations can also help prepare building occupants to respond to earthquake hazards to reduce the risk of injury. Building evaluation methods can facilitate a rapid damage assessment after the shaking stops. Finally, building evaluations can help speed the recovery process. Applications include identifying earthquake hazards through physical and social setting analysis, documenting occupant performance vis-à-vis building performance in specific building types and interior settings during and immediately after an earthquake, investigating building evacuation, the cause of earthquake injuries, and the behavior of people in collapse buildings, and in investigating the emotional impact of relocation on earthquake victims, as well as using techniques to assess the suitability of different relocation alternatives.

TIES TO CONVENTIONAL BUILDING ASSESSMENT

An important but often overlooked use of building evaluation is earthquake hazard reduction. Building assessment for hazards is similar to building assessment for occupant performance, satisfaction, and well-being: both types of evaluation focus on

Figure 1. Partially collapsed, multi-story building following 1985 Mexico City earthquake.

how buildings affect the well-being and productivity of people; both use building-specific methods to make such assessments. However, they differ in important areas: they are based on different criteria, concentrate on different topics, and are employed in a different time frame.

Earthquake evaluations apply the key principles of building evaluation: the emphasis on building purpose, the use of agreed-upon criteria, the employment of tested methods, and an increasing depth of study. Building purpose, for both, is highlighted in the relationship between design and activities. Both earthquake and conventional assessments emphasize that design is a prediction that certain activities will take place and both test to see if indeed that prediction was correct. Furthermore, agreed-upon criteria govern the investigations for both. Both earthquake and conventional assessments address health and safety issues, and both try to integrate these issues with other concerns such as functional requirements, permeability and interference, and flexibility. In addition, employment of tested methods typifies both kinds of assessment. Both earthquake and conventional assessments employ instruments that have been developed and refined over time. Finally, increasing depth of study characterizes both kinds of assessment in that they both seek to integrate concerns into a developing framework. Both seek to prioritize concerns, and to perform different types of evaluations within changing circumstances and at different times.

However, key differences between conventional and earthquake evaluations need to be considered. These include differences in emphasis, in specific topics, and in time-frames.

The first difference lies in emphasis. Whereas conventional building assessments underscore building purpose, earthquake hazard evaluations feature generic concerns for life safety, building integrity, loss of property, and the viability of the business or organization after a disaster. Conventional building evaluations emphasize topic categories so that privacy, interaction and symbolism are at the top, and health and safety concerns are at the bottom (with categories such as physical comfort and functional requirements falling in the middle). Earthquake building evaluations flip these priorities, making health and safety issues of primary concern. They do, however, strive to integrate health and safety issues with conventional evaluation concerns. They thus emphasize the implementation of hazard reduction strategies in such a way that they address such issues as privacy and social status. For instance, nonstructural hazards can be reduced in a way that ensures ongoing functional performance or privacy or social status. Framed diplomas dangerously positioned over the head of a business executive might either be taken down, moved or secured. The decision would be based on the fitting of health and safety measures with other priorities and concerns.

The difference in emphasis results in a second difference, a difference in topics: earthquake evaluation topics feature health and safety issues and all necessarily reformulate the notion of building functions. Earthquake evaluations predict or evaluate a building's performance in a disaster and predict or assess the building's effect on individuals' health and safety in specific settings. They assess actual and complete failings of particular functions, not anticipated inadequacies or partial failings. As a result, topics are less likely to include amenities that over time enhance function in a non-disaster situation, than topics such as physical elements that ensure functions in the immediate aftermath of a disaster. Indeed, whereas most conventional building assessments assume functions to be relatively stable over time, earthquake assessments assume disaster situations in which specific functions have suddenly been eliminated and others changed or modified.

Examples of these differences in topics can be seen in the differences between conventional vs. earthquake evaluations of hospitals. Whereas conventional hospital evaluations provide a detailed organizational analysis of normal functions, including department functions, department resource, needed adjacency requirements, equipment, and supplies, earthquake evaluations differentiate between normal and emergency functions. For example, for each hospital department, the evaluator compares emergency and routine functions, characterizing how existing functions might be adapted in an emergency and what new ones are necessary.

The difference in topics is tied to a third key difference, a difference in time frames. Conventional building evaluations assume a long time frame for building change: a building is designed, constructed, occupied and changed as building use changes; assessments can be integrated into each stage. However, earthquake evaluations assume a sudden and irrevocable change: a building collapses or is seriously damaged, key personnel are injured, and building functions cease; assessments cannot be integrated into a slow evolution but are clearly divided between pre- and post-earthquake evaluations. Furthermore, earthquake evaluations focus on particular time frames such as the period of shaking or the immediate period following the earthquake, with some studies now focusing on a longer time period, namely the recovery period following the quake.

SUMMARY OF THE NATURE AND SCOPE OF EARTHQUAKE BUILDING ASSESSMENTS

Using physical/social setting connections as a jumping-off point, one can begin defining the nature and scope of earthquake assessments in terms of three kinds of activities they address. In the first place, they address response activities to the quake. These responses include individual occupant activities before, during, and after the event. In the second place, these assessments address emergency function activities, namely the activities that in an emergency have to substitute for normal activities. For example, the kitchen staff of a hospital might assist in evacuating patients from the nursing units. In the third place, earthquake evaluations address normal activities that change during the recovery process. For example, a family might have to relocate its sleeping activities from a heavily damaged bedroom into a lightly damaged living room; or hospitals might have to relocate surgery activities to other parts of the building.

These three types of activities are clearly tied to building design. Earthquake assessments, like conventional assessments, attempt to determine the successes and failures of a design according to the activities they enhance or impede. Also, each design predicts that certain activities will take place. The evaluation "tests whether that prediction was correct" (Bechtel et al, 1975). The building design, or physical setting as it is called in earthquake research, affects all three types of activities. It clearly affects individuals' response activities before, during, and after a tremor. Before an earthquake, individuals engage in training and in identifying hazards in their immediate work settings. During an earthquake, individuals no longer find themselves in a stable environment: floors buckle, bookcases fall, elevators fail. After an earthquake, individuals have to participate in the recovery process of the business or organization. The physical setting also clearly affects the transformation of normal organizational activities into emergency activities. The failure of elevators or the entrapment of other personnel may require hospital kitchen staff to help with evacuation of patients; major damage to the facility may require that a reception area be transformed into an emergency care center. Finally, the physical setting also affects the transformation of normal activities during the long-term recovery process. Damage to settings, equipment, and supplies, as well as injury to personnel, affects the ongoing activities of individuals and organizations. Hospitals may need to relocate essential services to another floor within the hospital; businesses may temporarily have to operate out of a warehouse.

The building design elements that come to the forefront in earthquake assessments include building type, building configuration, building layout, building materials, structural elements, type and placement of nonstructural elements, and building contents. Earthquake evaluations determine whether or not these design factors enhance or impede the activities essential to the protection of life and property and to the viability of key functions. While the protection of property may seem trivial compared to protection of life, it ensures the survival of the organization. Earthquake evaluations are primarily concerned with the specific effects of earthquakes on buildings and people, at specific time periods. Building features, including physical performance in an earthquake, certainly stimulate immediate occupant response and result in injuries and entrapment. They also result in damage and disruption extending for long time periods after the event.

The major concern of such evaluations is to identify hazards, predict the performance of physical settings, select among alternative solutions, and implement those solutions.

WHAT ARE THE AVAILABLE PROCESS MODELS FOR APPLYING BUILDING EVALUATION TO EARTHQUAKE HAZARD REDUCTION?

We can apply building evaluation methods to earthquake hazard reduction both before and after an earthquake. Like design and construction, both the pre-earthquake and post-earthquake hazard reduction processes include the generic problem-solving steps of identifying the problem, developing alternative solutions, choosing from alternatives, and implementing the chosen solution. Earthquake building evaluations performed before a quake are used first to assess physical setting hazards; second, to estimate the potential earthquake impact; third, to assist the selection of appropriate hazard reduction strategies from among alternatives; and fourth, to implement the selected approach. The post-earthquake evaluation to assess damage can compare the impact estimates made in the pre-earthquake hazard identification with the post-earthquake reality as documented by the damage assessment. A second function of the damage assessment is to further estimate both immediate and longer-term consequences of the damage actually sustained on future operations. Pre-earthquake hazard reduction solutions aim at minimizing earthquake damage, while post-earthquake solutions strive to minimize damage impact.

WHAT ARE THE APPLICATIONS OF PRE-EARTHQUAKE BUILDING EVALUATION?

Assessing Building Hazards

The time before the earthquake is the best time for private and public sector organizations and individuals to assess facilities and take action to reduce earthquake hazards. Such assessments identify primarily life safety threats to building occupants, posed by structural damage, damage to architectural elements, equipment and furnishings, and by occupant actions. Organizations also increasingly use this time to identify hazards with respect to loss and recovery.

Existing buildings are frequently assessed to identify physical setting aspects dangerous to occupants. Such assessments can be very sophisticated, requiring experts of several disciplines and in-depth engineering studies; or they can be simple applications, requiring only that building occupants assess their own work areas. Many businesses and institutions hire architects and engineers to inspect facilities and identify potential hazards, based on expected building performance. The team normally recommends specific hazard abatement activities. The background for this assessment derives from research on similar physical settings that have undergone earthquakes, and on actions of similar populations exposed to earthquake threats.

Earthquake building evaluations seek to identify factors contributing to casualties, as well as loss and recovery problems. Thus they provide a promising research base for these assessments by focusing on a) the incidence of different kinds of damage – structural, nonstructural, and to building contents; and b) damage impact of direct costs and longer-term losses predominantly on the recovery process.

Researchers are investigating the connection between injury, occupant actions, selected physical setting aspects, and geologic performance. (Durkin, 1985c). Examples include the type and range of injuries that have occurred in a given earthquake (Durkin et al., 1983), as well as in specific building types like residential buildings (Durkin et al., 1988a) and office buildings (Durkin and Ohashi-Murakami, 1988c). The major research focus is on the building and occupant response patterns responsible for different types of injury.

Our current studies are analyzing occupant actions with respect to hazards posed by the seismic performance of specific building types and setting types. This research is documenting the behavior of building occupants in earthquakes, which include spontaneous action, action based on prior training, and organized response. Figure 3 illustrates current earthquake injury research concerns. Investigations documenting functional loss on commercial and institutional organizations are also underway (Durkin, 1988b).

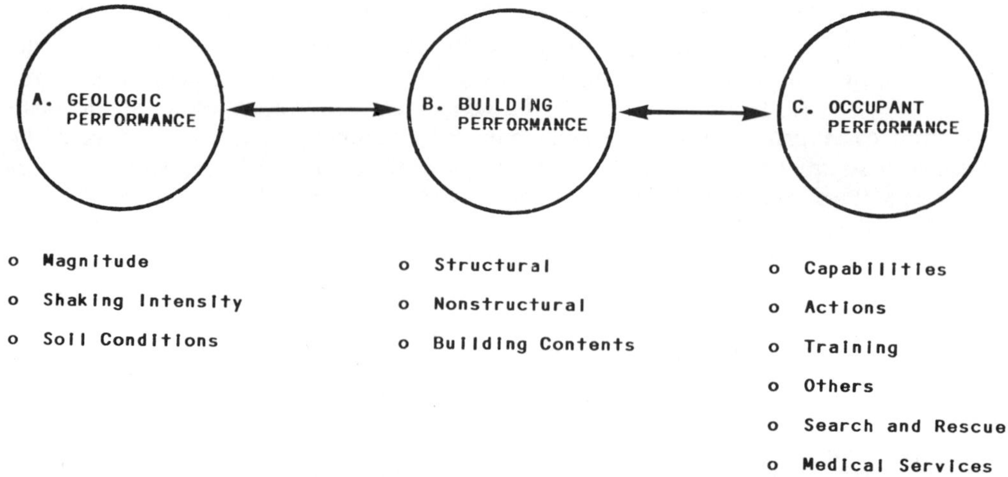

Figure 2. Geologic performance, building performance, and occupant performance. Three key components in earthquake hazard reduction research.

Renovation/Relocation

Earthquake damage can change an existing physical setting, change the relationship between an existing physical setting and other related settings, or necessitate creation of a new replacement setting (figures 4 and 5).

Post-earthquake relocation research is addressing residential, business, institutional, community, and individual needs. Topics include, at the basic level, 1) the number, sequence, and duration of each move, 2) type of alternative facility selected, and 3) the reasons for selecting one facility type over another.

For example, our analysis of household relocation following the 1983 Coalinga earthquake showed a strong tendency for those displaced to remain on their own property, sleeping in a tent, in an RV, or in the open. Many of the other families stayed with relatives or friends. There was little use of emergency shelter provided by volunteer organizations, e.g., only one survey respondent stayed in a Red Cross shelter (Durkin and Burnam, 1986e).

Initial housing research was geared to documating community-wide rates and suggesting potential associations between post-earthquake relocation and mental health disorders, such as PTSD, major depression, and acute anxiety. For example, the post-earthquake, overall period prevalence of major depression was found to be 17.4 per 100 population in the Coalinga, California community (Durkin and Burnam, 1986e) and 18.4 per 100 in a Chile housing project (Durkin, 1987a). We have since expanded our approach to document physical setting modifications and to compare the suitability of pre- and post-earthquake settings along relevant dimensions.

Estimating The Impact

Pre-earthquake assessments are next concerned with estimating the impact of an earthquake on buildings and people. Building owners and organizations need to know what kind of casualties and losses they can expect; they want to know what the potential

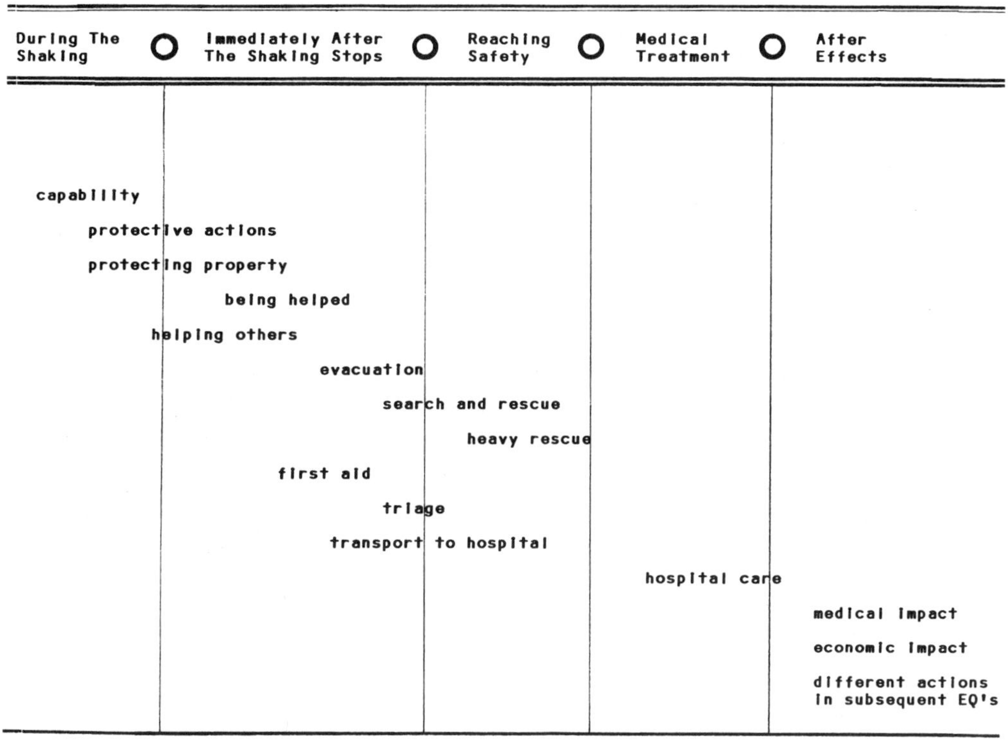

Figure 3. Critical time frame of current earthquake injury research, from onset of shaking to long-term injury impacts.

costs of recovery will be. These assessments are based on research in the areas of casualties, damage and loss, and recovery.

Research in these areas is moving from descriptive to analytic in certain areas, while remaining still generally descriptive in others. Studies in the key areas of casualties, damage and loss, and recovery have become increasingly analytic. For example, earthquake injury research first attempted to identify and describe the factors responsible for injury (risk factors). Next it attempted to quantify these impacts. Finally, it used epidemiological approaches to investigate the susceptibility of different types of persons to specific types of physical setting damage.

Problems remain in both the descriptive and analytic research on impacts. While paucity of data impedes progress from descriptive to analytic studies, the availability of quantitative methods encourages acceptance of gross analytic findings. For instance, data is lacking for earthquake impacts on specific building types (for example, high-rise office buildings). In addition, where good data exists, as in the area of injuries, one sees a rush toward quantitative results.

Selecting Appropriate Hazard Reduction Strategies

A number of hazard reduction strategies exist. These can be either physical or programmatic or a combination of both. Pre-earthquake physical setting intervention applies both to the design and construction of new buildings and to the renovation of

Figure 4. Major commercial street in Coalinga, CA central business district following 1983 earthquake.

existing ones. Techniques include strengthening and/or securing the building structure, architectural elements, and building contents, and reprogramming and/or relocating building functions. Pre-earthquake social setting intervention includes developing disaster response and contingency plans and training building occupants (Durkin, 1984c).

One fairly obvious application of building evaluations is in the creation and revision of seismic building codes and other provisions for both new and existing construction. Such regulations currently govern certain facility types in California and elsewhere. Existing regulations apply not only to structural elements but also to "nonstructural" elements, including cladding, glazing, interior systems, equipment, and furniture.

In Los Angeles peak occupancy is one basis for determining the scheduled structural retrofit of pre-1932 unreinforced masonry buildings. Since the uses and functions of old buildings change over time, occupancy classifications are not based on actual usage features but on code definitions of legal occupancy loads. Therefore, building evaluations can assist these efforts by providing more detailed actual occupancy information to guide structural retrofit approaches. Since many U.S. communities have many older, hazardous buildings but few seismic regulations, good building evaluations can document where people spend most of their time during a typical day. Designers can then use this information as a critical element of a vulnerability assessment, enabling selective strengthening of building areas having the highest occupancy loads or the most vulnerable occupant groups.

Selections of hazard reduction strategies have in the past been made by building codes. Many recent campaigns are more comprehensive and multifaceted, implementing architectural, administrative, and managerial hazard reduction techniques in concert with structural retrofit approaches. Seismic alternatives hazard reduction techniques have been incrementally employed, on an interim basis and where complete structural upgrading is impossible (Durkin, 1984).

In addition to those affected by the codes, building owners and organizations who have identified hazards and assessed impacts are usually concerned with selecting an appropriate strategy for hazard mitigation (Durkin, 1986d). Here, building evaluations can help provide a cost benefit analysis to evaluate the efficiency of different approaches for a given set of circumstances.

Figure 5. Coalinga commercial street following reconstruction.

Occupant training provides a third hazard reduction strategy. Directed at actions prior to an earthquake, such training could sensitize building occupants to risks and possible actions. However, to be truly effective, earthquake training measures must be based on a realistic appraisal of occupants' capabilities and actions during earthquakes (Durkin, 1985). Our studies suggest that many overly general beliefs about appropriate response, such as unconditionally evacuating a building, standing in a doorway, or getting under a desk, can endanger rather than protect building occupants.

Occupant training for the period during the earthquake can result in more effective occupant actions. Building occupants may find themselves in the midst of a distorted corridor or toppling stair tower. Earthquake training directed at the immediate shaking may prevent panic and foolhardy actions. Occupants will be trained in taking immediate cover, finding alternative escape routes, and avoiding injury from falling debris (Durkin, 1985d). Assuming that many spatial design, management, and utilization decisions are made by laymen, an additional benefit of a earthquake building evaluation is to recommend training measures that would improve the ability of building occupants to prepare for an earthquake.

Programming for Emergency Operations

Many organizations must continue to operate following a major disaster. Hospitals, for example, must stay operational to treat earthquake injuries and to handle normal community health needs (Durkin, 1986b). Many essential organizations will temporarily curtail non-essential services in favor of maintaining essential functions. The expanded scope of the earthquake building evaluation permits examination of the physical setting's suitability for supporting normal functions, and occupants, but also emergency ones.

Implementation

Market research. For a limited number of building types, researchers have employed market research to identify the specific information needs of diverse target audiences, to determine the specific kinds of information that each audience needed, and to discover the best way to provide this information. The recent development of small business earthquake preparedness materials (Durkin, 1985e) was preceded by a systematic review of existing research and the conduct of new research on earthquake impacts on small businesses (Durkin, 1987c; Durkin, 1985f), surveys of both information providers and users, and focus groups, to review preliminary material (Durkin, 1985; Durkin, 1986c).

Occupant training. In all buildings, occupants have a certain sphere of control over large portions of the physical setting, usually around individual work settings. Occupants can take numerous actions to reduce localized earthquake risks, actions such as locating physical hazards, removing or securing them, and determining safe locations in which to take immediate shelter. Furthermore, occupants can be trained in expected building performance, making them more likely to initiate and undertake both personal and collective hazard reduction actions. For instance, companies can devise earthquake specific evacuation plans or designate safe areas within buildings. In addition, staff can be trained for alternative roles, preparing them to assume different or additional responsibilities in an emergency situation. For example, hospital clerks can be taught how to transport evacuating inpatients. Finally, volunteers can be trained in how to take over emergency functions.

Occupant training for the pre-earthquake period is especially valuable to organizations and physical settings with a rapid turnover rate. Such training could help managers quickly acquaint people with the ways that the building is intended to operate. For one, the original design intentions for the way a building should function in an emergency are frequently not communicated to, or forgotten by, building occupants. For another, the building layout, including exit routes, is frequently unknown to new personnel.

Assessing hazard reduction approaches. We can use earthquake building evaluations to periodically assess and monitor earthquake hazard reduction programs that a particular organization might be employing. Such evaluations can pinpoint opportunities for achieving earthquake hazard reduction as part of an ongoing building maintenance process, determine which approaches are being successfully used and which are not, and aid selection from among alternative approaches. With some adaptation of traditional methods, we are teaching building managers and maintenance personnel to assess their physical settings on a continuous basis.

WHAT ARE THE APPLICATIONS OF POST-EARTHQUAKE BUILDING EVALUATION?

The main goal of immediate post-earthquake assessment is to ensure that the building is safe to remain in. The initial damage assessment step, therefore, is whether or not evacuation is necessary (for any or all of the buildings). The secondary damage assessment step determines whether or not a physical setting, group of settings, or building can still support its normal activities and/or its emergency ones.

If a setting no longer supports its intended functions, then a subsequent building evaluation focuses on the adequacy of the "back-up" settings that were identified prior to the earthquake. If any of these settings are found inadequate, or if an extensive relocation is advised, building evaluation methods may be employed to find suitable alternative settings somewhere else in the community. Next, alternative facility assessment, step number three, assesses particular rooms or settings to determine if they can still accommodate the contingent functions.

Immediately following the earthquake. Building evaluation can assist the selection of suitable facilities for short and interim time periods. Quick and dirty facility screening is useful for designating emergency facilities. More systematic building evaluation is useful for longer-term relocation.

Programming a permanent replacement facility. Although a "quick and dirty" building evaluation can address the previous uses, the recovery process requires more traditional building evaluation techniques with their earthquake embellishments. The recovery phase relies on the more traditional building activities of renovation, repair, and relocation.

Organizations and individuals can employ post-occupancy evaluation techniques either before or after a damaging earthquake to plan, program, and design a new replacement facility. The building evaluation adds the promise of the new facility supporting evolving needs and requirements. In addition earthquake damage can provide the opportunity for both facility and organizational "pruning" by causing the replacement of outmoded facilities, equipment, and organizational procedures.

Building evaluation techniques are employed to assist relocation at several different levels. They reveal what design solutions support the needs of different occupant groups, different activity patterns, and different organizational requirements. At a more detailed level, they pinpoint what specific aspects are supportive and what ones are not.

We are applying knowledge from past earthquake building evaluations to create new settings. In California, and elsewhere, lessons from post-earthquake building evaluations are leading to codes, standards, and design and operational criteria influencing the construction and operation of new hospitals, schools, and homes in these and in other areas facing similar circumstances.

WHAT IS THE FUTURE OF DISASTER EVALUATION?

Since current building evaluation provides proven ways to systematically study physical settings, it has numerous advantages for future application.

First, it can provide research specificity. Most earthquake hazard reduction information is not specific enough to the physical setting. As more studies are completed and more physical settings evaluated, researchers can tailor more information to specific circumstances and needs.

Second, standard evaluation techniques currently enable us to compare and contrast findings of other studies dealing with similar physical settings.

Third, like traditional evaluation, earthquake evaluation methods are evolving in an iterative manner. As more disaster building evaluations are completed, researchers learn more about methods that work and those that don't. We can therefore adjust existing methods, and select more appropriate ones for future projects. One future effort is to broaden the applicability of earthquake building evaluation to other building types by adapting existing methods and developing new ones for ready inclusion in the existing decision-making processes.

Fourth, earthquake building evaluation is evolving toward more integration, more comprehensive frameworks, and a better knowledge of complex relationships. We've broadened our conceptual framework and made it more comprehensive. Early studies were limited by such factors as gaps in shaking intensity, building types involved, and time between earthquake and study onset. Limited by available settings and earthquakes, we looked at the effects of damage and loss in whatever situations we could find them. After starting with available information, we simultaneously broadened and added detail to our research focus. For example, we've expanded our focus to study earthquake effects on entire health care systems (Durkin, 1986a; Durkin 1987a). We've also refined our focus to view in more detail the workings of specific hospital departments. Additional disaster building evaluations promise to improve our understanding of the complex relationships between building performance, damage to structure, nonstructural components, building contents (including equipment and essential supplies), and operational loss.

Fifth, earthquake building evaluations increasingly involves interdisciplinary teams rather than representatives of a single specialization. Thus a range of previously neglected issues is now covered. We have learned that collaboration can take place in a variety of forms, and that the collaborative process might have to be individually tailored to specific design projects. Many factors interact to determine the dynamics of a particular effort. We are currently exploring this dimension. Future work in earthquake building evaluation will better reflect the iterative nature of the collaborative process.

Sixth, researchers and practitioners from the social sciences and design disciplines are becoming more skilled in interdisciplinary collaboration. A small but growing number engages in training, research, and practice in both the physical and social setting dimensions of earthquake hazard reduction. This background enables them to translate the information and thereby serve as synthesizer, communicator, and intermediary between both disciplines.

ACKNOWLEDGMENTS

This chapter was completed with assistance from the National Science Foundation (current grant number ECE-8610890). Dr. William A. Anderson, of NSF, provided timely insights and encouragement. The California Department of Mental Health and the National Institute of Mental Health provided contract funds to investigate the Coalinga earthquake's emotional impact. Wilma O'Callahan and Dr. Paul Gorman managed the contract for CDMH. Drs. Mary Lystad and Susan Solomon directed NIMH's Center for Disaster Studies at the time of this contract. Rapid response grants from the Natural Hazards Research and Applications Center enabled travel and some data collection following the 1985 Chile and Mexico City earthquake. Dr. William Reibsame and Ms. Susan Tubbesing directed the Center's activities during this period. The Southern California Earthquake Preparedness Project (California Office of Emergency Services), Bay Area Regional Earthquake Preparedness Project (California Seismic Safety Commission), and the Federal Emergency Management Agency supported development of earthquake hazard reduction applications.

REFERENCES

Durkin, M., Coulson, A., Medina, M. H., Kraus, J., Ohashi, H., Romo, R. C., March 1988a, La Supervivencia De Personas En Edificios Derrumbados, in: *Cronicas De Desastres: Terremoto en Mexico, Septiembre 19 y 20, 1985*, Washington, D.C., Pan American Health Organization.

Durkin, M., April 1988b, *Impact of the October 1, 1987 Whittier Narrows Earthquake On Businesses, Building Owners, and Managers: Early Results of the BOMA/USC Survey*, Woodland Hills, CA, Michael E. Durkin and Associates.

Durkin, M., and Ohashi-Murakami, H., August 1988c, Casualities, Survival, and Entrapment in Heavily Damaged Buildings, in: *Ninth World Conference On Earthquake Engineering, Tokyo-Kyoto, Japan* (forthcoming).

Durkin, M., July 1987a, *The Emotional Impact of Post-Disaster Relocation: The Chile Earthquake Experience*, Boulder, Colorado, Natural Hazards Research and Applications Information Center.

Durkin, M., August 1987b, Casualties, Search and Rescue, and Response of the Health Care System, in The San Salvador Earthquake of October 10, 1986, *Earthquake Spectra*, Vol. 3, No. 3.

Durkin, M., and Hopkins, J., August 1987c, Architecture and Planning in The San Salvador Earthquake of October 10, 1986, *Earthquake Spectra*, Vol. 3, No. 3.

Durkin, M., Winter 1987d, Recent Quakes Teach Tough Lessons, in *Networks: Earthquake Preparedness News*, Oakland, CA, Bay Area Regional Earthquake Preparedness Project, Volume 2, Number 2.

Durkin, M., February 1986a, Casualties, Hospital Response and Response ofthe Health Care System, in The Chilean Earthquake of March 3, 1985, *Earthquake Spectra*, Vol. 2, Number 2.

Durkin, M., April 1986b, The Response of the Health Care System, in P. Bonelli (Ed.), *Proceedings of the Fourth Chilean Conference on Seismology and Earthquake Engineering and International Seminar on the Chilean March 3, 1985 Earthquake*, Vina Del Mar, Chile, Santa Maria University.

Durkin, M., September 1986c, *Facilitating Minority Business Involvement in Earthquake Preparedness and Integrating Local Business in Community Preparedness*, Los Angeles, CA, Southern California Earthquake Preparedness Project.

Durkin, M., 1986d, Reducing Earthquake Hazards in Existing Buildings in J. Ziony and W. Kockelman (Eds.), *Future Directions in Evaluating Earthquake Hazards of Southern California*, Menlo Park, CA, U.S. Geological Survey, Open-File Report 86-401.

Durkin, M. and Burnam, M. A., October 1986e, *The Emotional Impact of Post-Disaster Relocation: The Coalinga Earthquake Experience*, Woodland Hills, CA, Michael E. Durkin and Associates.

Durkin, M., February 1985b, Occupant Behavior and Organizational Response in Earthquakes: A Preliminary Research Agenda, in J. Pauls (Ed.), *The Proceedings of the International Conference on Building Use and Safety Technology - 1984*, Ottawa, National Research Council of Canada.

Durkin, M., February 1985c, Occupant Behavior in Earthquakes: A Comparative Analysis, in J. Pauls (Ed.), *The Proceedings of the International Conference on Building Use and Safety Technology - 1984*, Ottawa, National Research Council of Canada.

Durkin, M., February 1985c, The Behavior of Building Occupants in Earthquakes, *Earthquake Spectra*, Vol. 1, Number 2.

Durkin, M., May 8, 1985d, Earthquake Impact and Hospital Evacuation Requirements: Interaction of Design and Occupant Behavior. Paper delivered at the Hospital Evacuation Technical Session of the American Association for Medical Instrumentation Annual Conference, Boston, Mass.

Durkin, M., September 1985e, *Earthquake Preparedness: A Key to Small Business Survival*, Los Angeles, CA, Southern California Earthquake Preparedness Project. Also published by the Bay Area Regional Earthquake Preparedness Project for distribution in northern California, and recently by the Federal Emergency Management Agency for national distribution.

Durkin, M., September 1985f, *Facilitating Small Business Involvement in Earthquake Preparedness*, Los Angeles, CA, Southern California Earthquake Preparedness Project.

Durkin, M., 1985g, The Economic Recovery of Small Businesses After Earthquakes: The

Coalinga Earthquake Experience, in D. Brotherson (Ed.), *The Proceedings of the International Conference on Natural Hazards Mitigation Research and Practice: Small Buildings and Community Development*, Champaign/Urbana, IL, University of Illinois.

Durkin, M., April 1984a, Seismic Performance and Hazard Reduction in Unreinforced Masonry Buildings, paper given at conference on The Seismic Retrofit of Historic Buildings, Sacramento, California.

Durkin, M., July 1984b, Alternative Methods for Hazard Reduction in Unreinforced Masonry Buildings, *The Proceedings of the Eighth World Congress on Earthquake Engineering,* New York Prentice Hall. A longer version of this report is available from the National Technical Information Service, Springfield, VA, Report Number PB83-186858.

Durkin, M., Nov./Dec. 1984c, Improving Seismic Safety in Unreinforced Masonry Buildings, *Ekistics*, 309.

Durkin, M., Aroni, S., and Coulson, A., November 1983, Injuries in the Coalinga Earthquake, in *The Coalinga Earthquake of May 2,1983: EERI Report*, Berkeley, Earthquake Engineering Research Institute.

Durkin, M., 1981, *Methods for Hazard Reduction in Office Buildings in Earthquakes*, Woodland Hills, CA, Michael E. Durkin and Associates.

Green, M. and Durkin, M., January 1983, *Seismic Safety and Existing Buildings: A Research Agenda*, El Segundo, CA, Melvyn Green and Associates.

CHAPTER 7

EVALUATION OF ANIMAL HABITABILITY IN FARM STRUCTURES

Bengt Gustafsson

Swedish University of Agricultural Sciences
Department of Farm Buildings
Lund, Sweden

INTRODUCTION

The value or worth of an item is an opinion (Suter, 1974). That opinion may be a personal one in which the person making the evaluation largely ignores the real world and bases the value on sentimental and subjective judgements. In that instance, value is the worth of the item to one individual alone. The determination of the value of an item may, on the other hand, be based on certain assumptions and analyses and arrived at in an objective fashion. In that instance the value represents the worth of the item to all persons who might be owners of that item (Murray, 1969).

The value of a farm structure can mean many different things. In most textbooks in farm management or farm real estate appraisal, the following five values of buildings are defined and discussed.

1. *Book value*: The actual capital outlay spent on the farm building. The original cost is divided by the estimated life of the building to determine a depreciation schedule. The book value is used for record-keeping and income tax purposes.
2. *Structural value*: The present-day reproduction cost less the building's physical depreciation ("wear and tear") to date. In this case, the estimated life equals the building's economic life and is based on its physical condition and probable use.
3. *Contributory value*: The amount of money that the building adds to the market value of the farm. It is based on replacement cost less depreciation. Here, depreciation reflects physical deterioration, as well as functional inadequacy and economic obsolescence. The value is based on the typical operator concept.
4. *Value in use*: The present-day value of the building in the particular farming operation of the present owner or operator ("the utility value"). It reflects the extent to which a given building would be replaced if it were completely destroyed due to fire, wind, or other reasons. It typically equals the insured value.
5. *Salvage value*: The value of a given farm building, or some or all of its parts, when the building or its component parts are relegated to their next-best

alternative use or location. This implies that the building's present use is discontinued or that a change in the location of the building is required.

This list of common values of farm buildings illustrates that the purpose of, and the reason for, farm building valuation determines the value concept used. It also indicates that different methods of evaluation must be applied, and that different sets of data are required, depending on the purpose of the evaluation. The book value may easily be found in the farm records, while, for example, the determination of the contributory value of a building often requires a professional appraisal based on thorough inspection and expert judgement.

In the following we will deal with a special farm building valuation problem: the situation when a range of competing designs or options of (livestock) buildings are presented, which are to be evaluated and compared to help the decision maker to pick "the best" design. We will, in other words, deal with the evaluation of the relative merits of livestock buildings. This is becoming an increasingly important issue, especially in countries where today a farm livestock building project must meet a set of sophisticated legislative requirements regarding such factors as animal welfare, security, work environment, structural design, and environmental protection.

The complexity of the modern livestock building requires its programming to be team work. The owner/operator, in consultation with agriculturists, building experts, veterinarians, materials handling specialists, manufacturers of interior equipment and machinery, working environment experts, economists, and others, must select one of many alternatives. Therefore, the qualities of different options must be assessed by some means and the costs compared. At the present time, no established method is available for comparison of the various characteristics which must be considered and for reducing them to a common denominator. The method presented here has been developed and tested in a research project supported by the Swedish Council for Forestry and Agricultural Research and carried out at the Department of Farm Buildings of the Swedish University of Agricultural Sciences (Gustafsson, 1978). It is characterized by a quantitative comparison of the function embodied in various designs for a livestock building. Primary importance is attached to the evaluation of the structural design itself and to the function of the building with respect to how well it will meet the requirements concerning animal welfare, working environment, and pollution control (external environment). The method is based upon:

1. Listing an appropriate number of characteristics (or factors) of the design, working environment, etc., which should be taken into consideration. (Table 1.)
2. Checklists for evaluating and quantifying how well the various specifications are met. (Table 2.)
3. A diagram for determining the points for each factor (figure 1).
4. An evaluation scheme for weighting and subsequently adding together the points obtained from each characteristic or factor that meets the stipulated requirements, in order to obtain a total score (function index) for the respective alternative being evaluated (figure 2).
5. Diagram for converting building costs into points to be added to or compared with the functional scores obtained in step 4 (figure 3).

The principal features of the method will be generally described. Special attention will be paid to the results of an attempt to assess in quantitative terms the relative importance of a set of main factors considered in the evaluation. The outcome

TABLE 1
List of Factors for Evaluating
The Functions of an Animal House (Dairy Barn)

Main Section	Factors	Subfactors
1. Design	1.1 Available Space	
	1.2 Floor plan (arrangement)	
	1.3 Transport aisles	
	1.4 Adaptability of design	
	1.5 Operational safety	
	1.6 Other	
2. Animal environment	2.1 Ambient environment	2.1.1 Drinking water
		2.1.2 Hygiene, cleanliness, animal protection
	2.2 Design (1)	
	2.3 Climatic conditions	
	2.4 Noise and lighting	
	2.5 Flooring	
3. Personnel environment	3.1 Working environment	3.1.1 Working conditions
		3.1.2 Design (1)
		3.1.3 Ambient environment (2.1)
		3.1.4 Climatic conditions
		3.1.5 Noise, lighting
		3.1.6 Safety and security
		3.1.7 Flooring
	3.2 Personnel rooms	
4. Outside environment	4.1 The building's design in relation to the farm as a whole	4.1.1 Transport
		4.1.2 Possibility of additions to the building
		4.1.3 Other
	4.2 Adaptability of the building to the existing buildings	
	4.3 Waste handling (manure)	
	4.4 Water	
	4.5 Drainage system	

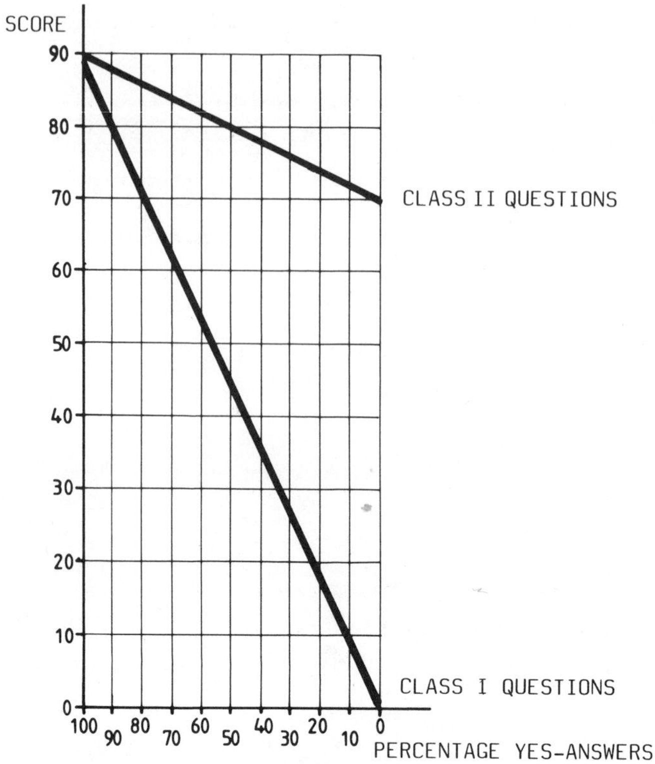

Figure 1

of a field test comprising the post-occupancy evaluation of different livestock buildings in southern Sweden will be commented upon.

SPECIFICATION OF REQUIREMENTS AND CHECKLISTS

When evaluating the quality or value of a specific function of a livestock building, it must first be decided which properties or characteristics (factors) should be assessed, and which requirements, according to present knowledge or information, and in respect to the client's wishes, should be fulfilled by the various building and equipment details. For a dairy barn, the division of building and environmental factors described in table 1 is considered to be adequate. It was developed by Montelius et al., 1981.

The method contains requirements or checklists for each factor, in the form of simple Yes / No questions set forth for various types of buildings (primarily for cattle and swine houses, table 2).

The questions are divided into two groups, according to the degree of desirability. Class I contains questions whose satisfactory solution is of greatest importance. Class II refers to items of lesser importance. Other ways of classifying the questions could be applied in special cases. The check lists are an aid to evaluate how well the requirements have been met by the building design. It is also possible to indicate by a plus or a minus if the stipulated requirement is fulfilled in a way that is considered above (+) or below (-) average standard.

TABLE 2
Excerpt From the Checklist for a Dairy Barn Where the Animals Are Kept in Group Pens

MAIN SECTION	FACTOR		
1. Design	1.1 Available space		
	Class		Remarks
Specification of requirements	I	II	+/-
Is there a sufficient amount of room for			
1. milk cows:	Yes		
2. dry cows:	Yes		
3. pregnant animals?	Yes		+
4. replacement animals, over and above the calculated number of animal places?	Yes		
5. Does the collection pen hold either the total number of cows which pass through the milking area, or for milking in groups, for the group that has the most cows?	Yes		
6. Is there at least 1 isolation pen for sick animals per 25 suckling calves?	Yes		
7. Is the number of places in the treatment section sufficient for the calving season?	No		
Number of questions	6	1	
Number of Yes answers	6	0	
Percent Yes answers	100	0	
Points (according to Fig 1)	90	70	
Net number of +/-			1
Sum of points divided by the number of classes used	160/2		
Points (functional scores for factor 1.1 to be inserted into the evaluation form in figure 2)	80		

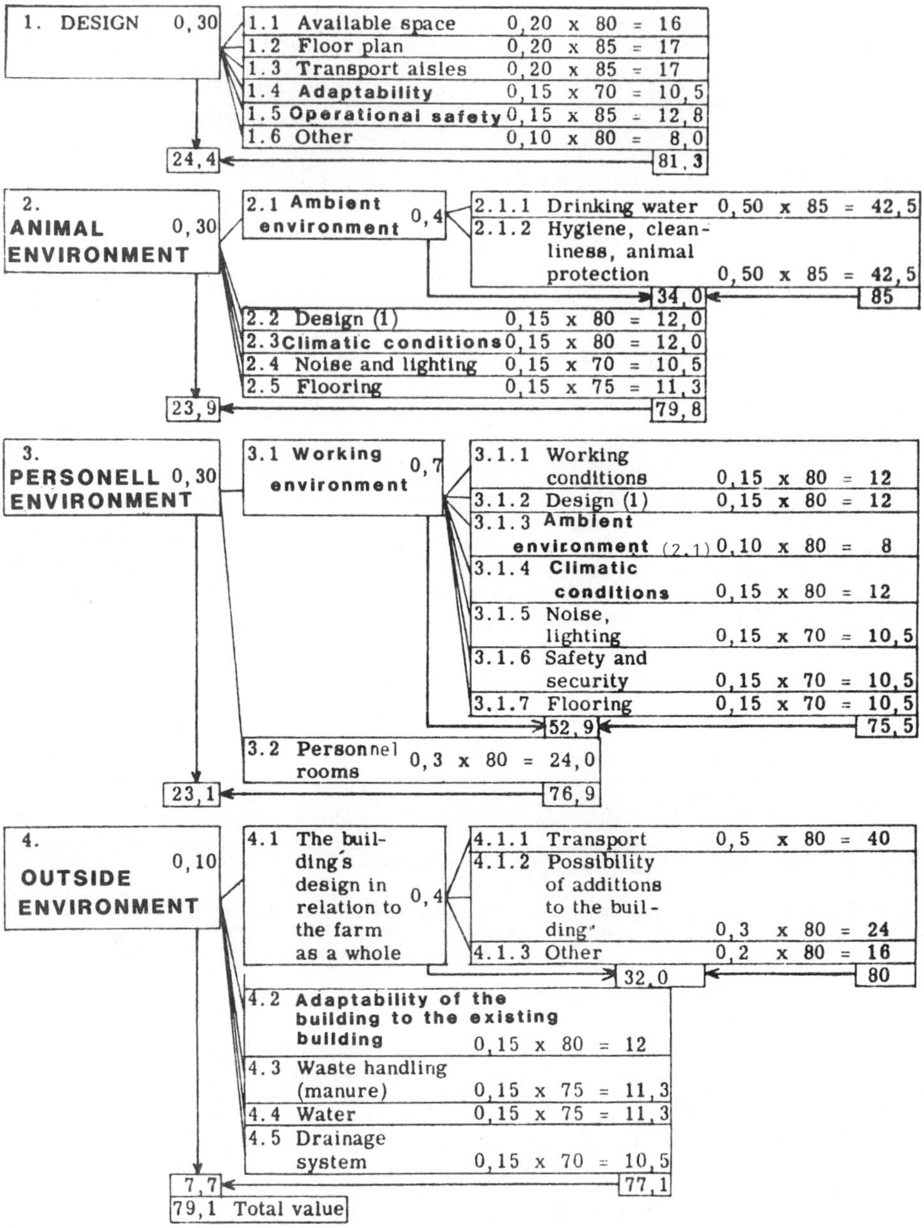

Figure 2

The degree of fulfillment of the stipulated requirements is expressed for each factor in points ranging between 70 and 90, whereby 90 is intended for the solution which embodies the highest imaginable degree of fulfillment, 80 is given for an average acceptable solution, and 70 is given for the solution which is most unsatisfactory. The selection of grading between 70 and 90 is a rather arbitrary suggestion originally made by a group of experts (Montelius op. cit.) and seems to serve its purpose well. To obtain the score for each factor, the percentage of Yes answers is calculated and entered into the suggested diagram in figure 1. Various principles can be used to combine the Class I and Class II Yes answers into one score. In order for a specific factor to obtain a score of 85 points, it may, for example, require a Yes answer for all Class 1 questions, together with Yes answers for half of Class II questions. The questions have been formulated so that a "Yes" answer is always positive, and a "No" always negative.

After all the factors have been assigned points which describe how well the listed requirements have been satisfied, a total score may be calculated by assigning the factors weights, according to the importance of the respective factors. The assigned weight is then multiplied by the calculated points for that factor. By adding all these products, a total score for the building design can be reached (see figure 2). The determination of the factor weights, which is discussed in more detail in the next section, is one of the most crucial problems in the whole evaluation.

Quantitative Assessment of the Relative Importance of Factors Determining the Function of a Livestock Building

How, then, should a weighting system be set up? A more or less "objective" distribution of weights is difficult to achieve. Therefore, it is natural to let the distribution of these weights be determined by the circumstances, so that the group making the evaluation should first discuss and then agree on the weight distribution for each separate case after they have agreed as to the characteristics or factors to be evaluated.

In the following, the results are presented of three experiments performed by the author to obtain subjective judgements of the importance of factors determining the function of a livestock building (Gustafsson, 1988, op. cit.).

The different factors to be considered were selected by the investigator in cooperation with a group of farm building experts. Factors were chosen within the following four main areas: general building design, animal environment, working environment, and outside environment. The factors chosen (see table 3) should make it possible to characterize major functions of a livestock building.

To obtain subjective judgements of the importance of each of these factors, a psychophysical scaling technique of magnitude estimation was used without any fixed standard. The technique was originally developed by S. S. Stevens (1975) as a method of measuring subjective sensations of the magnitudes of various types of stimuli. It has been applied in other sciences to scale subjective opinions about the importance of non-metric stimuli. Kvålseth (1980), for example, has presented such an application with reference to ergonometrics. The following draws on Kvålseth's general format.

A total of 23 factors were chosen. They are listed and briefly described in table 3. Among the about 60 staff members at a farm building research department, 13 persons were asked to participate in an experiment to investigate if it was possible to obtain reasonable magnitude estimates for the perceived relative importance of the 23 factors. The subjects were all in one way or another involved in the design and planning of farm buildings. Their ages ranged from 25 to 63 (mean: 42 ± 13 years), and they had worked with farm buildings from 1 to 27 years (mean: 11 ± 7 years). Three of them were females. Six had college degrees in engineering, and seven of them were trained at agricultural colleges.

All subjects were asked by the author to come at a certain time to a lecture hall within the department "to participate in an experiment." There, the group was given further written instructions and details about the experiment and told that they could withdraw if they now felt that they did not want to participate. Each person was given a booklet with each of the factors typed and explained on a separate page. The participants were asked to rate each factor at a time in terms of how important they felt each factor was in determining the general function of a livestock building (dairy barn for tied cows). They were given 15 to 20 minutes to complete the rating. The written instruction which was patterned after Kvålseth's format was as follows:

"On the following pages you will find some factors which are generally considered to be of importance when determining the function of a livestock building. The factors are arranged in a random order with each factor on a separate page.

"You are asked to assess in quantitative terms the relative importance that each factor has for determining the function of a livestock building. To the first factor (i.e., the one given on the first page in this booklet) assign any number that seems appropriate for your subjective feeling about the importance that this factor has for the general function of a dairy barn. For the following factors listed on the next pages, assign those numbers that you feel represent your impression of the factors' importance.

"For example, if you feel that the second factor is twice as important as the first factor, assign it a number that is twice as large as the first; if the second factor seems to be one-fourth as important as the first, assign it a number one-fourth as large, etc.

"Write each number at the bottom of each page. You may use whole numbers, decimals, or fractions, as well as a mixture of these. You may not use the number zero or negative numbers.

"Take one page at a time. Try to avoid looking back on earlier pages and avoid changing previous judgement. You have 15 to 20 minutes to complete the rating."

The mean and variation of the ratings of the various factors by the individuals were calculated. The geometric rather than the arithmetic mean was used since the data obtained from magnitude estimation generally appear to be log-normally distributed (Stevens, 1975, pp 279-280). The geometric mean, G, as defined as the nth root of the product of the items, X, in a series of observations, or:

$$G = \sqrt[n]{X_1 \bullet X_2 \bullet \ldots\ldots \bullet X_n}$$

The rating variations were following the procedure suggested by Stevens (op. cit.) determined in terms of the semi-interquartile range, Q, or the quartile deviation as:

$$\frac{Q_3 - Q_1}{2}$$

(In statistics, the concept quartile is used to characterize the dispersion of a frequency distribution. There are three quartiles, Q_1, Q_2, and Q_3, which divide the distribution into four equal parts. Q_1 is the lowest quartile, and Q_3 is the highest.)

It should be pointed out that since no standard was used, the data were transformed such that each subject's geometric mean rating across all factors coincided with the grand mean before the semi-interquartile range values were calculated. This means, that the difference between (a) the grand mean of the importance ratings of all factors by all individuals and (b) the mean of each subject's ratings for all factors was added to that subject's individual ratings.

The same experiment and procedure was repeated with two other groups of participants. In the first case, 39 people – 30 farm building extension service experts and 9 farmers – were used (average age 45 ± 8 years, average work experience 17 ± 6 years).

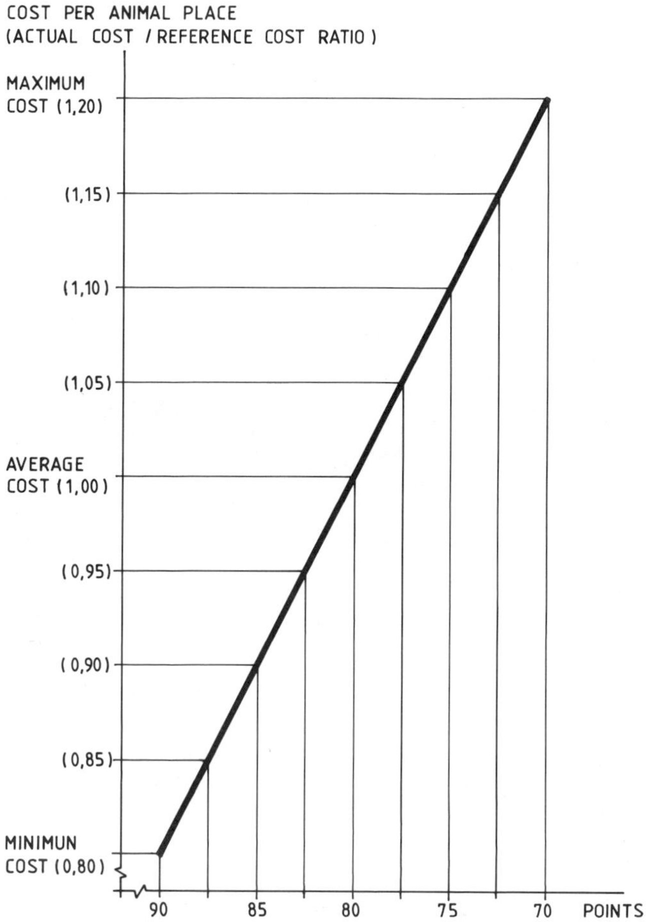

Figure 3

In another experiment in another part of the country a population of 25, 16 farm building experts and 9 farmers, participated (average age 43 ± 11 years, average farm building experience 13 ± 8 years). A statistical test of the ranking of factors (Snedecor's F-test, Moroney, 1951), revealed that there was a statistically significant degree of agreement in the three experiments. The three data sets were therefore combined; the importance ranking for all 77 individuals is presented in table 3. The results of the experiments should be interpreted with caution. It seems clear, however, that the psychophysical scaling technique of magnitude estimation is one possible method for ratio scaling of subjective perceptions of factors determining the function of a farm building. The mean values show a rather low variation which is somewhat surprising. On the other hand, it is quite reasonable to expect that farm building specialists with similar experiences and training should tend to have a consistent perception about the importance of different factors that are selected to determine the function of a livestock building.

For practical reasons, the mean values for different factors according to table 3 should be rearranged and transformed to make them add up to 1.0 within each main section and for each set of factors and subfactors before they are entered into the evaluation form as illustrated in figure 2. Other arrangements of factors and subfactors than are presented in tables 1 and 3 may be more appropriate under special circumstances. The character of the determining factors may change over time and with modifications in federal legislation, farm policy, regional regulations, building practices, and the like. As an example of this, it can be mentioned that in 1988 the Swedish Parliament passed a new Animal Protection Act for domestic animals and animals held in captivity. It contains very stringent requirements with regard to building space, environmental control, management systems, etc. and will certainly call for a revised set of attributes to be used when evaluating animal habitability in the future.

After the total score has been calculated, by finally adding the sums of the different factors and subfactors in the evaluation form (figure 2), the appraisal may be considered finished. However, in most cases it is also of interest to include the cost factor in the scheme and to carry out the functional evaluation in relation to and in combination with an estimation of the building costs. How this can be done will be further discussed in the following section.

THE TRANSFORMATION OF BUILDING COSTS INTO FUNCTIONAL SCORES

In an initial attempt to include in the evaluation scheme the costs of the various building alternatives considered, the following procedure has been employed:

1. Calculation of the "normal" capital investment cost for the kind of building to be evaluated. Standard building cost data, often produced normally by government agencies such as the Swedish National Agricultural Board, can be used to obtain a point of reference for the investment cost of a building specified with respect to essential features, such as type of livestock, size, general layout, etc. The evaluation task may, for example, refer to a free-stall system for 40 silage-fed dairy cows, with a liquid manure handling system, or a 2x4-meter walk-through milking parlor. In the case of a livestock building, it is practical to express the reference cost, C, in terms of investment cost per animal place.
2. Determination of "the lowest conceivable investment cost" for the particular building. For practical reasons the amount 0.80C is suggested.
3. Estimation of an upper limit for the investment cost per animal place over which a profit of the livestock business cannot be expected. Here, the value 1.20C has been considered a reasonable ratio under present Scandinavian conditions.
4. Making a graph according to figure 3 for the transformation of the building costs, assigning the higher investment cost (1.20C) a value of 70 points and the lower cost (0.80C) one of 90 points. Again, the selection of the maximum and minimum values of 90 and 70 is rather arbitrary, but it results in a simple and manageable system that will probably be refined when more experience is accumulated. The general idea is to "reward" with a high score a cheap but technically satisfactory building and to "charge" an expensive one by giving it a low cost score.
5. Marking the actual calculated investment costs per animal place on the graph on figure 3 and reading off the points. The actual costs are determined by employing the more detailed calculations normally carried out by farm building contractors.

TABLE 3
Mean and variation (semi-interquartile range) of magnitude estimates for the perceived importance of the various factors (N=77)

Factor No.	Factor Description	Importance Mean	Importance Variation
1.	Design, available space, floor plan (arrangement)	29.35	2.54
2.	Animal environment, hygiene, animal welfare, housing of calves, design of stalls	29.34	3.51
3.	Personnel environment, working environment with respect to work posture, carrying, lifting	29.09	3.73
4.	Personnel environment, working environment with respect to railings for the fencing of floor openings, gangways	28.87	2.21
5.	Outside environment, the location of the barn on the farm, distance from house to barn	28.56	1.72
6.	Design, transport aisles, width, floor level	28.30	1.70
7.	Animal environment, climate control	27.46	5.94
8.	Outside environment, access to water, water quality	27.10	2.79
9.	Design, reliability of machinery, alternative milking, feeding, and manure handling methods at power breakdowns	25.08	2.43
10.	Outside environment, transports, roads for (bulk collection) trucks	25.07	2.21
11.	Personnel environment, working environment, climate control, temperature, draught	25.04	1.83
12.	Animal environment, watering system location, accessibility	24.37	1.64
13.	Animal environment, floors and stalls, non-slip surface, cleaning	23.94	1.66
14.	Personnel environment, working environment, noise attenuation, lighting	21.84	1.54
15.	Personnel environment, working environment, floors and walkways, non-slip surface, cleaning	21.80	1.66
16.	Outside environment, manure storage water pollution	21.03	1.30
17.	Animal environment, noise attenuation, lighting	20.07	1.31
18.	Outside environment, the barn's design, size, and shape in relation to other buildings on the farm	19.82	2.46
19.	Design, flexibility, adaptability to other alternative purposes	19.19	2.46
20.	Design, construction of floor, walls, ceiling, etc., with respect to structure, resistance, strength, insulation	18.32	2.00
21.	Outside environment, the location of the barn, interference with future barn expansion, soil conditions	17.37	2.29
22.	Outside environment, drainage, hard top surface areas	16.48	2.64
23.	Personnel environment, working environment, personnel service room, design and facilities	16.15	2.37

TABLE 4
Examples of results of field evaluations of selected livestock building alternatives using the proposed method

Type of building evaluated	Actual capital investment cost per animal place as a fraction of reference cost	Building scores	Functional scores	Total scores
Dairy barn:				
Loose housing				
- Alt. 1, 50 cows	0.94	83	77	80
- Alt. 2, 50 cows	0.80	90	78	84
Tie stall				
- Alt. 1, 30 cows	0.95	82	82	82
- Alt. 2, 30 cows	1.08	79	79	79
Sow house:				
- Alt. 1, 46 sows	1.04	78	83	81
- Alt. 2, 46 sows	.80	90	77	84
- Alt. 3, 46 sows	.80	90	72	81
Hog house:				
- Alt. 1, 300 hogs	1.12	74	82	78
- Alt. 2, 300 hogs	.97	81	82	82
- Alt. 3, 300 hogs	.94	83	81	82
- Alt. 4, 300 hogs	1.00	80	87	84

The points determined for the actual investment costs may then be assigned a weight and the resulting figure added to the total functional score in figure 2. Using our practical field tests as a convention, the same weight has been assigned to the points read from the building cost graph in figure 3 as to the functional scores generated in the evaluation form in figure 2. After that, it is possible to make comparisons between either complete buildings or parts of buildings using just one quantitative measure for each object. For the final selection of the best option, the individual or group of people performing the evaluation has to take a position on the question of how great the difference in points must be in order to be decisive.

ASSESSING THE THE APPLICABILITY OF THE METHOD IN THE FIELD

The ideas and evaluation procedure developed in the previous sections have been tested in a limited number of field studies (Jonsson, 1982; Gustafsson, 1988). These studies have in most cases been carried out in collaboration with the Regional Agricultural Board in southern Sweden and included buildings for the most common livestock, i.e., dairy cows, sows, and hogs. The field assessments have been comprehensive and covered determination of both the functional value and the building cost score. Some examples of final results produced during field tests of the applicability of the method are presented in table 4.

The overall experience of the field investigations was that the proposed methodology is applicable in real-world situations. This refers to post-occupancy evaluations as well as assessments of plans and other building program documents for farm structures to be erected. The empirical results are promising, but call for certain refinements of the procedures employed. Some practical problems and advantages will be commented upon.

CONCLUSION

Magnitude estimation of subjective opinion using rating scales appeared to be an appropriate technique for the assessment of various factors and attributes chosen to determine the merits of livestock buildings. Many other principles for establishing a weighting system are feasible, such as cost, number of questions on the check-list related a given attribute, and the position of the attribute within the evaluation hierarchy (table 1, main section, factor or subfactor).

When selecting the sets of factors or attributes to be considered in the evaluation, one must make sure that all essential building features are accounted for. All the members of the evaluation team must agree upon the principles and procedures to be employed.

The method we used involved too many check-lists, forms, and questions for them to be handy under field conditions. There are, for example, 290 questions to be answered when evaluating a building for tied cows or sows, and 370 in the case of a barn for loose-housed dairy cows. The whole procedure has therefore been computerized and transformed into a question-and-answer dialogue between a personal computer and the evaluating team.

The result of an evaluation is naturally greatly dependent upon the extent to which the chosen weighting and point systems result in meaningful differences between alternatives. A position must be taken by the evaluating group regarding the question of how great these differences must be in order to be decisive. As a rule of thumb, a difference of at least 5 to 6 points should be required.

The advantage of performing a quantitative evaluation of complex built environments such as animal habitability in farm structures is that it results in coordinated, joint measurement and a common denominator for a series of completely different entities. It also forces the members of the evaluation team to make a systematic review of the object they are evaluating, together with presenting an open account of how the evaluation was made.

Scientific studies of animal behavior are nowadays more and more used in the design of housing specifically adapted to the animal concerned. The animal and its physical environment and the animal in relation to other animals and to man are studied by gathering data on health, physical welfare, levels of comfort, production level, social tension among animals, and human safety and working conditions.

Therefore, future evaluation methodologies will probably be much more sophisticated than the one presented here and will be tied into a computerized system, employing CAD (computer-aided design) and the like for the development of drawings and other documents. New data, experiences, and results from current scientific studies and field tests will continuously be brought into the system to keep it updated and refined.

REFERENCES

Gustafsson, B., 1988, unpublished material.

Gustafsson, B., and Montelius, J., 1978, *Evaluation of the performance of farm buildings*, The Swedish University of Agricultural Sciences, Allmant or 12, Uppsala. (In Swedish)

Jonsson, I., 1982, Valuation of farm buildings and dwellings 1974-1981, Special Report No. 122, Department of Farm Buildings, Lund. (In Swedish)

Kvålseth, T. O., 1980, Factors influencing the implementation of ergonomics: An empirical study based on a psychophysical scaling technique, *Ergonomics*, Vol. 23, pp. 821-826.

Montelius, B., Kemnefeldt, S., Jonsson, T. and Gustafsson, B., 1981, Evaluating the function of farm buildings, Progress report, Department of Farm Buildings, Lund. (In Swedish)

Moroney, M. J., 1951, *Facts from figures*, Penguin Books, Harmondsworth, Middlesex.

Murray, W. G., 1969, *Farm appraisal and valuation*, The Iowa State University Press, Ames, Iowa.

Stevens, S. S., 1975. *Psychophysics*, John Wiley, New York.

Suter, H. C., 1974, *The appraisal of farm real estate*, The Interstate Printers and Publishers, Inc., Danville, Illinois.

CHAPTER 8

TOWARDS AN ACTIVATION PROCESS MODEL

Wolfgang F. E. Preiser, John P. Petronis, and Ingrid B. Vigil

**Architectural Research Consultants, Inc.
Albuquerque, New Mexico**

ABSTRACT

Activation is the process of preparing people and a new or renovated facility for move-in and operation. It is a complex and little-understood process, particularly for hospitals, where the health, safety, and care of patients is the highest priority and where systems must be operational and fault-free on move day to safeguard a cost-efficient and smooth transition (especially when multi-phased construction and moves are involved). Activation includes the selection and ordering of equipment and supplies, the hiring of new staff, staff training, move planning, facility readiness, culminating in the move itself and facility operation. The challenge of the project described here was to clarify the meaning of activation, to develop an incipient "activation process model" and to demonstrate that activation brings a new focus to and expands the scope of the Post-Occupancy Evaluation (POE) process. This model describes the significant phases, steps, decisions, interdependencies, resources, policy and procedural guidance and time expenditure that are involved in activation.

Research for this project focused on in-depth study of a large facility which had recently activated. After the data gathering and document analysis phase, workshops were conducted with project and move coordinators from four similar facilities throughout the United States, two of which had recently been activated and two of which will activate new facilities within the next two years. These workshops provided a unique opportunity for "knowledge transfer" and action research by applying the lessons learned in facility activations to those future facilities that will benefit most from them. More importantly, the newly developed process model was critiqued by those who had undergone activation. The process model will be tested in facilities of varying complexity and scale around the country. The innovative process model views activation from the management perspective. With further development, it will evolve into a more detailed model that can be adapted and applied to the activation of any large and complex facility.

The post occupancy study of a facility's construction and operation should be expanded to include the activation process. A successful activation is a key determinant of long-term operational success. The value of transmitting activation knowledge for organizations will be in the activation of new facilities, with improved training of personnel, better and more timely facility readiness, improved morale, and last but not least, significant savings in personnel, time and overall cost.

INTRODUCTION

This chapter describes key findings and recommendations that resulted from the evaluation of the activation of a new and major facility in Albuquerque, New Mexico.

The study focused primarily on the local activation experience of a new facility that is part of a national organization. Representatives from two other locations that will soon activate a new facility, and two locations that had recently activated major and new facilities, and over twenty representatives from the organization's headquarters also participated. The study's major findings, issues, and recommendations, were presented at a workshop and received their consensual validation.

This section summarizes the findings and recommendations that addressed the principal purposes of the evaluation, which were to:

- Develop a generally accepted definition for Activation.
- Develop a structured Activation Process Model for use and development at other facilities.
- Identify key activation issues that affect facility staff and users.
- Develop and test an expanded scope of Post-Occupancy Evaluation.

During the course of the study, several other key issues emerged that involve and affect the organization's headquarters policies and procedures. They are briefly described below.

METHODOLOGY

Various methods were used to evaluate the facility's activation experience. They included interviews with people involved in activation, analysis of the literature on activation, analysis of general information and facility-specific documents on activation, administration of an extensive questionnaire to all departments, and finally, workshops to clarify, refine and focus information.

Interviews

Many interviews and meetings were held with members of the facility's staff who had been a part of activation, including multiple meetings with the move coordinator and the chief engineer and his staff. These people, in particular, helped in the understanding of the activation process, and assisted with the refinement of the Activation Process Model. They also provided access to activation documents.

Document Analysis

One of the first steps taken in conducting this evaluation was a search for existing literature on activation. The field, as one for formal study, is rather new. Little has been

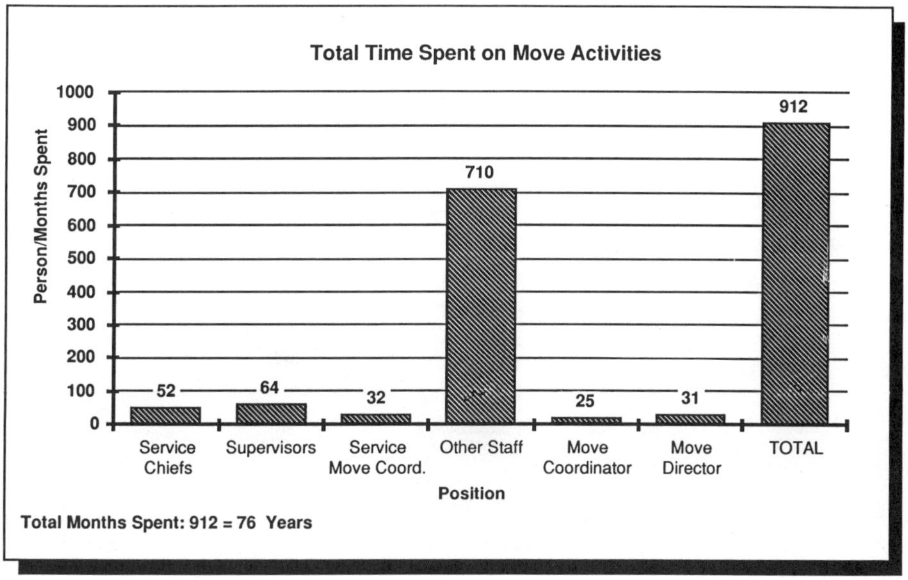

Figure 1

written on activation and even less is readily available in journals. The articles found on the topic tended to be general in nature, lacking substance that might be of assistance for an activating facility.

Documents from the organization's headquarters and the facility were identified and reviewed. Headquarters documents primarily involved existing policies and legislation. Documents provided by the facility were useful in understanding the activation process and its timing. They included four years of agendas from the facility's activation steering committee. The agendas helped to clarify the time sequencing and activities that took place during activation. A formal move and transition plan explained the planning for the actual start-up of the new facility. Departmental move plans illustrated the tasks and scheduling planned by each department to carry out the transition to the new building. Documents such as "showstopper" lists, critical item lists, "to do" lists, and completion item lists indicated the types and complexity of facility-related items to be completed. The document search also yielded copies of forms created for use during activation that could be useful to other facilities.

Survey Questionnaire

A survey questionnaire was prepared and distributed to all departments. Approximately 90% of departments completed the questionnaire. All departments that were heavily involved in activation responded. Departments were asked to quantify activation in terms of time expended, personnel, and dollar costs involved. They were also asked for opinions on issues such as adequacy of planning input, quality of training, and guidance materials, as well as the success of the move itself. In the following, some highlights of the survey data are presented:

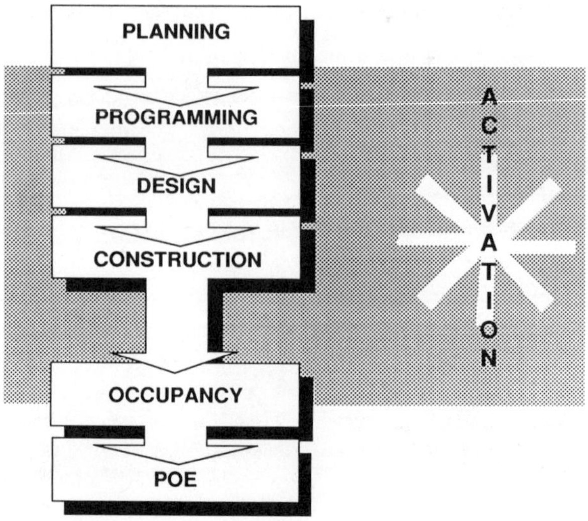

Figure 2. Activation spans aspects of the building delivery process.

FINDINGS

The experience of replacing an aging 250,000-square-foot facility with a new 550,000-square-foot facility employing 1200 provided a unique opportunity to examine the Activation process. The data gathering process quantified some significant information on complexities involved:

Time Spent on Activation – Questionnaire results indicate that an extremely large amount of time is necessary for activation. The total amount of time spent by department chiefs, supervisors and staff on activation of this facility was reported as 76 person/years. If equated to dollars, this totals approximately $3 million (based on an average cost of $40,000/person/year, including salary and benefits). This amount represents approximately 3% of the total project cost, i.e., construction and equipment expenditures. The majority of the cost/time spent on activation was absorbed into regular operating budgets by the facility. Departments reporting the most time spent (person/months) on activation were: Building Management – 228 person/months, Move Director's staff – 144 person/months, and Engineering Maintenance/Repair – 214 person/months.

Not surprisingly, 57 out of 76 person/years (75%) were spent on the lengthy pre-Start-Up planning and tasks. 7.5 person/years were spent on the brief Start-up phase and the remaining 10.6 person/years were expended during the Operation Phase.

59 out of 76 (78%) person/years spent on activation were by staff members, with the remaining 22% or about 17 person/years spent by department heads, supervisors, and Move Coordinators.

Training and Orientation – Training and orientation were important factors in activation preparation. Departments with large amounts of training were those with heavy responsibilities for facility operation, or departments with large staffs and new high-tech equipment. Departments were asked to rate training and orientation quality from the various providers. The rating scale used ranged from 1 to 5, with 1=poor and 5=excellent. The quality of equipment training was generally rated below average, with a rating of 1.76 for contractor-provided training and 2.5 for vendor training. Training provided by the facility's organization was rated at 2.21, while in-house orientation was received quite favorably and was rated at 3.03.

Activation Planning and Execution – Departments were also asked to rate the quality of the move, again on a scale from 1=poor to 5=excellent. The quality of overall activation planning for the entire facility was rated quite favorably at 3.25. Specific department moves rated their own move plans at 3.16, with the entire move execution receiving a rating of 3.38.

Temporary Staff – A total of about 60-65 temporary staff were reported hired by various departments, with Engineering Department/ Maintenance and Repair, Building Management Department, the Move Director and staff requiring the bulk of temporary labor. The need for temporary staff was greatest during the people/facility readiness phase – or six months before Move Day. The total cost for departments using significant numbers of temporary staff were: Building Management Department, $400,000; Engineering Maintenance & Repair, $372,000; Move Director's Staff, $200,000.

Workshops

Three workshops were held with personnel from the activated facility and headquarters to report on information gathered and to present the activation process model for critique and clarification. The major and final workshop was held in Albuquerque to present initial study findings to a broader audience. There were 28 attendees plus the project team.

As outlined above, several key findings were identified from the questionnaires, interviews and focus sessions that were conducted with the management and staff of the Albuquerque facility, based upon their experience during the activation of that new facility. These findings warrant the special attention of managers and staff who will be planning the activation of new facilities in the future.

THE FACILITY ACTIVATION PROCESS MODEL

The Activation Process Model describes the Albuquerque activation experience, as interpreted by the project team. It is a project management tool, outlining the logic and sequence of decisions, tasks, durations and dependencies. The Activation Process Model is an outline that identifies tasks without detailed description. The future challenge is to test this model at other facilities, to fill in information gaps and ultimately to refine the model from its present preliminary state to a more generalized model. In this evaluation, the activation process used by the Albuquerque facility has been documented and explained for the first time, defining its distinct phases and tasks. Furthermore, the process model presented here is a combination of what actually happened and recommendations as to what should happen during the activation process.

The Activation Process Model consists of the five phases described below.

PHASE 1: INITIATION

Purpose

The Initiation Phase identifies and requests activation funding in general, and plans and coordinates the specific funding and ordering of equipment and furnishings.

Timing

The activation process starts during the programming and preliminary design process (possibly more than 5 years prior to Move Day, depending upon facility size and complexity) and ends about 12-18 months before Move Day. Some aspects of equipment and furniture ordering and receiving may extend past Move Day.

Resources

In the Initiation phase, top management at the facility is involved in preliminary funding decisions and long-range allocation of staff. A formal Activation Steering Committee may be functioning but actual planning for the move has not yet begun. Initiation Phase activities are carried out by the facility and headquarters staff responsible for the construction project.

Major Tasks

1. Budgeting for Activation.

During this step, initial activation and impact funds are estimated. Activation and impact funding issues shall be considered, such as:

- Commercial movers
- Visits to other facilities
- Project staff and special personnel
- Warehouse costs
- Overtime
- Contract personnel
- Training costs

2. Working with the Design Team to determine basic Equipment/Furniture needs and preliminary specifications.

A significant lead time is usual between design and activation. The inevitable technological changes that may occur affecting equipment and furniture specification should be anticipated.

3. Equipment Identification and Ordering.

This process includes policies and guidelines related to equipment and furniture requirements for major and minor construction and renovation projects.

4. Furniture Identification and Ordering.

The process for furniture requests is similar to that for equipment requests.

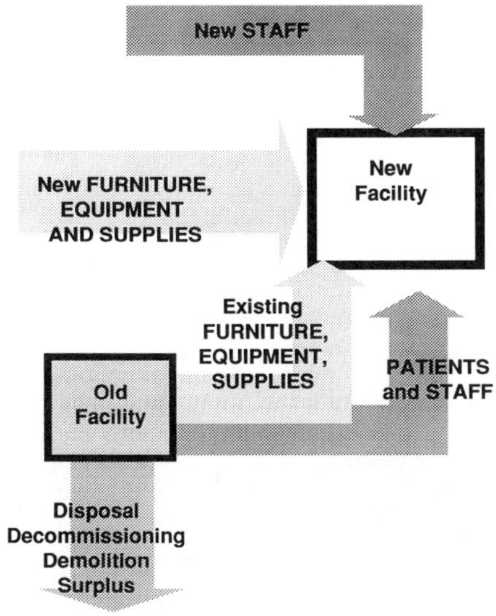

Figure 3. Activation is complicated when a new building replaces an old one.

5. Receiving and Warehousing Equipment and Furniture.

Equipment and furniture will start to be received any time from this point to Patient Move Day. It is important to anticipate the major impact that activation planning will have on the Supply Department including:

- Budgeting for use of temporary storage space, either on-site or off-site.
- Organizing the warehouse for efficient transfer of stored items.
- Allowing adequate lead time for equipment orders, especially for complex or one-of-a-kind items.

6. Impact Funds Identification and Requests.

PHASE II: ORGANIZATION/PLANNING

Purpose

The purpose of this phase is to mobilize facility management for organizing people, equipment, and resources through an activation plan, to create detailed procedures and planned actions in preparation for the move.

Timing

This phase starts 12-18 months prior to patient move (lasts up to 9 months) and ends 3-6 months prior to Move Day. It is important not to start this phase too early, risking premature dissipation of enthusiasm for the activation process.

Resources

- Facility Top Management
- Move Steering Committee
- Move Director and Dedicated Staff
- Construction Project Manager and Staff Departments
- Headquarters

Major Tasks

1. Appoint Activation Steering Committee

The appointment of a Steering Committee is the first visible activation activity. The Steering Committee represents the interests of top facility management as well as those of all of the major departments. The purpose of this committee is to clearly define activation roles and authority.

2. Organize Move Committee

The Move Committee is the implementation instrument for activation and oversees the entire process. The Move Committee monitors two distinct functions: the "people" move and the moving of furniture, equipment and supplies. Each of these functions should have a coordinator, supervised by the Move Director. These are key activation positions; it is vital that they be filled by individuals with proper qualifications and experience. Professional and operational experience is particularly important for the Move Coordinator, and Supply Department, Engineering Department and Building Management experience is useful for the Move Coordinator.

An important responsibility of the Move Committee is to create effective, frequent communication with all persons and groups involved in activation (patients and their families, facility staff, headquarters personnel, and the public) to create enthusiasm and "esprit de corps." Professional staff involvement is particularly important in the process.

3. Determine Move Approach

The Move Committee must determine an overall move approach. Key elements of the Albuquerque facility's approach were to:

- Delegate move responsibilities to individual departments
- Hire commercial movers for the majority of equipment and furniture to be moved to the new facility

Other approaches may be available, including delegating the majority of the process to commercial movers. Other move approach elements of the facility activation experience include:

- Identifying move priorities, including:
  - Moving those with overall preparatory or security functions at the earliest date.
  - Non-essential activities.
  - Essential activities.
- Compressing move into a few weeks to minimize disruption.
- Lowering the workload in critical areas to the maximum practical extent.

- Keeping life support room and other essential activities functional at all times.
- Retaining adequate telecommunications.

4. Delegate Responsibilities

Responsibilities for department moves are formally delegated to individual department chiefs.

5. Prepare and Disseminate Move Plan Guidance

Comprehensive guidance is developed for department moves to assure highest quality performance. Department plans include:

- Identify the who, what, where, and how of department actions.
- Describe department operations including:
  - How will the department operate (before, during, after move).
  - Key relationships to other departments
  - Backup requirements (duplicate supplies, additional personnel, etc.).
  - Staffing levels and overtime required.
  - Outside resources required.
  - Needed training programs.
  - Vacation rescheduling.
- How will space be used including location of furniture and equipment and space assignments.
- Supplies (what, how much, where placed, what to get rid of).
- Equipment (What testing is required, training and maintenance is required).
- Personnel roles.
- Telephones, paging system, computers and mail requirements.
- Transportation of people/equipment.
- Security (egress/ingress, type and security, type of doors that are locked).
- Parking and location Directions.
- Work/Activity/Material Flows.

6. Prepare and Present Department Move Plans

Move plans are prepared by individual departments, incorporating the items described above.

7. Approve (Revise) Move Plans

The Move Committee with facility management reviews and approves department move plans.

PHASE III: PEOPLE/FACILITY READINESS

Purpose

The People/Facility Readiness Phase is used to prepare the building and its new occupants for the move. Equipment and furnishings will be moved into the building, checked and calibrated. Orientation to the building and training for equipment operation and maintenance are conducted.

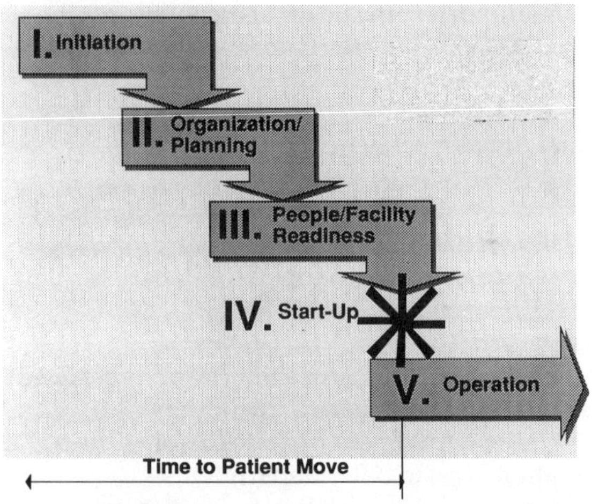

Figure 4. Activation process.

Timing

3 to 6 months prior to move.

Resources

- Move Director
- Resident Engineer
- Education Department
- Engineering Department
- Supply Department
- Building Management
- Affected Administrative and Clinical Departments
- Equipment Vendors/Contractors

Major Tasks

1. Identify and Prepare Orientation and Training Materials.

A lead training group is chosen to coordinate orientation and training of facility staff (The Education Department is a logical choice for this role). This group develops and implements a Training Master Plan which establishes specifications and expectations for facility training and orientation. The plan budgets adequate resources for training, including potential out-of-town equipment training sessions, and includes clear specifications for contractor and vendor training components.

2. Departments conduct Tours and prepare "To Do" lists.

As soon as is practically feasible, individual departments should examine the new facility, both through study of the architects' drawings and/or model and by visits during construction. Prior to beneficial occupancy of the facility, departments should be provided ample opportunity to check their individual areas in the new facility, accompanied by Engineering staff. Detailed "to do" lists are generated for items that need to be completed prior to acceptance and occupancy.

3. Initiate Orientation and Training.
Orientation and training in the new facility is conducted as soon as possible after beneficial occupancy. Use of "mock-up" clinical ward areas – such as patient rooms in a hospital, for example – are very useful for orientation and training.

PHASE IV: START-UP

Purpose

The Start-Up phase assures a smooth move-in and transition with new and reused furniture and equipment, causing minimum disruption to facility operations. The "public" population is reduced. Move activities are phased according to priority, culminating in Move Day.

Duration

Begins up to 4 months prior to Move Day. Culminates in moving building occupants.

Resources

- Top Management
- Move Director and Staff
- Departments
- Commercial movers
- Temporary/Auxiliary Department Providers

Major Tasks

1. Estimate Move Requirements.
The Move Committee and each department must realistically estimate the amount of equipment, furniture, supplies and personal belongings to be moved and budget adequate time for department moves.

2. Identify and Hire Mover.
A reputable commercial mover with experience and resources to perform the move in the required time period is retained. Use of commercial movers or using available facility staff for portions of the move are options that should be considered.

3. Move New Furniture and Equipment.
As soon as is practically possible, new furniture and equipment are moved into the new facility. Anticipate and provide adequate staff to move equipment from the warehouse to the new building. Special contracts to move certain equipment should be considered, as well as the need to supervise commercial movers.

4. Move Departments by Priority.

Departments are moved according to priority established in the Move Plan. All furniture, equipment and supplies that will be reused in the new facility are moved at this time.

5. Move Staff.

The Staff Move is the single most significant milestone of the activation process. The move closely follows a previously prepared plan (Phase II: Organization/Planning). While some preparations occur up to 4 weeks prior to the move, the Move itself occurs in the shortest time safely feasible. Avoid move delays since they can be very costly, with significant loss of revenue and additional direct expenses for movers and temporary staff.

Preparations for the move include:

- Reducing the facility population census by:
  - Temporarily discontinuing non-essential activities
  - Limiting public visits.
- Ensuring that all signage is installed by the time of department moves.
- Providing adequate security at both old and new facilities.
- Assuring that the telephone system, intercom and computers are working.
- Preparing and identifying staff for every activity, identifying and using volunteer help and providing proper orientation.
- Preparing staff including:
  - Obtaining staff consent for pictures and interviews
  - Identifying records and personal belongings.

Strict organization and discipline are required for the Move. This includes designating move dispatchers and receivers for each ward. Various forms, checklists and logistical procedures aid in the effort.

PHASE V: OPERATION

Purpose

The Operation phase completes all move activities and the remaining "to dc" list items (Engineering Department); remaining punchlist items dating from building acceptance, and any other minor adjustments that may be required as departments begin to operate in the new facility. The old facility, if any, is deactivated and essential adjustments are made to the new facility.

Duration

From Move Day to up to one year after move.

Resources

- Project Director and Staff
- Move Director and Staff

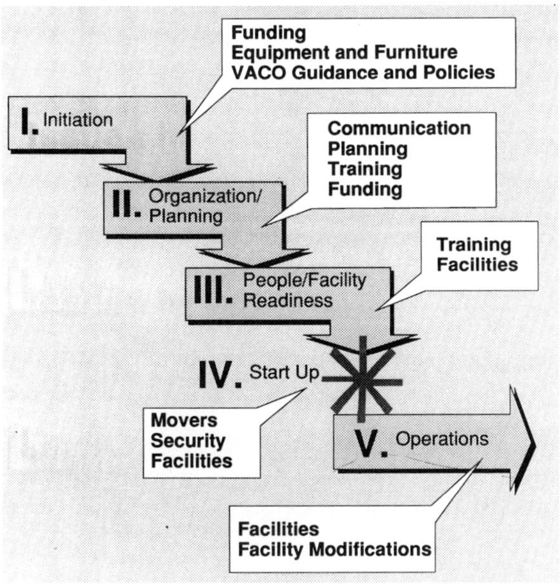

Figure 5

- Supply Department
- Building Management
- All Affected Departments
- Engineering Department

Major Tasks

1. Secure Old Facility.

Secure the "old" facility to prevent loss of equipment and furniture that may be left behind.

2. Excess Furniture and Equipment.

Furniture and equipment that is no longer needed is disposed of. Provide adequate time for departments to identify and review the excess material.

3. Decommission Old Facility.

4. Demolish Old Facility.

5. Identify and Budget One-Year Modifications.

During facility operation, significant items requiring adjustment are identified. In some organizations, modifications requiring major construction have a moratorium of one year from building acceptance before they can be funded.

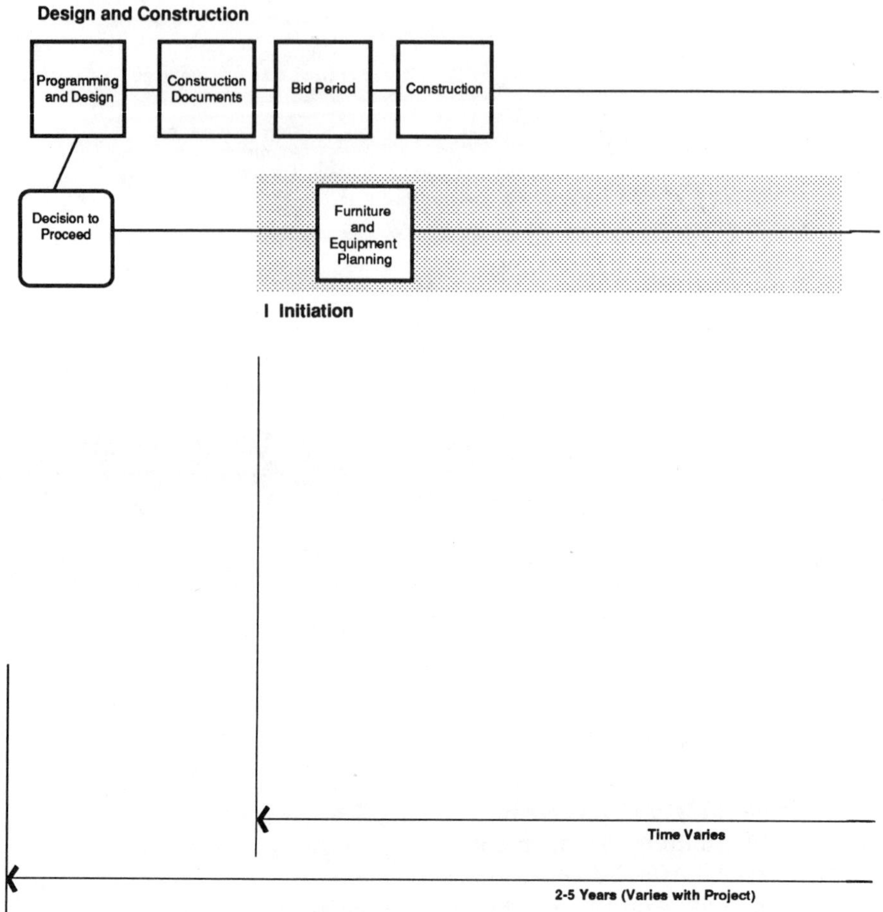

Figure 6

CONCLUSIONS

The Evaluation of the Recent Facility Activation Successfully Demonstrated the New POE Approach

There was little written guidance material available to the Albuquerque facility for their activation effort. As other facilities before them had done, the Albuquerque facility contacted other staffs and organized on their own to identify and resolve many unfamiliar problems. In turn, after they had successfully activated their new facility, other facilities contacted them for help, continuing the process of "reinventing the wheel."

The Chief Engineer at the Albuquerque facility and key members of his staff recognized the need to evaluate and to document their experience, and to develop an Activation Process Model that could be used, and continue to be developed, by other facility staffs. A study was proposed, using the recently developed Post-Occupancy Evaluation approach, that included the following features:

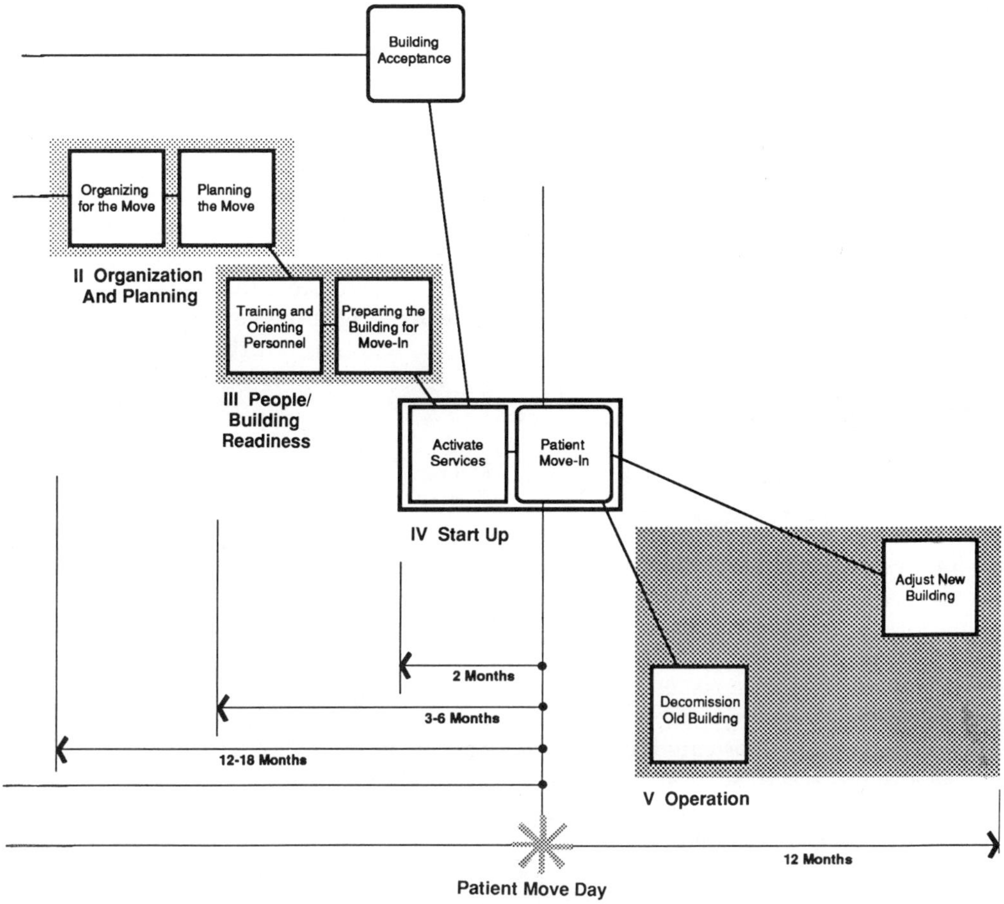

Figure 6 (continued)

- Sponsorship by a direct user-facility.
- Orientation toward a specific issue that directly affects the building delivery process.
- Participation of four other facility staffs. Two had recently activated major new facilities. Two would be activating major new facilities.
- Involvement of the Regional and Central Offices of the sponsoring organization as well as its program and construction staff representatives.
- Evaluation and description of activation problems and alternative approaches.
- Identification of key activation issues and proposals for resolution.
- Development of a generalized Activation Process Model for immediate use by other facilities.
- Use of available documents, data, questionnaires, focus groups and workshops.
- Presentation of a final report for use by other facility staffs and program officials.

Evaluation of the Facility Activation Process

The evaluation redefined Post-Occupancy Evaluation (POE) to include the evaluation of the process of facility activation and development of a process model for activating other new facilities.

A More Focused and Proactive POE Approach Was Used to Evaluate Recent Facility Activations

The sponsoring organization recognized that the outcome and effectiveness of conventional Post-Occupancy Evaluation (POE) activities needed to be improved. Traditional POE methods, including activation, have not produced the expected results. By focusing in on the process of activation and extending it onward by sharing and clarifying procedures, information, issues and lessons learned, the organization expects to help new facilities through the complexities of activation, saving time, effort and money.

The Activation Process Evaluation Achieved its Purposes

The evaluation of the activation process used in the start-up of the new building in Albuquerque, NM succeeded in meeting the principal objectives of the study:

- An Activation Process Model is now developed for immediate use, and for further testing and development, at other facilities.
- Key activation issues are now identified and documented for immediate use.
- Specific task-oriented recommendations are now available, to organize and manage activation activities at new facilities.
- Major components of the current approach to conducting POE studies are further validated.

As a direct result of this evaluation study, many of the recommended improvements continue to be made in programming, design, construction and project management practices, to directly improve both the activation process itself, and activation activities at new facilities. Other workshops are now being planned to continue communication and to further develop the Activation Process Model.

A Task Group is addressing activation process issues, organizational roles and responsibilities that directly involve headquarters activities identified during the study. A computer program is now being used to manage activation funds, and equipment requests and approvals. A policy initiative has been proposed to permit the organization to more effectively budget and to provide funds earlier for the activation of new facilities.

Most importantly, attention is now focused directly on the activation process and on addressing key issues that often determine the effectiveness, success, and acceptability of new facilities.

NOTES

1. The authors wish to thank Mr. Ron Richter, Chief, Engineering Service and Mr. Tom Casper, Project Coordinator, VAMC Albuquerque, as well as all participating staff of the organization for their excellent collaboration in this project. Furthermore, we wish

to thank Mr. Irwin Axelrod and Mr. Dick Kelly of the Veterans Administration Central Office and Regional Office, respectively, for their foresight and guidance in this project.

2. Reports and further information on the project can be obtained from: Architectural Research Consultants, Inc., P.O. Box 1158, Albuquerque, NM, 87103. Phone: (505) 842-1254. Fax: (505) 293-0100.

BIBLIOGRAPHY

Architectural Research Consultants, 1988,*VAMC Albuquerque, New Mexico Activation Process Evaluation,* Albuquerque, NM, ARC, Inc.

Boucher, M. P., and Hobbins, A. J., 1986, "Thorough Planning Creates Smooth Move for Parkwood Residents, *Dimensions,* November 1986, pp. 55-57.

Bunning, R. L., 1982, "A Moving Orientation," *Hospital Topics,* May/June 1982, pp. 26-29.

Bunning, R. L., 1984, "Coping With Large-Scale Change," *Hospital Forum,* May/June 1984, pp. 71, 72, 74.

Galvagni, W., 1976, "Planning Facilities Moving Two Hospitals Into One New Center," *Hospitals, J.A.H.A.,* February 16, 1976, pp. 61-64.

Handel, D. J., Hilling, L. N., and Lingo, K. H., 1983, "Transition Planning – An Integrated Approach," *HCM Review,* Fall 1983, pp. 61-67.

Hanlon, M. E., 1976, "Hospital Follows Two-Phase Plan for Move into New Building," *Hospitals, J.A.H.A.,* March 1, 1976, pp. 69-72.

James, M. V., 1985, "How Pacific Bell Made Its Enormous Relocation Work," *Office Administration and Automation,* June 1985, pp. 56-58, 60.

Mark, K., and Rowe, C., 1982, "Minimizing the Problems of Hospital Relocation," *Dimensions,* June 1982, pp. 22-23.

National Institute of Building Sciences, 1985, *Hospital User Manual Series Guidelines, Volume 1: VA Facility Description and Documentation Manual,* working document, August 1985.

National Instutite of Building Sciences, 1985, *Hospital User Manual Series Guidelines, Volume 2: Systems Start-Up Manual,* working document, August 1985.

National Instutite of Building Sciences, 1987, *Hospital User Manual Series Guidelines, Volume 3: Operations Manual,* draft model, September 1987.

National Instutite of Building Sciences, 1987, *Hospital User Manual Series Guidelines, Volume 4: Maintenance Manual,* draft model, September 1987.

Roth, B. G., 1985, "Moving a Medical Center Library," *Special Libraries,* Winter 1985, pp. 31-34.

Saint Vincent Hospital, Santa Fe, New Mexico, 1977, Move-Related Materials.

Vanderbilt University Medical Center, Nashville, Tennessee, 1980, Move-Related Materials.

CHAPTER 9

POST-OCCUPANCY EVALUATION AND IMPLICIT THEORY: AN OVERVIEW

Craig M. Zimring

College of Architecture
Georgia Institute of Technology
Atlanta, Georgia

INTRODUCTION

The chapters in this section address the problem of *usable knowledge*. The ultimate goal of most building evaluations is to produce better buildings by supporting decisions about planning, design, construction or management of buildings. The seven chapters tackle these problems by arguing that we should refine theory and develop better, more standardized, methods. The authors draw on examples from several successful POE programs affecting billions of dollars of construction annually.

In my chapter, I map the kinds of issues that arise in POE and argue that POE can effectively advance environmental design research and support decisions by attending to the implicit theories guiding evaluations, such as the assumed role of evaluators and the assumed ways organizations make decisions and learn.

Becker proposes that a distinction be made between POE and environment-behavior research. He suggests that POE be reserved for relatively simple practitioner-driven studies that could be carried out by facilities managers as part of their everyday activities. As in medicine, these can serve as clinical experience that aids day-to-day decision-making and can provide a basis for more rigorous evaluation research.

Machado outlines a more holistic, interdisciplinary, and integrative approach to the analysis of the man-made environment as it affects occupants, whether in industrial or institutional settings, such as tropical hospitals in Brazil's Amazon region. This approach goes beyond that of traditional architectural programming and evaluation to include risk management and disciplines like human factors and environmental health concerns.

Loftness, Hartkopf, and Mill write that POE must be viewed in an integrated framework that considers a full spectrum of human needs, building systems, and research approaches over the life cycle of a building. Focusing particularly on high tech offices, they suggest that the relationship between buildings and human needs can be defined in terms of functional/spatial quality, thermal quality, air quality, acoustic quality, visual quality, and building integrity. Each building system needs to be designed and evaluated based on their impacts on these attributes.

	Theory	Technique
Practice	Roles of evaluators (Connell & Sanford, 1986; Saegert, 1988; Shibley & Schneekloth, 1988)	"How to do" POE (Friedmann, Zimring & Zube, 1978; Bechtel, Marans & Michelson, 1987; Preiser, Rabinowitz & White, 1988) Development of standardized methods (Zeisel, this issue)
Context	Role of "strategic design choices" about building form in establishing "social encounter fields" (Hillier & Hanson, 1984) Role of time in the life cycle of settings (Stokols, in press; Wicker, 1987)	Analyses of decision-making by large organizations (Shibley, 1985)

Figure 1. Some sample issues in POE divided by the distinction between context and practice and technique and between theory and technique.

In his chapter on standardizing POEs, Zeisel makes a call for sharing and standardizing research approaches, instruments and scales. He argues that several elements of the Hospital POE Project of the Canadian Department of Health and Welfare can be generalized, such as basing POE on building purpose and critical performance criteria, standardizing multiple methods, employing a range of depths of study, predesigning data handling and archiving results in a usable data-base.

Francescato, Weidemann and Anderson provide a cognitively-based integrative model of housing satisfaction. They argue that generaliziblity will only result from a theory that captures the multiple aspects of housing experience such as objective measures, user needs, hedonic tendencies, and residents' satisfaction. In reviewing housing research over the past 50 years, they suggest that recent developments in cognitive social psychology provide the basis of a useful model of housing satisfaction.

POST-OCCUPANCY EVALUATION AND IMPLICIT THEORY

As Kurt Lewin said: "There's nothing as practical as a good theory." However, few discussions of post-occupancy evaluation (POE), including my own, have addressed theory. Most books and articles POE have discussed POE as a set of techniques and methods rather than as an approach toward better theoretical understanding of buildings. The purpose of this chapter is to consider the functions theory presently serves, and can serve, in light of the organizational and institutional relationships that operate in post-occupancy evaluation.

I will first map the kinds of issues that have arisen in POE and will argue that most attention has been given to what I call "practice-technique" issues: concerns with "how to do" POE. I will then explore the context within which POE operates and examine three evaluator roles emerging from this context. Finally, I will discuss a detailed case study that shows the relative impacts of these different roles.

The Context of POE

A map of issues in POE can be understood in terms of the two distinctions illustrated in figure 1: between issues related to context and those related to practice; and, issues related to technique and those related to theory. For these purposes, "context" means the setting, organization or system being evaluated, and if different, the client. This represents a slight departure from other discussion of POE. Friedmann, Zimring and Zube (1978), for example, talked about the setting, the social-historical context, and the proximate environmental context. In this chapter, "context" refers to all to these.

Context issues consist of many different kinds of concerns; they might include, for example, theoretical constructs about the impact of characteristics of the setting on certain user groups, or mechanics of how the client makes decisions or procures buildings. By "practice" I mean issues related to the conduct of POE, including methods and procedures, but also theories and values about the role of POE and evaluator in society and decision-making, ideas about the relative importance of different issues, etc.

The distinction between technique and theory is between the means and methods of achieving ends and the development of a coherent system of propositions. "Technique" can refer to the client's or setting's methods or means of achieving ends (this is "technology" in the sense it is used in management: the ways of achieving outcomes; March and Olsen, 1982). Or, it may be the evaluation's means and methods ("practice"). Similarly, "theory" can be about the context or about POE itself. POE theory can, of course, be about the technique of evaluation, but this seems to be conceptually separable from the "how-to-do-it" procedural rules that make up technique. This division of evaluation issues into context-theory, context-technique, practice-theory, and practice-technique, along with some sample issues, was shown in Figure 1, above.

However, this division is only a very abstract mapping of issues; the specifics of practice, context, theory and technique for any evaluation derive from a complex system of evaluator peers, clients, evaluator himself or herself, and the specifics of the setting: who performs POE, who pays for it, and what the intended use of the information is. This system as it has evolved in present-day POE has resulted in an emphasis on practice-technique issues. These influences are illustrated in figure 2.

Although early surveys of the field found that most POEs were performed by academics for academic or government research funding sources, the emphasis in POE from the outset has been to be useful in design decision-making rather than to create theory (Bechtel and Srivastava, 1978. The intention has been to produce better buildings by rationalizing the design process for designers by providing feedback about building performance (see for example, Friedmann, Zimring and Zube, 1978). As such, the context-technique issues were often taken as givens. For instance, these include assumptions that architects were the most appropriate point at which to affect the building development and management process, that architects would use such information if it was provided them, and so on.

Interestingly, context-theory issues were also not explored. Although academics have dominated the field, with their pressures for peer-reviewed publications and tenure, the emphasis on being of instrumental value seems to have pervaded the field. The theory-building that has occurred has been in the environmental social sciences. The recent Handbook of Environmental Psychology contains several chapters relating to theory, but few are related to POE explicitly (Stokols and Altman, 1988). Similarly, most theory-practice issues have been assumed or unexplored. For example, many evaluators have taken a scientific or technical role where they attempt to take an objective noninvolved role in decision-making, and there has been relatively little discussion of

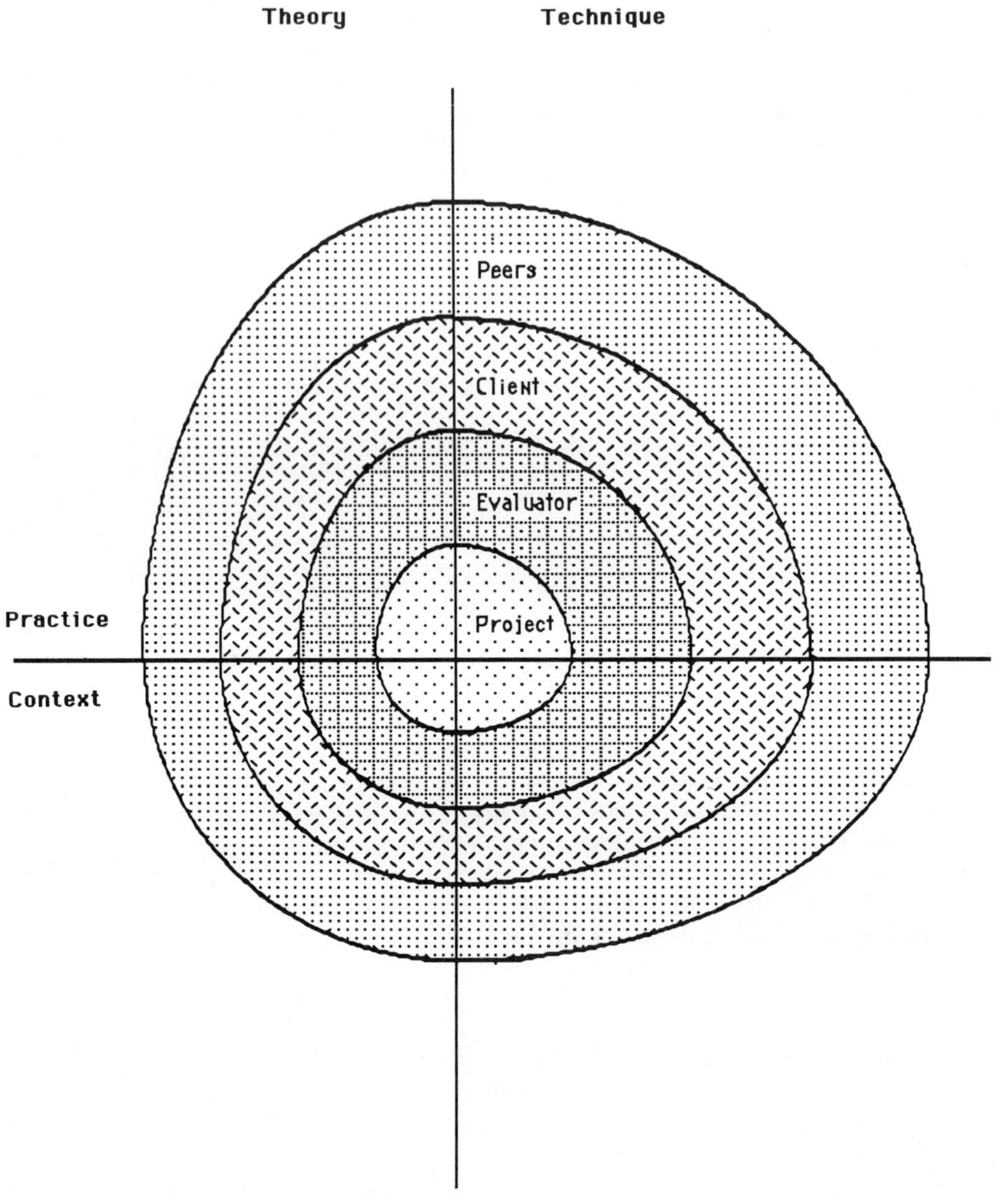

Figure 2. Because of the variety of issues impinging on a POE, most consideration has been given to practice-technique issues, the largest area on this chart.

alternatives (see Shibley and Schneekloth, 1988 for a discussion of the technical role versus one based on critical theory).

In his chapter in this volume, Farbstein argues that there is a shift underway in POE from an academic orientation to evaluations by consultants and inhouse decision-makers for large building delivery organizations. Whereas this shift is encouraging – POE now affects billions of dollars of construction annually – it suggests that the emphasis on practice-techniques issues may continue.

Roles in POE

Although the multiple influences on an evaluator are potentially quite complex, evaluators often simplify their choices by adopting a role based on the situation, client, and peers, and his or her world view and philosophy, interests, background and professional socialization. This role includes a set of rules that at least loosely establishes who defines the issues, what the evaluator's relationship should be to the client and how the results should be used (Connell and Sanford, 1986; Saegert, 1987; Shibley and Schneekloth, 1988). Although there can be considerable overlap between roles, and evaluators may choose different roles in different situations, evaluators choose at least three distinct roles: the inquirer, the performer, and the changer. Many other roles are possible, but these seem to have dominated in the first 20 years of POE practice. My purpose in discussing these roles is to describe what seems to be common choices rather than suggesting that these are the most effective roles or that they are necessarily internally consistent.

The inquirer. The inquirer is interested in learning about issues and questions that he or she defines out of personal interest or professional socialization or are defined by the academic community. Of the three roles he or she is most likely to address context-theory issues. However, whereas such a role would seemingly allow a wide range of theories to be investigated, an early emphasis by many evaluators on POE as a way to empower the "nonpaying" client has resulted in a focus on exploring influences on the satisfaction of building users (Zeisel, 1975). Recently the inquirer role has broadened to include wayfinding in buildings (Weisman, 1981), or the role of the environment for work groups that relocate (Stokols, Personal communication 1988). Such inquiry questions may also be methodological, and as such may be concerned with practice (see, for example, Moos and his colleagues' development of scales to assess housing, correctional and treatment settings, Moos, 1974, 1975).

The inquirer typically comes from a scientific discipline such as environmental psychology or sociology; these fields have traditions of practice that the inquirer should be a passive observer in most situations, and should not intervene in the situations being studied. This is reflected in the research tradition of most inquiry-based POEs, which are field studies rather than action research or other interventionist strategies. The evaluator seldom has much control over the setting.

Although a wide range of data gathering methods have been used, inquirers have often employed social science field methods such as self-report (interviews, questionnaires, etc.), observation of physical traces and ambient qualities in the setting or observation of behavior (see Zimring, 1987 for a discussion of these methods). The POE inquirer often seems to have multiple goals: he or she wants to influence decision-making (that is, to influence context-technique) but also hopes to advance knowledge and have his or her findings appear in journals, books or in the professional press, or to present them at professional meetings (Zimring and Reizenstein, 1980).

The performer. Similar to what Saegert (1987) describes as the "technician," this role takes as its charge providing answers to questions raised by some specific group (typically clients). This is one of the roles that is often referred to as "applied" and evaluators often strive to produce "action-oriented" studies in that they aim to affect the decision-making of clients. The performer's role is to develop and carry out effective research technologies. Performers bring a wide range of backgrounds and approaches to evaluations, although social scientists and social science methods seem to predominate.

These studies tend to focus on issues that impact organizational effectiveness – context-technique issues – such as original and life cycle cost, maintenance, efficiency of space utilization, and ability to carry out critical tasks such as supervise prisoners or provide post office lobbies attractive to the public. Because of the interest in supporting organizational decision-making, this role has the potential to use action research where the evaluator participates in implementation of recommendations and monitors their impacts, rather than more passive scientific models. This seldom seems to happen, but the increase in in-house POE programs noted by Farbstein may help make action research a more accepted alternative.

The performer hopes to be useful to the client or other decision-maker, and typically presents findings as inhouse reports or presentations, although he or she may develop design guides or other publications for broader distribution.

The changer. The changer has a view of what ought to be, and uses the POE to attempt to bring this about. This may be achieved by representing one set of views, being a mediator between different positions in a setting, such as workers and management, or by attempting to instigate new positions not necessarily held by any present actor (Knight and Campbell, 1981). The positions changers have taken seem to focus on context-technique issues, and particularly on developing equitable decision making processes. For example, he or she may be interested in providing public housing tenants a role in design decisions and uses POE to identify their needs and hence give them power by providing a voice to "nonpaying users" who are not typically consulted (see Francescato, Weidemann, and Anderson, this volume; Zeisel, 1975). Or, the evaluator may have a view of how organizational learning should take place (Argryis, Putnam, and Smith, 1985), or that occupants should feel more powerful in influencing decisions about buildings (Shibley and Schneekloth, 1988).

There may also be positions based on theories about the context. For example, Bill Hillier and his colleagues have argued that certain "syntatic" qualities of buildings and urban settings are desireable and an evaluation can advance such a position (Hillier and Hanson, 1984; Peponis, 1985). For instance, Hillier argues that buildings should generally have a well-defined core where there is high probability of encountering others; a building can be analyzed as to the presence of such a core and the results presented to decision-makers (Hillier, Hanson, and Peponis, 1984).

Most change-oriented studies are typical field studies but may include action research or other approaches. Although the evaluator may depend on remote observation of the setting or users, changers seem to have been more likely to choose self report or participatory methods such as interviews, user walkthoughs, gaming, or task forces. For example, Mark Francis and his landscape architecture class used group interviews and other methods to assess a playground that they had previously designed (Francis, 1982).

Changers have picked from a wide range of dissemination strategies: inhouse reports, public press articles, etc. The changer may also participate in lobbying or as an expert witness in court cases, although to date such approaches are rare.

The Importance of Role

The choice of role is useful because it makes subsequent decision-making easier. For instance, the choice of the "performer" role not only sets the overall goal of the POE – utility for the client – but probably, or at least ideally, establishes about how the study resources will be distributed: initial investigation about how to be useful to the client, preparation of inhouse reports, and so on.

Like all such roles, however, these establish systems of constraints and opportunities. Some of the constraints are contractual or are inherent in the situation, such as when time or budget does not allow a time-series research design. Other constraints result from tradition or from the training of evaluators (see Shibley and Schneekloth, 1988, for a discussion of the impact of professional education on the consultant-client relationship).

In the following example, the evaluators adopted a performer role intended to affect organizational decision-making, yet because the evaluation did not adequately explore context-technique issues, it was limited in its effectiveness.

IMPLICIT THEORY IN A PRISON POE PROGRAM

The California Department of Corrections/Kitchell Capital Expenditure Managers Post-Occupancy Evaluation Program

In 1980, in response to seriously overcrowded prisons and burgeoning state population growth, the California Department of Corrections (CDC) initiated a prison construction program to more than double California's prison capacity by 1991. At a cost that is expected to exceed $4.5 billion, this is one of the world's largest public construction programs. In order to provide a large number of prisons quickly, the CDC and their program planner, Kitchell Capital Expenditure Managers (Kitchell CEM), developed a prototype system where standard building designs and building components would be refined and repeated up to 100 times around California.

In 1983 the CDC and Kitchell CEM initiated a POE program to learn about the performance of entire prisons and of prison components. It was intended that this would serve as input into the programs and designs of future prisons and as an aid to fitting out new prisons as they were finished. The program is jointly conducted by planners from Kitchell CEM, correctional administrators from the CDC, and by me, Zimring, as POE consultant. Data is collected by both the Corrections Department and Kitchell, as well as by staff of the facility being evaluated.

A broad range of issues has been studied in POEs of several California prisons, such as efficiency of prototype warehouse designs, communications between inmates and officers, control room functioning, durability and safety of cell furnishings, kitchen operations, and use of common day room space. This program has evolved into a multi-level program that includes: "profile assessments," rapid overviews of operating facilities; "issues studies," that focus on particular concerns of decision makers; and, "comprehensive studies," that examine the interrelated functioning of multiple departments in a prison. (This division is similar to that used in the Public Works Canada POE program and others; see the chapters by Harvey and by Zeisel in this volume.)

A typical study begins by attempting to develop evaluation criteria based on

Figure 3

available programs and policy documents and by interviews with decision-makers. Data-gathering methods include structured methods based on these criteria such as questionnaires with inmates and staff, direct observation of inmate and staff activity, analysis of disciplinary reports, complaints and other records, and observation and measurement of the physical setting and ambient conditions. More flexible participatory methods are usually also included; these allow the viewpoints of participants in the setting to emerge. Such participatory methods include interviews and walkthroughs with administrators, staff, and inmates. The program has been quite well accepted by the CDC and Kitchell CEM. Its scale and complexity has increased and Kitchell's former Director of Planning and Programming, Mark Goldman, has identified over 100 specific programmatic and design changes that can be directly traced to POE recommendations.

The Role of Implicit Theories of Organizational Decision-Making

We have adopted a performer role in the CDC POE program and have taken a synchronic perspective on problems. We have generally seen our role as supporting decision makers in making better, more rational decisions by objectively studying how well completed prisons fit the expressed goals of decision-makers and by suggesting solutions to improve this fit.

In one POE, for example, it was discovered that epoxy paint is not a good surface for shower room walls and floors; although initially less expensive than ceramic tile, epoxy paint requires frequent repainting and has a much higher long-term cost. Specification of shower surface material was clearly within the authority of a group of prison construction program managers from Kitchell and the CDC, who met weekly to decide such issues.

Moreover, this finding raised little controversy about values. There was general agreement among decision-makers that showers should be reasonably clean and pleasant and that life-cycle costs should be reduced (where initial cost is not too much higher). The POE recommendation was adopted immediately in a single meeting; all new prisons in California will have ceramic tile showers.

However, sometimes the decision-making process has been more dispersed, changeable and complex. Some decisions have been made by several groups or actors with different goals. In addition, goals often changed, or were not widely held, or the goals that were expressed were not the goals-in-action (see Argyris & Schon, 1978, for a discussion of this distinction). This meant that we could not rely on the goals established at the beginning of the program to form the basis of evaluation criteria.

For example, many functions of California medium-security prisons are controlled by officers in a control room that is separate from the inmate area. Although the control room was originally planned to have been raised only a few feet off the day room floor and to be surrounded by bars, it was decided to glaze the windows and further raise the floor to improve surveillance and to enable the control room to be cooled by mechanical air conditioning rather than by the evaporative coolers used in the inmate areas. These changes allowed a further set of possibilities: the control room could be darkened so that inmates could not easily see in from the day room and tell what the officers were doing.

Because of the role of the control room in the prison design, the final design represented a very complex decision-making process, with several different groups making decisions. Top corrections department decision-makers had a role in deciding the general concept of the control room, which dictated that some officers should be partially separated from inmates and placed in a control room, whereas other guards would patrol the floor of the housing unit. (This represents what the CDC views as a "mixed model" of supervision. Some officers have central control of the housing unit from a control room and other officers get to know inmates by spending time in the housing unit and hence can diffuse problems before they get serious.) Facility programmers specified the glass, mechanical systems and general control room layout; staff in the prisons reduced the lighting.

The original control room design was seen as part of a transition to a "direct supervision" model of incarceration which emphasized the role of continuous and easy interaction between inmates and officers. The final design bears a remarkable resemblance to the panopticon, Jeremy Bentham's 1794 prototypic prison design that has been used by Foucault as an example of the spatialization of power (Foucault, 1979). The control room design became a solution to the problem of how to separate inmates and officers. This was not a conspiracy to change policy. Each group operated from premises that were reasonable given the pressures facing it. Top decision-makers set the broad policy direction, such as establishing the mixed supervision model as a transition to an approach that requires less central control. Facilities programmers raised the control room and glazed it to help make the staff more comfortable and to aid visual surveillance of the housing unit. The officers felt uncomfortably "on display" in the brightly-lit control room, so they reduced the lighting. None of these decisions involved a conscious decision to subvert the original intention to move toward more interaction between inmates and officers, but they all had that effect. Top decision-makers and some mid-level administrators had moved into their jobs from other social service programs in California, and they tended to support a model of corrections that emphasized interaction between officers and inmates and presumably supported a less separate control room. However, there seemed to be an implicit agreement by many of the other actors about the appropriate technologies of control by officers: separation, distant visual supervision, a focus on aid coming from outside the housing unit for officers in trouble, etc.

The new solution and its assignment to new problems posed a problem for us as evaluators: We couldn't use the original goals or understanding of the problem as the basis the evaluation. How could we have been most helpful in supporting future decisions about control rooms? Should the control room be evaluated using the original criteria of allowing easy and immediate communication between officers and inmates? Should it be approached from the perspective of providing as complete isolation as possible for officers? Our POE focused on technical aspects of the control room on which there was agreement, such as that the design of the control panel should allow error-free control of cell doors. We made suggestions for improvement of the control panel, which were accepted.

The CDC/Kitchell CEM POE program is primarily staffed by mid-level managers, who did not see it within the program's purview to challenge major program directions. We very cautiously made our observations that the final control room design seemed to reflect a different policy than had been originally proposed. These observations did not figure prominently in any policy debate, and in fact the primary effect of the POE program has been on specific aspects of the design or facilities program rather than on broader policy.

In the shower tile example, the values of reasonably pleasant shower rooms and reduction of life-cycle costs were widely shared and unchanging, and decisions about the outcome was clearly assigned to one group who saw no trouble in making the decision. These seem to be the most likely circumstances under which an evaluators' performer role, as it is usually defined, can influence decision-making. In the second example, decision-making was dispersed and there were at least implicit disagreements about the goals of the control room design. As a result, we evaluators had little impact on policy.

The Role of POE in Institutional Learning

Rather than simply supporting decision-making, much POE, including the California Department of Corrections POE program, has been justified on the basis that it provides more efficient learning processes for organizations and architects. That is, that POE not only supports specific decisions but it helps decisions improve over time: organizations "learn how to learn." For example, Dennis Dunne, then California's Deputy Director of Corrections for Planning and Construction, said, "Post-Occupancy Evaluation allows us to get it right on the second or third prison rather than on the seventh or eighth." As Mark Goldman's list of 100 positive POE-generated impacts attests, we have had a cumulative effect on California prisons. However, the role we adopted is primarily directed at what has been called "single-loop learning" (Argyris, 1976), focusing on specific questions of whether outcomes achieve goals, rather than on reflective "double-loop" questions of whether goals and objectives are appropriate, or whether goals-in-action or policies-in-action fit espoused goals or policies.

In the shower tile example, we provided information about the performance of shower surfaces with respect to management's criteria of pleasantness and low maintenance. Because there was relatively little disagreement in values, and decision-making was concentrated in one group, our results had direct impact. In the example of the control room design that apparently signalled a conflict between policy-in-action and espoused policy, we were less successful in affecting learning because at least some of the questions were reflective kinds of questions that we were not well equipped to

Figure 4

address. I am not suggesting that environmental design consultants suddenly attempt to sell ourselves as management consultants. However, when consultants provide feedback to the organization about issues that arise during the course of a POE (or programming or other activities) they may help the organization to function in ways that go beyond the technical questions the evaluator was originally asked to address.

For example, it may be of long term importance to the CDC to understand that the way decision-making is distributed between the various internal groups may result in a solution that no one particularly intended to produce. In addition, there were important organizational political relationships that surfaced in the development of the control room that could help the organization manage its intergroup relationships in a positive way. For instance, because of the rapidly growing inmate population, the CDC had to open new prisons without the final security perimeter being completed. As a result they felt some special pressure to satisfy the officers' union. Although I have no direct evidence to support this contention, it seems likely that some people who had concerns about the directions that the control room design was going might have not pressed such concerns. The process that produced the final control design was not necessarily bad, but had the evaluation team established better ways of providing such feedback, information about the role of internal politics in design decision-making might allow the CDC to use politics in a positive way.

Both single-loop and double-loop learning can be quite useful. As in the shower example, it is often very important to learn the consequences of decisions that are likely to be repeated with the same criteria. In this sense we have helped produce better prisons than probably would have been produced otherwise. But our theory of decision-making has been flawed, or at least incomplete, in that we have been of less help to the California Department of Corrections in helping them maintain consistent and flexible policy-in-action and to reflect on whether policies need to change to accommodate changed conditions.

CONCLUSIONS

The context within which POE operates and the roles that evaluators have chosen have caused the emphasis of most POE discussion to be on technique-practice issues, and the three other cells in the issue map have been assumed or ignored. The California Department of Corrections example suggests that the impact of POE was limited by adopting a performer role while assuming that single loop learning is appropriate – that is, by not exploring technique-context issues. In addition, by looking at the relationship of a "strategic design decision" (Peponis, personal communication) about control room design to theories about architecture's role in control, we potentially could have clarified for the client a critical and expensive decision.

This analysis opens up a much larger, and to me more interesting, role for POE, yet also requires greater expertise and self awareness on the part of evaluators. Theories of practice need to be explored and debated. Theories of the context may be necessary if we are going to propose strategic changes in programs. This analysis also suggests that evaluators who hope to impact decision-making should explore how decision-making operates both generally and in a given organization. Perhaps we must finally retire the highly rational model of decision-making that has dominated POE: we cannot simply provide feedback on how well criteria resulting from policy have been met; we must also consider if the policies themselves need refinement or change.

NOTE

1. For their comments on previous drafts or detailed discussion about the ideas and issues in this chapter I am very grateful to Jan Carpman, Yoon Choi, Michael Elliott, Sandra Howell, John Peponis, Wolfgang Preiser, Lynda Schneekloth, Donald Schon, and Fred Zimring.

REFERENCES

Argyris, C, 1976, Single-Loop and Double-Loop Models in Research on Decision Making, *Administrative Science Quarterly*, 21:362-375.

Argyris, C., Putnam, R., and Smith, D, 1985, *Action Science*, San Francisco, Josey-Bass.

Argyris, C. and Schon, D., 1978, *Organizational Learning: A Theory of Action Perspective*, Reading, Massachusetts, Addison-Wesley.

Bechtel, R. B., Marans, R.W., and Michelson, W., (Eds.), 1987, *Methods in Environmental and Behavioral Research*, New York, Van Nostrand Reinhold.

Bechtel, R. B. and Srivastava, R. J., 1978, *Post-occupancy evaluation of housing*, report submitted to the U.S. Department of Housing and Urban Development.

Connell, B. R. and Sanford, J., 1987, Workshop presentation in the 18th Annual Conference of the Environmental Design Research Association, Pomona, California.

Foucault, M., 1979, *Discipline and Punish*, New York, Vintage.

Francis, M., 1982, Designing landscapes with community participation and behavioral research, *Landscape Architectural Forum*, Spring:15.

Friedmann, A., Zimring, C., and Zube, E., 1978, *Environmental Design Evaluation*, New York, Plenum.

Hillier, B. and Hanson, J., 1984, *The Social Logic of Space*, Cambridge, Cambridge Press.

Hillier, B., Hanson, J., and Peponis, J., 1984, What do we mean by building function?, in: J.A. Powell, I. Cooper, and S. Lera (Eds.), *Designing for Building Utilization*, London, Spon.

Knight, R. C. and Campbell, D. E., 1980, Environmental Evaluation Research: Evaluator Roles and Inherent Social Commitments, *Environment and Behavior*, 12, 533-541.

March, J. G. and Olsen, J. P., 1982, *Ambiguity and Choice in Organizations*, Bergen, Universitetsforlaget.

Moos, R., 1974, *Evaluating Treatment Environments: A Social Ecological Approach*, New York, Wiley.

Moos, R., 1975, *Evaluating Correctional and Community Settings*, New York, Wiley.

Peponis, J., 1985, The Spatial Culture of Factories, *Human Relations*, 38, 357-390.

Preiser, W. F. E., Rabinowitz, H., and White, E. T., 1988, *Post-Occupancy Evaluation*, New York, Van Nostrand Reinhold.

Saegert, S., 1987, Environmental Psychology and Social Change, in: D. Stokols and I. Altman (Eds.), *Handbook of Environmental Psychology*, New York, John Wiley.

Shibley, R. G., 1985, Building Evaluation in the Mainstream, *Environment and Behavior*, 17(1), 7-25.

Shibley, R. and Schneekloth, L., 1988, Risking Collaboration: Professional Dilemmas in Evaluation and Design, *Journal of Architectural and Planning Research*, 5(4), 304-321.

Stokols, D., 1988, personal communication.

Stokols, D., in press, Transformational Processes in Person-Environment Relations, in: J. McGrath (Ed.), *The Social Psychology of Time: New Perspectives*, Newbury Park, California, Sage Publications.

Stokols, D. and Altman, I. (Eds.), 1987, *Handbook of Environmental Psychology*, New York, Wiley.

Weisman, G. D., 1981, Evaluating Architectural Legibilty: Wayfinding in the Built Environment, *Environment and Behavior*, 13, 189-204.

Wicker, A. W., 1987, Behavior Settings Reconsidered: Temporal Stages, Internal Dynamics, Context, in: D. Stokols and I. Altman (Eds.), Handbook of Environmental Psychology, New York, Wiley.

Zeisel, J,, 1975, *Sociology and Architectural Design* (Russell Sage Foundation Science Frontiers, Series No. 6), New York, Free Press.

Zimring, C. M., 1987, Evaluation of Designed Environments: Methods for Post-Occupancy Evaluation, in: R.B. Bechtel, R.W. Marans, and W. Michelson (Eds.), *Methods in Environmental and Behavioral Research*, New York, Van Nostrand Reinhold.

Zimring, C. M. and Reizenstein, J. E, 1980, Post-Occupancy Evaluation: An Overview, *Environment and Behavior*, 12, 429-451.

Zimring, C. M., Weitzer, W. H., and Knight, R. C., 1982, Opportunity for control and the built environment: The case of an institution for the developmentally disabled, in: A. Baum and J. Singer (Eds.), *Advances in environmental psychology, Volume IV: Environment and health*, Hillsdale, N.J., Lawrence Erlbaum.

CHAPTER 10

POST-OCCUPANCY EVALUATION: RESEARCH PARADIGM OR DIAGNOSTIC TOOL

Franklin Becker

Department of Design and Environmental Analysis
Cornell University

ABSTRACT

POE (Post-Occupancy Evaluation) has existed, in some form, since people began occupying buildings. Its association with relatively systematic assessment of how well a building performs on explicit criteria is more recent, but has existed for at least 25 years. This chapter argues that the role of POE in improving building performance has been inadvertently undermined by trying to make POE an academically acceptable form of evaluation research. POE as a diagnostic tool (essentially a clinical technique) and environment-behavior research (intended to help develop a solid research tradition to guide architecture and interior design practice) are both useful. They share common concerns, but they are not synonymous. The development of facility management, a client-based professional discipline whose foremost concern is for buildings-in-use, has shifted attention from the architecture and design (as well as academic) communities as direct beneficiaries of POE results to the organization paying for and occupying the building. In doing so it has created a set of conditions that is beginning to make POE an acceptable management tool with potential for significantly improving building performance.

INTRODUCTION

Post-occupancy evaluation (POE) developed over the past 20 to 25 years as a technique by which design practitioners could learn from their past, mistakes and successes alike. The intent was simple: to avoid continually reinventing the wheel. By doing so, presumably the cost of design development could be lessened, occupant satisfaction, comfort, and performance could be enhanced, and organizations could get better value for money from their facilities.

The need for some more deliberate, systematic effort at assessing how well design actually worked for the occupant was articulated in a number of widely-read books and articles (Sommer, 1969; Perin, 1969). The picture which began to emerge of architecture and design was that of a profession gone awry, meeting its own and others' professional and personal images of proper housing, schools, dormitories, hospitals or pedestrian malls, while remaining ignorant of or dismissing the living, work, and play patterns of persons actually occupying and using the designed environment (Sommer, 1983).

To their credit, members of the design professions participated both in exposing these kinds of environmental dysfunctions and in proposing alternative approaches which more seriously took into consideration what Zeisel (1976) called the "non-paying client," the person who occupies a facility without necessarily owning or paying for it (as in the case of prisoners, students, employees, etc.). Often, some form of formalized POE was a part of these calls for reform.

Social Scientists

POE was the bridge over which social scientists walked on their way to becoming involved with the design professions. If architecture was going awry because architects did not understand the values, behavior patterns, and attitudes of the non-paying client or end-user, then the social sciences seemed a logical professional ally, since investigations of such matters was their balliwick.

POE became one of the means of identifying how well buildings worked from their occupants' viewpoints (Becker, 1974; Cooper, 1975). This information was taken as a measure of user needs or requirements. The intent was to feed it forward into the next project of a comparable nature. It thus became a part of the programming process.

Both social scientists and design professionals conducted these kinds of studies. Such POEs had a number of characteristics (see table 1). They tended to be serendipitous; that is, they were selected more because of convenience and proximity than because they provided the opportunity to test some specific idea or hypothesis. Many were quantitative, relying on surveys of occupants, but (with the exception of some academic-oriented studies) rigorous population sampling procedures were rarely followed, and statistical analysis of results was often minimal or non-existent. Clearly formulated research designs, particularly those involving comparisons among groups experiencing different environmental constraints and opportunities were extremely rare. Single-site studies were the rule, but they differed from more traditional case studies, which typically provide in-depth analyses of a single situation in great detail, with considerable focus on the social, psychological, and organizational processes underlying observed behavior patterns.

From a practitioner's viewpoint, these are *positive* characteristics. They make POE accessible to persons without formal social science training, and cause POE to better fit the time and resource availability of most organizations.

Broad Focus

The focus of most POE was extremely broad. It included users' self-reported satisfaction with dozens of aspects of the built environment, ranging from site planning and design to the interior arrangements and size of rooms, lighting, signage, storage, and privacy (cf. Becker, 1974). Rarely were behavior patterns described in detail, or were outcomes besides user satisfaction measured. Little attention was paid to contextual factors, such as how long the occupants had been using the space, the circumstances under which they came to use it (e.g., compulsory or voluntary), or the nature of their relations to those running it or using it with them. The desire to be "relevant" and to provide information of real practical value quickly to designers provided a rationale for this kind of approach. These kinds of studies were useful, for if nothing else they recorded in case after case user dissatisfaction with the designed environment, often despite good intentions by architect and owner. More to the point, they were accessible in terms of required time and cost to the design practitioner.

While attributing blame was not their (primary) intention, unintentionally or otherwise POE studies often seemed to point the finger at the architect, who presumably (though this is a simplistic viewpoint) was considered responsible for the form the built environment took . The failure to explicitly consider the full range of players and other contextual factors in most POEs, unlike the typical practice in more traditional case studies, contributed to the discrediting of the design professions when, ironically, in many cases the POEs were conducted or sanctioned by design professionals who themselves wanted to reform their profession.

Environment-Behavior Research

At the same time as POEs were being recognized as a useful technique for assessing how well buildings were performing for their occupants, environment-behavior research was also being conducted by social scientists with largely the same goals; namely, to improve the quality of the built environment by increasing our understanding of how physical form affected behavior and attitudes. Sommer's research on small group ecology (see *Personal Space*, 1969) and Altman's work with human territoriality among Navy crewmen (1966) were good early examples of this type of research.

These research endeavors had a quite different set of characteristics (see table 1) than those associated with POEs. In general, they tended to be more scientifically rigorous. Most had clearly formulated research designs and followed accepted canons of population sampling. Statistical analyses of data were used to test for differences among groups or treatments and to test specific research questions and hypotheses. The research focused on one or two variables selected for theoretical or other deliberate reasons. Behavioral measures (e.g., actual seating locations, communication patterns, avoidance behavior) as well as interviews and questionnaires were used to assess the effect of environmental variables. Little emphasis was placed on self-reported satisfaction with specific features of the built environment. In between these two approaches

TABLE 1
Characteristics of POE and
Environment-Behavior Research

Environment-Behavior	**POE**
• semi-systematic	• systematic
• sampling casual	• sampling rigorous
• minimal research design	• explicit research design
• broad focus	• narrow focus
• satisfaction key measure	• attitude & behavior
• serendipitous with respect to time and place	• deliberate selection of time and place
• minimal data analysis	• elaborate data analysis
• single-site case study	• comparative analysis

fell rigorous, systematic POEs conducted by academic researchers (Becker, 1974; Cooper 1975; Marans and Spreckelmeyer, 1982).

Why Distinguish Between POE and Other Environment-Behavior Research?

It is worth distinguishing and then clarifying the relationship between POE and other types of environment-behavior research, for several reasons. It seems to me that considering POE as a form of scholarship or "scientific" research does all concerned an injustice. To the extent that POEs are conducted by academics or others outside practice who wish to publish or otherwise make public their results, the POE is under pressure to become more academic; that is, to take on more of the characteristics listed under "environment-behavior" research in table 1. This form of research generally takes more time and more trained skill in research design and data analysis than is available in even relatively sophisticated design practices or large organizations.

It thus discourages the practitioner from becoming involved in briefer, simpler, and less scientific – but still useful – efforts that they may conduct for their own immediate internal edification and which, at some point, could stimulate more rigorous, academic research efforts. If POE is viewed as a distinct research paradigm, such efforts are likely to be considered as second class, if not second-rate, research in comparison to more rigorous methodologies. As such, they are likely to have minimum credibility with either academic researchers, designers, or facility managers.

Facility Management and POE

The design community was viewed originally as the primary audience for POE. This is changing. Over the past eight years, facility planning and management has evolved as an increasingly influential client-centered function responsible for coordinating all aspects of building planning, design, and operation (Becker, 1988a; 1988b; 1988c; Becker and Sims, 1988). For facility managers, the concept of building-in-use is central to their professional activities, in a way that was never true for the architecture and design community.

Facility management responsibilities range from real estate acquisition and leasing and major new design, construction, and renovation, to planning and managing small scale reconfiguration of space and furniture, developing space and furniture allocation policies, and cleaning and maintaining the building. The facility management function integrates the fields of real estate, architecture and design, engineering, space planning, business and general management, human factors, and what Becker (1981) and Steele (1986) have called "organizational ecology." This is the study of the interdependency between organizations and the physical settings which they create and occupy.

Fundamentally, the facility management function is responsible for the planning, design, and management of the total work environment. The facility management function's responsibility for the built environment starts before and ends after the responsibility of the architectural and design professions does. It is deeper as well as broader, since it embraces development and implementation of planning processes, management policy concerning allocation and use of space and facilities after occupancy, maintenance, and cost control, as well as design and construction. The facility management "audience" is an internal one: management. Charged with insuring that the organization receives value for money in its facility-related expenditures, management is the natural audience for the kind of information generated by POEs.

While facility management as a new field is at the moment largely associated with the the workplace and corporate offices, as a function it is bounded by no particular type of setting. The function exists and is developing in all complex settings, from housing, hospitals, and health care facilities, to schools, hotels, research laboratories, airports, and convention centers.

POE serves several key functions for facility managers. They are under pressure to improve the level of services provided while containing costs (Becker, 1988d), and they are being held accountable by management for the quality of facilities they provide to the organization (Becker, 1988b). POE provides them with a way of assessing the level of professional service provided by architects and designers, either as consultants or as in-house staff. Over time, such feedback forms a valuable database for benchmarking the relation between cost of renovating or building new facilities and their performance.

Second, POE is extremely useful as a diagnostic tool for pinpointing specific aspects of the facility that need to be improved, or should be preserved, in environmental improvement programs. Thus it helps in resource allocation decisions, especially in justifying to top management why funds are being spent as they are.

Third, interviews, observations, focus groups and other data collection techniques are a form of staff involvement appreciated by a more educated and demanding staff. Such face-to-face contact also allows the facility management team to educate staff and management about their functions, which are typically poorly understood within the organization (Becker, 1988d).

Fourth, POEs represent a kind-specific management tool that can contribute to enhancing the image and credibility of the facility management function within the organization. Data integrated into presentations to management from even very simple surveys or interviews, especially in graphic form, conveys a more professional and technical image than one based only on personal experience or judgement.

Delimiting the POE Role

Considering POE as a diagnostic tool intended largely for internal use by practitioners can help establish more reasonable expectations for its role in organizational and design practice. As a clinical technique, POE is not meant to have scientific reliability and validity, nor is it meant to be widely generalizable. Its primary intent is to provide immediate feedback to the practitioner about occupant response to a new or renovated facility. As a simple, low-cost technique with minimal statistical analysis, focused on a single case selected because those conducting the POE are interested in the results for their own professional purposes, POEs are more likely to be widely carried out. While the data will rarely meet scientific canons of research, such an approach is far better than no evaluation or haphazard and deliberately skewed "guesstimates" about how well a facility is working for occupants. More sophisticated facility managers and design practitioners will undoubtedly begin to look for and analyze patterns in internal data bases developed from what are essentially individual case studies. The more typical practitioner, limited at least initially by experience, training, time, and cost restrictions, will tend to focus on the most simple level of POE, which Preiser, Rabinowitz, and White (1988) call the "diagnostic" POE.

These simple POEs might even be somewhat standardized in format (perhaps developed by a group from a professional association such as the AIA or IFMA [International Facility Management Association]). An academic researcher could then collect them and, using these individual records as a form of archival data, search for patterns that are not clear from individual studies but which may emerge from analysis of many cases. The individual cases could be published on a regular basis as a column in professional journals. This would provide a steady source of research topics to more academic researchers, as well as keeping practitioners better informed about the consequences of different design solutions. This is akin to a medical model of research in which the clinician reports on what seem like patterns emerging from clinical observations, or on problems they cannot solve, with those patterns or problems then taken up by medical researchers for more focused and rigorous study.

There is a significant glitch in this proposal, of course. How many architects or facility managers want to make public "failures" in their own designs, even to reduce the likelihood of similar mistakes by their fellow professionals and thus improve the overall quality and reputation of the profession (!) ?

Making POE More Professionally Acceptable

This is a very real problem, but not necessarily an intractable one. Diagnostic POEs, largely for consumption inside the organization, are likely to encourage the facility manager and architect to provide a fuller account of the project because of the direct responsibility they had in planning and implementing it. In particular, there would be an incentive to provide more contextual information about the decision-making process and about the basis (evidence used) for different kinds of decisions. Did the facility manager or architect propose the room arrangement, or was it demanded by the client – and if so, for what reason? What information was presented to the client to help make the decision? Did the client sign off on the design after reviewing it? Did the client provide the facility manager or architect with sufficient information about the organization, its values, work patterns, culture and philosophy, political structures, group dynamics, and so on, to allow the architect to make informed choices as they sought and proposed various design solutions? To avoid "buck passing" and and organizational form of "blaming the victim," facilities decisions would need to be seen, from the outset, as a collaborative effort involving both facility managers and a range of users.

A practitioner-initiated POE would inevitably push back the focus of inquiry to the programming process and even further to the organization's decision-making process and the role of different players in it. In this case, concern for legal liability might actually improve professional practice, rather than merely creating a burdensome bureaucratic layer on top of an already complex set of activities. This assumes, of course, a sincere interest on the part of facility manager, designer, and client in improving the quality of design by obtaining feedback about how well the design actually works.

Collaboration on the purposes and conduct of the POE by all concerned is necessary to make it useful. The rapid change and ambitious building and/or renovating programs of many organizations today creates an incentive for the organization, not just the architect or designer, to obtain this type of information. Initiating this type of evaluation from the beginning should also provide an additional incentive for all parties to clarify their roles and expectations.

In this approach the POE remains a diagnostic technique, but it begins to be rooted in professional practice itself, with a self-interest motive for both client and designer. Such POEs are valuable even if unreported because of the direct learning opportunities inherent in them for practitioners and clients alike. If reported, they have the potential to stimulate environment-behavior research. Such information, in turn, can be fed forward into the programming process, to evaluate the suitability of proposed design alternatives, and as a basis for rethinking policies and procedures for planning and managing facilities.

Linking POE to Environment-Behavior Research

More formalized environment-behavior research, as noted earlier, shares with POE the goal of improving the quality of the built environment and of professional practice. It does so by providing a solid, credible research foundation from which design practitioners and clients can search for and evaluate potential design solutions. One would like, for example, to be able to propose a given number of conference rooms, located in particular parts of a building, based on some research evidence that these choices would actually provide the necessary support for a given type of organization engaged in specified activities. No such data now exists.

We do have data about architectural characteristics that create a sense of privacy (Brill, 1984; Sundstrom, 1980), or how furniture arrangements or open plan or distance between groups will affect communication patterns (Allen, 1976; Becker, 1981; Sommer, 1969; Sundstrom, 1986). Relatively rigorous studies with controlled research designs, appropriate statistical analysis, and accepted sampling procedures underlie these research efforts and enhance their reliability and validity. Such studies are woefully few in number, and they cover relatively few areas of design decision-making. They do, however, have a legitimacy based on their rigor that exceeds (as it should) the informal diagnostic POE. They should be complemented by less rigorous, more informal, yet still standardized POEs conducted by practitioners.

SUMMARY AND CONCLUSION

Given that the purpose of POE is to improve design practice, enhance the quality of the built environment, and improve individual and organizational effectiveness, it should be integrated as closely as possible into facility planning and management and design practice. For this to happen, it must be simple and quick to conduct, and its results must be meaningful to both practitioner and client. To be useful it does not have to meet academic canons of scientific respectability. In this form, the POE's contribution to a formal knowledge base is minimal. Its value to the practitioner is, however, real and immediate.

In the long term, a more formalized data base is critical to the development of a credible research tradition similar to that found in engineering and medicine. It is in large part this research tradition that has distinguished the medical and engineering professions from architecture and given them their far wider social influence and respectability.

By distinguishing clearly between what might be called diagnostic or formative POEs, with their greater emphasis on influencing the course of events in the particular situation in which data is collected, and more academic environment-behavior research, greater pressure can be brought to bear on those with the time, knowledge, and resources to conduct formalized research to actually do so. In this way the efforts of the practitioner and the academic build on and expand each other, rather than either diluting or driving each other out.

Thinking of POE as a relatively simple, *practitioner-driven* management technique, distinct from more formalized environment-behavior research or sophisticated evaluation studies, may help it gain credence within the organization, where it is most likely to be used on a regular basis and have some direct influence on the quality of the built environment.

REFERENCES

Allen, T. J., 1976, *The Flow of Technology*, Cambridge, MIT Press.

Altman, I., 1976, Privacy: a conceptual analysis. *Environment and Behavior*, 8:7-29.

Becker, F., 1974, *Design for Living, Residents View of Multifamily Housing*, Ithaca, New York, Program in Urban and Regional Studies.

Becker, F. , 1981, *Workspace, Creating Environments in Organizations*, New York, Praeger.

Becker, F., January 1988a, Managing facility management, *Premises Management*, 18-19.

Becker, F. , February 1988b, FM pressure points, *Premises Management*, 14-15.

Becker, F., 1988c, Defining quality, *Premises Management*, 20-21.

Becker, F., 1988d, *The Changing Facilities Organization*, Haverhill, England, PROJECT.

Becker, F. and Sims, W., In press, Facility management, in: F. Duffy, C. Cave, J. Worthington (Eds.), *Planning Office Space, 2nd. Edition*. London, Architectural Press.

Brill, M., 1984, *Using Office Design to Increase Productivity*, Buffalo, New York, Workplace Design and Productivity.

Cooper, C., 1975, *Easter Hill Village*, New York, Basic Books.

Marans, R. W., and Spreckelmeyer, K. F., 1982, Evaluating open and conventional office design, *Environment and Behavior*, 14, 333-351.

Perin, C., 1972, *With Man in Mind*, Cambridge, MIT Press,

Preiser, W. F. E., Rabinowitz, H. Z., and White, E. T., 1988, *Post-Occupancy Evaluation*, New York, Van Nostrand Reinhold.

Sommer, R., 1969, *Personal Space, The Behavioral Basis of Design*, Prentice-Hall, Englewood Cliffs, N.J., 1969.

Sommer, R., 1983, *Social Design*, Englewood Cliffs, N.J., Prentice-Hall, 1983.

Steele, F., 1986, *Making and Managing High Quality Workplaces*, New York, Columbia University Press.

Sundstrom, E., Burt, R., Kamp, D., 1980, Privacy at work, Architectural correlates of job satisfaction and job performance, *Academy of Management Journal*, 23, 101-117.

Sundstrom, E., 1986, *Workplaces, The Psychology of the Physical Environment in Offices and Factories*, NY, Cambridge University Press, 1986.

Zeisel, J. , 1974, Fundamental values in planning with the non-paying client, J. Lang, C. Burnette, W. Moleski, and D. Vachon (Eds.), *Designing for Human Behavior*, Stroudsburg, PA., Dowden Hutchinson & Ross.

CHAPTER 11

A HUMAN ECOLOGICAL APPROACH TO THE EVALUATION OF MAN-MADE ENVIRONMENTS

Paulo A. Machado

National Health Council
Brasilia, Brazil

INTRODUCTION

Human ecology is a new discipline devoted to studies of the interactions between Man and Environment. According to the definition proposed in the early seventies by a World Health Organization Expert Committee, it is "the study of Man in his circumstance." It seeks knowledge on a definite subject: interactions between humans and environment. It has its own peculiar interdisciplinary methodology. It seeks knowledge of facts and the organization of such knowledge. It is not just a descriptive activity. Understanding facts is the goal. As a new science, human ecology has been differently defined by different authors, and some time will elapse before a consensus is reached, but the basic concept of human ecology is implicit or explicit in every proposed definition: it is the study of humans where they exist and of the interactions between humans and their surroundings.

Modifying the environment in order to obtain safer and more comfortable shelter was one of the earliest organized activities of mankind. The cultural evolution of humans is marked by the increasing ability to build micro-environments according to the current physiological, religious, political and economic conventions. From the primitive hunter-gatherer shelter to cave space organization, that ability has increasingly improved. Building skills have improved impressively over time, as revealed by Sumerian, Egyptian, Chinese, Hindustani, Greek, Roman, and Byzantine architecture, Middle Age castles and cathedrals, Renaissance palaces, and modern metropoli with their skyscrapers, tunnels, causeways, highways, subways, railways, airports, and so on. Mankind is accumulating achievements and abilities in building micro-environments at unforeseen levels.

Humans are gradually concentrating in urban metropolitan areas, and with this the perception of natural phenomena, as inherited from primitive man, is fading out. Since the industrial revolution, human beings have progressively been transformed into an indoor species. Without underestimating the importance of interactions with the natural environment, which are essential to human life, it is obvious that for modern humans the man-made environment is their most important surrounding. Interactions between humans and the space they have built and organized for themselves dominate whenever quality of life is concerned.

It may sound odd to a few traditional ecologists, and it may be deplored as a sad reality, but the natural environment now must share its significance with man-made environments. Humans in most industrialized countries work 8 hours a day, enjoying 52 weekends, 6 holidays, and a 30 day vacation per year. They stay indoors 80% of the time. There are now special professions calling for outdoor and leisure activities; however, most "outdoor" activities are still carried out in man-made environments, such as streets, highways, golf and tennis courts, swimming pools, parks, and gardens.

Man-made environments have become the closest environment, and this situation affords cultural and economic benefits, but exposes humans to risks. Because of that, human ecology needs to devote increasing attention to man-made environments. It is primarily architects who are responsible for the near environment, so it is therefore fair to look for an increasing cooperation between human ecologists and architects. This should be a two-way cooperation, the result of which would be harmonizing physiological, psychological and cultural factors with aesthetics, building technology, economic factors, and social and political reality.

In evaluating any man-built space, human ecologists will pose two questions: How does the man-built environment fit in to the natural environment, and how do humans interact with the built environment? Building in harmony with the natural environment is desirable, and such harmony will not be achieved without comprehensive environmental assessments. The environment is not a random entity. Rather, it is the result of interacting factors, such as climate, soil stability, classical ecological factors, social and economic utilization, cultural values, and landscape, to name a few.

These factors should be objectively and precisely evaluated, and for some, such as climatology, geology, and classic ecology, precise, quantified data are indeed available. For other factors, quantitative methods are not yet available and other evaluation methods must be employed, keeping in mind that subjective impressions are often considered a threat. Landscape, for instance, may be appraised using questionable and subjective aesthetic preference scales. It can also be objectively appraised according to the cultural values prevailing within the respective communities being evaluated. In hot climates, for example, one frequently sees huge concrete and glass buildings, ill-considered replicas of styles used in temperate climates. These concrete monsters, all too efficient in concentrating and storing heat, have no connection to local climate and culture.

Nevertheless, wise strategies are to be pointed out where they exist. In Bar Harbor, Maine, for example, the landscape is dominated by the coastline. Exploring Bar Harbor, it is easy to recognize the Maine coast as something more than just a geographic accident or a beautiful landscape. History and economy are connected with the oceanfront. Fishermen, an impressive tourist industry, literature, private associations and clubs, art galleries, souvenir shops, cuisine, lobster ponds, restaurants, and yachting all have their roots in the marine landscape. Boundless ocean, peaceful bays where tiny waves roll over the sand, wild waves in other places pounding rocky borders, mysterious foggy mornings or bright sunny days, glorious sunsets and dark impenetrable nights, that is the oceanfront in Bar Harbor. It is not just an aesthetic value, but also an outstanding cultural value. In this context the College of the Atlantic is to be commended for its new Thorndike Library. Sitting on a hill and facing the coast, it is a rather large building that could have blocked the marine panorama. However, the architect wisely planned wide glass windows at both ends of the entrance hall. Even passers-by can look "through" the building and see the marine landscape. The architect perceived that such a landscape should not be concealed by a massive building. On the contrary, the large windows merely frame the beautiful scenery, and thus the aesthetic and cultural values of landscape were respected.

Man-made environments are expected to improve upon and not disturb human life. Homes, office buildings, apartments, shopping centers, and recreation complexes, besides fitting into the existing environment, should facilitate improving the quality of life, not prevent it. The planning of large buildings for hundreds or thousands of occupants is complex: parking/traffic amenities, sewage disposal, potable water and energy availability, and many other factors need to be considered by human ecologists. This may seem a naïve and obvious recommendation, but throughout the world it is easy to find many man-built monstrosities which introduce many costly problems and which erode the quality of life.

Man-built environments can endanger the quality of life itself in another way: by generating various forms of pollution. Among other issues, human ecology is concerned with biological and chemical pollution, but it is also concerned with vibration, radiation, noise, and light pollution. Recreation and industrial buildings should be evaluated in terms of these parameters, too. The importance of light as a pollutant is a new concept. For example, it is easy to understand why commercial interests call for larger and brighter light signs. Today, electronics allow the production of spectacular signs, which add prestige to a product or a brand-name. The best selling sign is the brightest and most outstanding one – which is to say, the one which breaks the existing harmony, disturbs everyone who lives nearby, and exposes people all night to intermittent blasts of light and color. Is it permissible for someone to disturb the resting time of people who have worked for a long day?

RISK MANAGEMENT AND CHANGE

A new attitude is being shaped, as technological advances develop an increasing variety and frequency of utilizing new forms of energy, new chemicals and new processes. Knowing the risks of such new ventures is scary: progress means added risks, but nobody is insane enough to stop progress. Mankind has to coexist with increasing risks; therefore, the wisest approach is to face the risks, control them, or, according to a modern expression, manage the risks. Organizing risk management is a new occupation, although for a long time the automotive industry, traffic authorities, land, sea, and air transportation organizations, and space sciences have been charged with reducing risks. Nevertheless, what is needed is an organization of existing knowledge and the subsequent consolidation of principles, strategies and tactics in reducing risks. Building up a theoretical framework for risk management should be considered, and it will certainly be a valuable discipline in the next century. Man-made environments, from the architectural blue-print to post-occupancy quality standards, have critical significance in risk management.

The home is the primary component of the man-made environment, since women and men spend at least 40% of their time at home. Interestingly, when discussing environments, authors seldom cite the home as an important environment, but in fact it is the micro-environment with which we interact most during our lifetimes.

As the population became denser, the building industry had to yield to economic pressures, and traditional homes – with an attic for momentos, and so on – were constructed less often. Standardized apartments, strictly impersonal, sprouted everywhere. Nevertheless, little by little, the dwellers introduced personal items and the apartments became customized, reflecting the dwellers' personalities. Territoriality and privacy are significant sources in the spontaneous drive towards personalizing the home. Cultural factors, physiological factors resulting from handicaps or health problems, and psychological factors will increase or decrease the urge for personalizing the indoor space. In any case, the home is the only space whose organization is still within reach of ordinary humans in big cities. After coming home and locking the door, the dweller feels free. To organize the home space is to enjoy freedom.

LEGAL ASPECTS

There are many legal requirements which regulate safety, hygiene, and even energy efficiency. Assuming that all these requirements have been met, what legal texts do not explicitly consider is respecting dwellers' need for freedom, territoriality and privacy. This is an interesting issue deserving discussion by human ecologists and architects. While careful analysis of user needs, preferences, and even extravagances usually precedes the planning of mansions, increasing numbers of huge apartment buildings are planned for sale or rent to anonymous consumers. There is no chance for hearing from future inhabitants of public speculative housing. Standardization and limitation of usable space are economic imperatives in this common scenario.

In the early years, the building industry did not care about human factors when planning low-cost apartments. Recently, though, planners unveiled a new trend. Never mind how small a low-cost studio might be; ingenious architects have been able to plan small, allowing for multiple choices regarding different decoration, furniture placement, and lighting. A small studio apartment can become a cozy, personal space. Past studios were planned in such a way that few possibilities were left for dwellers' imaginations. It is easy to guess how inadequate interactions are in such studios. Poorly interacting with their nearest environment, humans will be exposed to threats to their quality of life, as well as their physical and mental health. Neuroses, violence, and drug use are also stimulated by inadequate interactions between dweller and housing, inhibiting the free organization of indoor space.

Human beings spend about 20% of their lifetimes in work places: an office, a laboratory, a library, a restaurant, a hospital, a store, an industrial facility, and the like. While this is only one fifth of the yearly time, they have to perform to the best of their capability during this time to achieve professional success. The workplace is not merely a place where humans spend 8 hours per day. It is the place where they are continuously challenged, an arena where they can build prosperity and happiness – or merely fail. Physical fatigue and mental stress are inevitable, even in the age of ergonomics, automation, informatics and robotics. While modern technology reduces some muscular efforts, it increases visual strain, as well as manual and mental stress. Consequently, there is an increasing interest in reducing labor stress. This is the newest approach to occupational health and industrial safety. Equipment design, furniture redesign, and carefully planned layouts are developed with increasing interest in ergonomics. Legislation in some countries is very comprehensive, setting standards for safety and health. Occupational medicine and engineering have identified health hazards defining norms for protecting workers' health. Ecologists have identified many forms of pollution. Some national environmental protection agencies have been active in fighting pollution. Industry, as a whole, has been cooperative in reducing pollution and labor hazards. What is the role for human ecologists in such a well developed and researched field? Two case studies are reported in brief and by way of illustration, should inspire some answers.

CASE STUDIES

1. *Hospitals in the Amazon*

In the late fifties, in central Brazil, a number of modern hospitals were built to assist "primitive" Indian tribes. Everything was nice and clean, all white, following modern Western standards. The Indians' reaction was astonishing: they just ignored the modern hospitals. The author studied the problem after reading several reports on the Indians' rejection of the hospitals.

The reason for their behavior? As many intelligent persons do, Indians think hammocks are far superior to conventional beds in hot climates. (During cold nights, they like small fires burning under the hammocks.) In the white man's hospital they were given uncomfortable blankets limiting their movements at night.

Supported by some cultural reasons, we can endure hospital diets. Indians hated them too, but had no cultural reasons for accepting such diets. The Indian family does not abandon sick relatives, and the whole family wants to stay close to the person enjoying medical care. Modern hospitals could not possibly cope with that type of situation.

The Indians need the spiritual support of their priests; they must attend ritual ceremonies in order to get rid of bad spirits and heal. There is no opportunity for such ceremonies to occur in sterile, white, modern hospitals. An architect understood the problem. A project for a new hospital was laid out, intended to be in tune with the characteristics of the community to be served – one for which it will take a very long time to find to find out why they should adhere to our Western industrialized standards. There were two inexpensive buildings, with an open area in between, where ceremonial dances could be performed. (Does this sound silly? Many of the white man's hospitals have chapels. White people, in spite of trusting their medicine, also call for divine help!) One building is the guest house for lodging relatives, while the other building is the hospital. There are two rooms with two beds each, for special cases under medical supervision, and two other rooms which are wide enough to contain hammocks – local hammocks are almost twice as long as ordinary hospital beds. In each room it was possible to hang 40 hammocks.

The kitchen was redesigned and adapted to the Indians' own diets: mostly grilled or smoked food with no salt or sugar. Free access to the kitchen was provided in order to take advantage of relatives' own fishing abilities. The floor was designed to support the building of fires under the hammocks, as was the ventilation.

Successive administrations long discussed that "revolutionary" project. Finally, yielding to classic prejudices, they preferred to impose orthodox and civilized standards on those "inferior" people. The hospital was never built and "primitive" Indians still avoid modern hospitals. Nevertheless, there was an attempt to understand people, to study how to coexist without imposing what our civilization unilaterally decided to be the best.

2. *Manaus Research Campus*

Early in 1970, the author was called to plan and build the campus for the National Institute for Research in Manaus, capital of the Amazon region in Brazil. It was a challenging task: a beautiful plot, 25 hectares of rain forest, equatorial climate, poor electric power supply, no information about energy-efficient buildings in the tropics, high construction cost, scarce funds, and the owner's conservative prejudices to deal with.

Occupants included senior scientists training a new generation which was going to be the permanent staff of an institute devoted to Amazon Ecology and technology of Amazon products. The majority of senior scientists came from temperate climates, south of Brazil and industrialized countries: not the kind of people who are used to hot, humid climates. Electric power limitations did not allow for the air conditioning they needed. An innovative ecological model was badly needed. The author met Professor W. F. Brinkmann, who was the pioneering researcher on Amazon climatology for the Max Planck Institute, and posed two questions: "What are the characteristics of the Amazon heat?" and "What are the natural climate moderators?"

A fruitful interdisciplinary project ensued, involving climatologists, ecologists, chemists, architects, engineers and a human ecologist. Professor Brinkmann had accurately quantified direct, reflected, and scattered infra-red radiation. The significance of the gentle east breeze and the calories involved in plant transpiration were assessed. Valuable information was already available.

The next step was to find an open-minded architect who would feel free to forget orthodox styles and plan something new, according to available scientific data. Severiano Porto, a recent prize-winner who was emerging as an innovative architect for humid tropical climates, was called in and gladly joined the team.

The first approach was to abandon the project calling for a huge imposing building and use 30 inconspicuous scattered buildings instead. Scattering buildings would avoid blocking the east breeze. Conservative minds on the Board were shocked: expecting a single big outstanding building, they were eventually confronted with several buildings that were aligned in straight parallel and perpendicular streets crossing at right angles. It took some time to sell such a "disorderly" project. Since ecology was becoming popular, it was not difficult to convince the Board to maintain untouched forest between buildings.

In planning long rectangular buildings with the smallest wall areas facing east and west, the largest wall areas were not exposed to direct sunlight and direct infra-red, and the heated surfaces were reduced. The main direct heating problem was the roof. A polished reflecting ceiling material was needed. Such material should be efficient as a heat-conductor and practically incapable of storing heat. The obvious choice was aluminum. However, ecologists reported an unexpected problem: vulture feces corrode aluminum and there was a large population of vultures. Chemists suggested aluminum-chromium alloy, light plates adding chemical hardness to the physical properties of conventional aluminum ceilings. Heat would be partially reflected and partially conducted entering the space between ceiling and roof. That space might be an insulating cell, provided hot air could escape out, being replaced by fresh air drawn from the exterior by the movement of hot air. Hot air would go out through vents at the east and west ends of buildings. Fresh air, pulled by the hot air flow, would be admitted through broad wooden grilled eaves.

Direct radiation hitting southern and northern walls after 9:30 A.M. and before 4:30 P.M., was avoided by limiting the height of walls and extending the eaves out 5.5 feet. A reasonably wide sidewalk was shielded. Cross-ventilation proved to be a practical way for reducing scattered infra-red. Classic internal corridors are the constant obstacle to cross ventilation. Since wide shielded sidewalks could replace the internal corridors, they provided an innovative cost-reducing option coping with the requirements of human ecology.

It was learned too late that scattered heat and reflected heat are funneled indoors by modern windows that open swinging out their lower end. They stay at a 45 degree angle, just the perfect angle to direct indoors the ascending hot air and fumes from outside. Furthermore, cross-ventilation is not favored by such windows. The incidence of reflected infra-red could be significantly reduced by an 18-foot green band between the building and the road. Buildings would fit in, without standing out and arrogantly obfuscating the forest. Only the rain forest would dominate the landscape.

Other interdisciplinary inputs by human ecology were:

- In order to avoid the need for odd signs saying "Keep Off The Grass," it was decided to wait until users conveyed, by their footprints, where they preferred to walk;
- An aquatic birds' sanctuary was built facing the guest apartments, where a few wives who could not join the research work suffered nervous breakdowns after weeks of looking at the dense green wall of forest across the road;
- The laundry was a decorated, nice-looking small building, including a living room where people could play chess or cards, sit and read, or just chat without disturbing one another, while the machines did their job. What was, under 90-degree heat and 98% relative humidity, a tedious task, became bearable, a pleasant opportunity to get together and hear different languages.

Post-occupancy monitoring revealed that indoor temperatures were lower than the ones recorded at the very same time in other monitored buildings: classical Portuguese architecture, modern concrete buildings and slum huts. Actually slum huts displayed unique temperature charts: the highest point was higher than the highest recorded at the Institute's buildings but the high-temperature period was shorter. The difference was intriguing and a modified project was worked out for the last building: reducing the volume of concrete and eliminating glass surfaces, employing lots of good local timber, it was possible to reduce both - the peak and the length of the warmer indoors period.

These are completely diverse cases. In one case, the occupants were primitive Indian tribes, the building a hospital; in the other, modern climatic scientists in an extensive research campus. The scope of these two cases is intended to illustrate how complex the human ecological view is, as far as the evaluation of man-made environments is concerned.

The subject is complex and significant variables are plenty. Man-made environments have different purposes and are built in different natural environments to serve different types of occupants. The first step in the human ecological approach is to recognize such differences. Another important goal is harmony between buildings and users. There are different users in buildings, housing, recreation, health services, teaching, administration, business, banking, production plants, etc. Evaluating work places is a fascinating subject. Industry has been the first one to invest in human ecological advice and there is already some experience, albeit limited, in this special field. The human ecological approach studies each plant as a system. No plant exists by itself, but rather it is a dynamic system sustained by interactions. Humans, machinery, raw materials, products, building, methods and norms are some of the components of the plant system.

Before occupancy, it is hard to envision cooperation from human ecology as far as industrial facilities are concerned. Industrial development goes fast and updating of equipment, raw materials and methods is never-ending. Consequently, identifying all future variables just after checking a blueprint can be risky at best. Concerning industrial facilities, the human ecologists' experience seems to be limited to post-occupancy evaluation. In this process, two guidelines have been observed:

1. Treating the plant as a system, and;
2. Implementing feedback and follow-through.

As a system, plant performance depends upon harmonious interactions between parts. Harmonious interactions occur when all parts are compatible: whenever humans are involved, that means adaptation. Management, workers, the building, machinery, raw materials, products, norms, communication and supervision must be mutually fine-tuned within the whole system. In evaluating man-made environments in industrial facilities one needs to consider the adaptability of both building and equipment as permanent components. By "equipment," human ecologists mean production machinery, maintenance equipment and pollution control devices. Raw materials, processes, products, norms, and communication, all of which are subject to frequent modifications, are semi-permanent components. The humans who manage and operate the plant are the third component. Evaluating their role is a detective process preceded by the identification of different activities performed at the plant.

Feedback and follow-up are not mere spreadsheets, showing quantity or monitoring data. For human ecological purposes, those spreadsheets are noble tools. After statistical analysis, spreadsheets need to be condensed, according to the plant's systemic view, organized and reduced to a workable amount of data representing activities, rather than work or measurement points. The significance of follow-up needs to be emphasized. A still-picture taken in one single audit may be helpful whenever checking one specific item is the objective, and these may be enough for official controls. A motion picture, however, discloses tendencies and as such emerges as another tool for managers.

Adapting means first of all coping with official standards and available recommendations as issued by accredited specialized associations. Those standards are therefore the first item in an audit check-list. Previous interviews with management, to learn about specific higher standards to which the management aspired, are helpful and time-saving. There are general standards and specific standards. General standards focus on hygiene and safety and are applied to every workplace. Specific standards must be added whenever labor involves particular risks. They are applied when special raw materials, equipment or processes are utilized. Both classes of standards have been chiefly inspired by occupational medicine, a branch of knowledge as old as the Industrial Revolution, although continuously updated. Despite international agreements, not all countries adopt or enforce the very same standards. Some countries accept lower standards while others rigidly enforce higher ones. Human ecologists aspire to higher standards, even where national authorities formally adopt or usually tolerate lower performance. Fortunately, corporations hiring a human ecologist are quality-minded; they want the best even when lower standards are locally acceptable.

Human ecology is not a substitute for occupational medicine and safety engineering. It is, rather, an "add-on." Human ecology performs its role after safety standards have been implemented. Continuous adaptation of workers (and even managers) facing prevailing norms and higher standards, is human ecology's domain. Transition from lower accepted standards to higher requirements may be a painstaking and time-consuming job for both management and the human ecologist auditor. Changes must be planned aiming at a gradual improvement within the limitations of economic possibilities. Passionate, hurried perfectionism may lead to failure. Whenever health hazards and/or life risks are at stake, no procrastination will be tolerated.

Usually official and union controls will take care of such situations and human ecologists usually are not supposed to take any steps in this field. Nevertheless, in exceptional instances, they may be compelled to keep management informed. It is management's privilege to adopt goals and to approve time tables. Human ecologists will not cultivate Utopian perfectionism. They will be glad if every year the predefined goal is attained. It is to be kept in mind that across the table sits a businessman who wants improvements but who is aware that funding such improvements is an economic issue.

The original architectural design can facilitate or hamper improvements. Planning a building requires knowing the circulation and designation of each room. Notwithstanding, modern architects are aware of dynamic industrial changes and the consequent the need to create plans without imposing inflexible schemes. Improvements can become discouragingly costly when dealing with rigidly designed building layouts. A healthy work environment requires adequate lighting, ventilation, noise control, radiation control, reduced vibration and comfortable temperature. The air must be clean and humidity must be within the limits compatible with both human comfort and production process requirements. Human ecology, after checking all those parameters, will take neighboring areas into consideration by checking the destinations of solid, liquid, and gaseous wastes. Monitoring at regular intervals, by qualified technicians and working in accordance with accepted methods, is a basic tool for systematic feedback which reveals management's success in implementing worker safety and anti-pollution norms. Open-minded discussions with management and staff will lead to a realistic interpretation of the feedback data. Human ecological feedback and follow-through disclose activity performance, while occupational health stresses working conditions.

An activity may be performed under different working conditions The human ecological method for studying industrial facilities is comprehensive and supplies information on humans adapting to a plant's conditions. Usually, training and education will improve interactions. Human ecology can also facilitate the planning of education and training within industrial environments, whereby monitored data are not the only source of information. Labor accidents, turnover, absenteeism, incidence of classical occupational diseases and other diseases, individual and team productivity, rejections by quality control, are all data subject to statistical analysis. Beyond these data, satisfaction, participation, cooperation, rivalries, and so on may be just as valuable and indicative of human adaptation to a facility's environment. Evaluating such data will require a "strategy of choices," as Quinn Patton once said. Where quantification is feasible, statistical methods will be applied. But eventually, statistical methods may not be applicable and it may be wise to find appropriate ways of qualitative evaluation.

The reliability of evaluations by human ecologists depends upon the strategy of choices for each situation. Last, but not least, one needs to keep in mind that human ecology is an interdisciplinary science and that its implementation requires the art of bringing together different professionals, not merely for occasional advice, but for collaboration on the screening and ranking of variables, on evaluating both procedures and results, and on making objective interpretations of data. As such, evaluations by human ecologists cannot be a "one-man show."

BIBLIOGRAPHY

Clerc, J. M., 1985, *Introduction to Working Conditions and Environments*, Geneva, International Labor Office.

Danko, S., 1988, The practical applications of lighting research, in: R. J. Borden and Jacobs, J. (Eds.), *Human Ecology: Research and Applications*, College Park, Maryland, Society for Human Ecology.

Hinkle, Jr., L. E., 1973, Man's environment and health, in: *Proceedings of the International Conference on Environmental Health*, Belgrad, Yugoslavia.

Lessa, A., 1985, Preface to: Machado, Paulo A., *Ecologia Humana*, Saõ Paulo, Brazil, Cortez Edit.

Machado, P. A., 1973, The ecology of human settlements in the Amazon region, in: *Proceedings of the International Conference on Environmental Health*, Belgrad, Yugoslavia.

Malmberg, T., 1980, *Human Territoriality*, The Hague, The Netherlands, Mouton Publishing.

Mendes, R., 1980, *Medicina do Trabalho - doencas profissionais*, Saõ Paulo, Brazil, Servier.

Patton, Q., 1986, *Qualitative Evaluation Methods*, Beverly Hills, California, Sage Publications (7th printing).

Preiser, W. F. E., 1986, The Habitability Framework, in: R. J. Borden and alt, *Human Ecology: A Gathering of Perspectives*, College Park, Maryland, Society for Human Ecology.

Shaden, Egon, 1965, Aculturacaõ indigena, *Revista de Antropologia*, 13:1-315, Saõ Paulo, Brazil.

Schmid, P, 1988, New building concepts, in: R. J. Borden and Jacobs, J. (Eds.), *Human Ecology: Research and Applications*, College Park, Maryland, Society for Human Ecology.

Simmons, J. G., 1974, *The Ecology of Natural Resources*, Edward Arnold Publishing, London.

Stallones, R. A., 1973, Epidemiological Evidence of Effective of Man's Environments on Health, in *Proceedings of the International Conference on Environmental Health*, Belgrad, Yugoslavia.

Vayda, A. P., 1988, Actions and consequences as objects of explanation in human ecology, in: R. J. Borden and Jacobs, J. (Eds.), *Human Ecology: Research and Applications*, Society for Human Ecology, College Park, Maryland.

Westney, O., Brabble, E., and Edwards, C., 1988, Human ecology: concepts and perspectives, in: R. J. Borden and Jacobs, J. (Eds.), *Human Ecology: Research and Applications*, Society for Human Ecology, College Park, Maryland.

World Health Organization (WHO), *Health Hazards of the Human Environment*, Geneva, 1972.

World Health Organization, Technical report #714, *Identification and Control of Work-Related Diseases*, Geneva, 1985.

Young, Gerald, 1983, *The Origins of Human Ecology*, Stoudeberg Publishing, Pennsylvania.

BASIC METHODOLOGY FOR HUMAN ECOLOGICAL ASSESSMENT OF INDUSTRIAL FACILITIES

TABLE I

A. PRELIMINARY DATA

INPUTS	SOURCE	SPECIFIC TARGETS
Scope	Management's reasons for calling Human Ecologist	Productivity? Rejections by Quality Control? Labor Accidents? Occupational Hazards? Occupational Diseases? Poor Interactions? Others?
Facility	Corporation or Local Management	Chemical, Mechanical, Steel Electronics, Food, Toys, Pharmaceuticals, etc.
Flow Sheet	Local Management	Identifying and locating working positions, sequences and activities
Blueprint	Local Management	Building Characteristics
Standards	Local Management Corporation Management International Organisms Official Agencies	Autonomy, uniformity Policy, forecasted changes Compliance with International and National norms
Supervision and Controls	Local Management Corporation Management Official Agencies Internal Committees Unions	Frequency, effectiveness Frequency, stability Frequency, authority Workers' participation Participation, objectiveness
Equipment	Local Management and inspection Engineers Health team Chemists Personnel Administration Technicians Workers Ecologists Other	Manufacturing equipment Methods (generic) Raw Materials Internal Transportation Products Quality Control By-products Waste Maintenance Health Services Internal Communication Signs Equipment Internal Environment Protection Individual Protection External Environment Protection
Definition of Auditing Team	Management & Auditing Organization	Identifying needed and available specialists.

TABLE II

HUMAN ECOLOGICAL AUDIT

INPUT	SOURCE	TARGETS
B.1 HABITABILITY*		
Perceived habitability	Qualitative methods and measurements	Space Circulation Comfort
Measurable habitability	Measurements**	Temperature Humidity Lighting Noise Vibration Radiation Dust Microscopic airborne particles Chemical pollutants Other

Source: P. A. Machado, "Audit Routines" internal document, REV 1987.

*"Habitability defines the degree of fit between individuals or groups and their environment, both natural and manmade, in terms of an ecologically sound and humane built environment." W.F.E. Preiser, 1986.

**Measurements are to be performed by qualified professionals, using tested equipment and employing accepted methods. All data reflect the situation at the measurement time. Continuously monitoring some variables is already feasible. In the absence of continuous monitoring, results of intermittent measurements, after adequate statistical treatment, can supply results which represent the prevailing situation with acceptable accuracy. Blind cross-checking, conducted by independent experts, will add credibility to monitoring data.

TABLE III

INPUT	SOURCE	TARGETS
B.2 SYSTEM PERFORMANCE	B.2-a: quantified data, in Tables I and II	Compliance with norms Variation measurements on each work station during successive months or years Variation measurements on similar work station in other facilities of the same corporation Consolidated work station results to assess activities as performed at the plant and within the corporation.
	B.2-b: Qualitative methods; pertinent statistical analysis performed.	Perception of environment Understanding of norms and recommendations. Acceptance of devices for individual protection. Workers' participation. Effectiveness of training Understanding of monitoring. Cooperation with monitoring. Attitude facing health services available Attitude facing safety Attitude facing local managers Attitude facing corporation Cultural factors Internal tensions External tensions
B.3 SCANNING AND SCREENING	ABOVE DATA Absolute values for w.p. and activities. Dispersion band of measurement data.	Screening significant variables
B.4 SEARCHING CORRELATIONS	B.3 data + equipment description + qualitative data, B.2-b	Spotting possible correlations.
B.5 TESTING CORRELATIONS	Field testing effects of induced changes.	Identifying correlations.
B.6 CONCLUSIONS		

Source: P. A. Machado, "Audit Routines," internal document, REV 1987.

CHAPTER 12

CRITICAL FRAMEWORKS FOR BUILDING EVALUATION: TOTAL BUILDING PERFORMANCE, SYSTEMS INTEGRATION, AND LEVELS OF MEASUREMENT AND ASSESSMENT

Vivian Loftness, Volker Hartkopf, and Peter Mill

Center for Building Performance and Diagnostics
Department of Architecture
Carnegie-Mellon University
Pittsburgh, Pennsylvania

FRAMEWORK FOR EVALUATION 1: TOTAL BUILDING PERFORMANCE

Although there has been heated discussion over the past few years about the need for both objective and subjective field evaluation methods, there has been very little discussion about the need to complete those field evaluations in all performance areas simultaneously. One possible explanation of this void might be the difficulty that the design community has in defining total building performance, much less establishing limits of acceptability and testing for them. Notwithstanding, there have been some collective attempts at the definition of total building performance by the National Bureau of Standards (NBS, 1972), the International Standards Organization (ISO, 1972), and the Centre Internationale de Batîment (CIB, 1982). The authors of this chapter have built on these efforts, to develop a more manageable yet comprehensive list of six performance mandates for the built environment: spatial quality, thermal quality, acoustic quality, visual quality, air quality, and long-term building integrity against degradation (figure 1) (Hartkopf, Loftness, Mill, 1983).

A Definition of Total Building Performance

First, there has been a fundamental mandate over the centuries for building integrity – protection of the building's appearance and critical properties from degradation through moisture, temperature shifts, air movement, radiation, chemical and biological attack, human attack, and natural disasters (fire, flood, earthquake). Established by concerns for health, safety, and welfare, resource management (energy, materials, manpower, money) and image, the requirements for building integrity are bound by limits of "acceptable" degradation – ranging from slight decay (of the

I. FUNCTIONAL/SPATIAL QUALITY = satisfactory:
Based on knowledge of the building occupancies, occupancy functions, and organizational structures
   A. Individual Space Layout Quality
      Useable space, furnishings, layout efficiency, access, anthropometrics, ergonomics, image, flexibility/growth, occupancy controls
   B. Aggregated Space Layout Quality
      Proximities, access, compartmentalization, useable space, layout efficiency, image, amenities, flexibility/growth
   C. Building Siting Layout Quality
      Access, public interface/image, indoor-outdoor relationships, outdoor space layout, flexibility/growth
   D. Quality of Conveniences and Services
      Sanitary, fire safety, security, transportation, electrical, telephone, information technology; flexibility/growth

II. THERMAL QUALITY = satisfactory:
   A. Air Temperature
   B. Mean Radiant Temperature
   C. Humidity
   D. Air Speed
   E. Occupancy Factors and Controls

III. AIR QUALITY = satisfactory:
   A. Fresh Air
   B. Fresh Air Distribution
   C. Restriction of Mass Pollution — gases, vapors, micro-organisms, fumes, smokes, dusts
   D. Restriction of Energy Pollution — ionizing radiation, microwaves, radio waves, light waves, infrared
   E. Occupancy Factors and Controls

IV. ACOUSTIC QUALITY = satisfactory:
   A. Sound Source — Sound Pressure Levels and Frequency
   B. Sound Source — Background Noise
   C. Sound Path — Noise Isolation (air and structure-borne)
   D. Sound Path — Sound Distribution; absorption, reflection, uniformity, reverberation
   E. Occupancy Factors and Controls

V. VISUAL QUALITY = satisfactory:
   A. Ambient Light Levels — artificial and daylight
   B. Task Light Levels — artificial and daylight
   C. Contrast and Brightness Ratios
   D. Color Rendition
   E. View — visual information
   F. Occupancy Factors and Controls

VI. BUILDING INTEGRITY = satisfactory:
Based on knowledge of loads, moisture conditions, temperature shifts, air movement, radiation conditions, biological attack, man-made and natural disasters
   A. Quality of Mechanical/Structural Properties
      Compression, tension, shear, abuse
   B. Quality of Physical/Chemical Properties
      Watertightness, airtightness; transmission, reflection, absorption of heat, light and sound energy; fire safety
   C. Visible Properties
      Color, texture, finish, form, durability, maintainability

Figure 1

building's visual, mechanical, and physical properties), to debilitation in the ability to provide weathertightness or environmental conditioning for the function, to total devastation. Second, there are a series of mandates relating to interior occupancy requirements (human, animal, plant, artifact, machine) and their elemental needs for health, safety, and welfare (in human terms, comfort and protection of the five senses), i.e., thermal quality, acoustic quality, visual quality, air quality, and spatial quality.

Spatial quality problems	• Tripping, ducking cables and cable changes • Pile-ups of print-out and computer peripherals • Work surface now covered by equipment • Neck strain from screens on conventional desk • Tired arms, backache from keyboards on desk • Sitting long hours in "dining room" chairs
Visual quality problems	• Glare from windows without controls • Reflections of light fixtures/windows on screens • Coping with contrast ratios (from black-on-white text, to white-on-black screen) • Headaches, eyestrain from focusing on screen
Thermal quality problems	• Overheating from equipment, hotspots (500–1000 W each) • Radiant heat effect near equipment • No heat, air when high partitioning is introduced
Air quality problems	• Outgassing from some equipment and wiring • No control of fresh air, removed from windows (because of glare) • Need for more environmental change and control owing to reduced mobility
Acoustic quality problems	• Printers without acoustic covers (sold separately) • Annoyance from some keyboards (clicks) and disk drives (whirrs) • Modified mechanical systems with greater noise • Future voice-activated technologies
Building integrity problems	• Unanticipated weight of equipment and paper • Heightened need for dust-free environment

Figure 2

A Rationale for Comprehensive Performance Evaluation

The need for a manageable yet comprehensive list of performance mandates for designing or evaluating buildings is becoming more and more imperative as we review the failures in today's office environments. As the 1970's demonstrated, an emphasis on one performance area such as energy, without consideration for the range of performance areas in buildings, often results in failures in other performance areas, such as serious air quality and degradation failures. Yet building evaluations continue in singular performance areas (acoustic studies in factories, lighting in offices, heat loss in old buildings) with recommendations for action that will solve that performance problem – and create three more.

Today, with the emphasis on office automation, it is even more critical that a total building performance approach be introduced in building evaluations. Where conventional offices have already stretched the limits of humane working conditions (Hartkopf, Loftness, Mill, 1983), office automation will make conditions intolerable: vast, open floor areas; high-density layouts; inadequate surface and storage; an army of fluorescent light fixtures marching flush in an "acoustic" ceiling; and such unresponsive or untouchable environmental controls as zone thermostats, floor light switches, and fixed windows. Not only has the haphazard introduction of computer equipment made these environments more unbearable for the humans who must occupy them (figure 2), but the retrofit measures that have been recommended as solutions (acoustic equipment covers, glare screens, additional air conditioning) solve one performance problem at a time, in one component, without recognition of the implications for other performance areas and other components.

	Physiological Needs	Psychological Needs	Sociological Needs	Economic Needs
	Performance Criteria Specific to Certain Human Senses, in the Integrated System			
1 Spatial	Ergonomic Comfort Handicap Access Functional Servicing	Habitability Beauty, Calm; Excitement, View	Wayfinding, Functional Adjacencies	Space Conservation
2 Thermal	No Numbness, Frost-bite; No Drowsiness, Heat Stroke	Healthy Plants Sense of Warmth Individual Control	Flexibility to Dress with the Custom	Energy Conservation
3 Air	Air Purity; No Lung Problems; No Rashes, Cancers	Healthy Plants, Not Closed in, Stuffy No Synthetics	No Irritation From Neighbors; No Smoke, Smells	Energy Conservation
4 Acoustical	No Hearing Damage Music Enjoyment Speech Clarity	Quiet, Soothing; Activity, Excitement "Alive"	Privacy Communication	First-cost Conservation
5 Visual	No Glare, Good Task Illumination, Way-finding, No Fatigue	Orientation, Cheer-fulness, Calm, Inti-mate, Spacious, Alive	Status of Window Daylit Office "Sense of Territory"	Energy Conservation
6 Building Integrity	Fire Safety; Struct. Strength + Stability; Weathertightness, No Outgassing	Durability, Sense of Stability Image	Status/Appearance Quality of Const. "Craftsmanship"	Material/Labor Conservation
	Performance Criteria General to All Human Senses, in the Integrated System			
	Physical Comfort Health Safety Functional Appropriateness	Psych. Comfort Mental Health Psych. Safety Esthetics Delight	Privacy Security Community Image/Status	Space Conservation Material Conservation Time Conservation Energy Conservation Money/Investment Conservation

Figure 3

The resulting dictum can only be that the evaluating community must begin with a comprehensive outline of "total building performance" to be achieved (Building Research Advisory Board, 1985), which is finite enough to be manageable in the field, yet developed enough to represent that "integrated multi-sensory evaluator" known as a human being. Aesthetics, to the building occupant, is a thermal, air, acoustic, visual, and spatial experience. Although a building evaluation need not focus equally on all six performance areas (an acoustic emphasis or lighting emphasis may be critical), its construct and its recommendations must deal with all of the building performance areas in an integrated fashion.

Evaluating Against the Physiological, Psychological, Sociological and Economic Limits of Total Building Performance

The field evaluation techniques must also cover the six performance areas with adequate depth, measuring against physiological, as well as psychological, sociological and economic, limits of acceptability (figure 3). Research efforts, and consequently codes and standards, tend to focus on the physiological limits of acceptable performance, aimed at ensuring the physical health and safety of the building's occupants, sheltering basic bodily functions – sight, hearing, breathing, touch, and movement – from wear or destruction over time. Given the "hidden cancers" in today's working and living environments, there needs to be a more refined definition of physiological "comfort," and an additional emphasis on psychological and sociological "comfort." Psychological requirements would aim to support individual mental health through appropriate provisions for privacy, quiet, interactions, clarity, status, and change.

Sociological requirements would aim to support the well-being of the community within which the individuals act, relating the needs of the individual to the collective. Finally, economic requirements would aim to allocate resources (financial, technical, energy, material, and time) in the most effective manner, towards the overall goal of serving client and user needs, within the wider social context. The field evaluation expert not only has to establish physiological, psychological, sociological, and economic limits against which performance in each of the six areas will be measured, but has to take an active role in the refining of these "comfort zones" in the standard developing organizations.

Evaluating Total Building Performance Over a Given Time

In the evaluation of occupied environments, it is also necessary to establish the length of time during which the building or space must perform (Hartkopf, Loftness, Mill, 1985). Temporary offices, for example, need only be immediately suitable for occupancy, while corporate headquarters must not only be immediately suitable, but be reliable over the long term, and flexible for changing occupancies and functions. Suitability is a measure of the degree to which an office and its component parts serve user needs in the present and near future. Reliability is expressed as the probability that the building will continue to remain suitable throughout the life of the facility, given appropriate maintenance and operation. Flexibility, including adaptability, is a measure of the ability to accommodate changing occupancies and functions, along with a measure of the continuing effort and resources required during the building life-cycle to maintain suitability. The building evaluator must establish the level of suitability, reliability, and flexibility expected by the clients for the time they expect to remain in the spaces to be evaluated.

This framework for evaluating six performance mandates, and their physiological, psychological, sociological and economic acceptability over time, might be entitled "Field Evaluations for Total Building Performance."

FRAMEWORK FOR EVALUATION 2: INTEGRATED BUILDING SYSTEMS

The second imperative for building evaluation, is the need to study performance qualities in the integrated setting. For example, the focus of an evaluation cannot be the success (or failure) of office automation systems alone, but those systems within their integrated setting of interior systems, structural systems, enclosure systems, mechanical/servicing systems, and building occupancies. This implies that the building evaluator must have a working knowledge of the generic choices in each of those systems, and their effectiveness in relation to the office automation focus, for example. Alternatively, the building evaluator must be an expert team capable of covering those knowledge areas.

A Definition of the Integrated Setting

Just as a manageable list of performance mandates is needed, a manageable but comprehensive list of building systems and subsystems is needed (Hartkopf, Loftness, Mill, 1985). Four basic system groups must be discussed at a minimum: structural systems, envelope (or enclosure) systems, interior systems, and mechanical (or servicing) systems. Indeed, four disciplines have developed over the years specializing in

STRUCTURAL	spatial	thermal	air quality	acoustical	visual	building integrity
General System Type: Frame, Diaphragm, Tensile, Inflated	●	●		●	○	○
System Materials and Properties: Steel, Concrete, Wood, Plastics	●	●	●	●		●
Span, Bay Sizes, Column Spacing	●	○		○	●	●
Floor to Floor Height	○	●	○	○	●	
Cross-section of Structural Elements (height, width, depth)	○				○	○
Building Form: Plan, Section	○	○		●	○	
Expansion Capabilities (Vert. Horiz.)	●	○	○	○	○	○
Connections to/Accommodation of Other Structural Components	○	●	○	○		●
ENVELOPE						
Wall/Roof/Exterior Floor						
Exterior Surface, Material Prop.		○				●
Composite Materials, Thickness	●	○	○	○		●
Interior Surface	○	○	○	●	●	○
Form: Planar, Curved	●	○		●	○	○
Slope, Orientation	●	○		●	●	○
Module Size, Shape	●	○				●
Connection to Other Envelope Comp.	○	●	○	○		●
Windows/Openings						
Material Properties	●	●		●	●	●
Size, Shape, Spacing	●	●		○	●	
Orientation		●		○	●	○
Control Systems, Sunshading	○	●		○	●	●
Control Systems, Heat Loss		○				○
Control Systems, Security/Privacy	○			●	○	
Frame Connections, Plan/Section	○	○		○	●	●
Access, Visual and Physical		○	○		●	○
Expansion Potential (Vert. Horiz.)	○				○	
Change Potential for Access & Image	○				○	○
Color, Texture, Ornament	○		○		●	●
MECHANICAL						
HVAC (HEATING, VENTILATION, AIR CONDITIONING)						
Service Generators						
Size, Volume	●	●	●	○		
Form, Configuration	●	●	●	○		
Expansion Capability	●	●	●			
Material, Ornament	○					○
Service Conduits						
Thickness, Volume of Service	●	●	●	○		
Form, Shape	○	○		○	○	
Configuration, Distance, rise/run	○	●	○	○		
Interface/Expansion Capability	●	●	●	○		
Material, Ornament	○				○	○
Connection to Other Mechanical	●		○	○	○	○
Access		●	●	○		●
Service Terminals						
Planning Module	●	●	●	●	●	●
Number, Size, Capacity	●	●	●	●	○	●
Form, Material, Ornament	○			●	●	●
Interface/Expansion Capability	●	●	●	○		

●: Critical implications for the delivery of this performance mandate. ○: Some implications for the delivery of this performance mandate.

Figure 4. Examples of system design decisions affecting performance.

	spatial	thermal	air quality	acoustic	visual	building integrity
Relocation Capability	●	●	●	○	○	
Connection to Other Mechanical	○		○		●	○
Control Systems						
Central Management Systems	●	●	●	○	○	●
Local Management, Automatic/Manual	●	●	●	○	○	●
LIGHTING						
Service Generator—Size, Capacity	○				●	
Service Conduit						
Thickness, Volume of Service	○				●	
Interface, Expansion Capability	●				○	
Material, Ornament						●
Access					●	○
Service Terminals						
Planning Module	●	●		●	●	●
Size, Capacity	○	●			●	●
Form, Material, Ornament	●		●	●	●	●
Interface, Expansion Capability	●				●	
Relocation Capability	●					●
Connection to Other Mechanical	○	●		○		●
Control Systems						
Central Management	●	●		○	●	
Local Management, Automatic, Manual	●	●		○	●	
POWER, TELECOMMUNICATIONS, & SECURITY						
Service Generator—Size, Capacity	●					
Service Conduit						
Thickness, Volume of Service	●					
Interface/Expansion Capability	●					
Material, Ornament	●					●
Access						●
Service Terminals						
Planning Module	●			●		
Number, Size, Capacity	●	●				
Form, Ergonomics, Maneuverability						●
Material, Ornament	○				●	●
Interface/Expansion Capability	●					
Relocation Capability	●					
PLUMBING AND FIRE SAFETY						
Service Generator—Size, Capacity	●					
Service Conduit						
Thickness, Volume	●			○		
Configuration, Distance, rise/run	●			○		●
Interface/Expansion Capability	●					
Access						●
Material, Ornament						
Service Terminals						
Planning Module	●					
Number, Size, Capacity	●		○	○		●
Form, Material, Ornament						●
Interface, Expansion Capability	●					
Relocation Capability	●			○		
VERTICAL TRANSPORT						
Size, Volume of Service	●			●		●
Form, Configuration	●	○	●	○		○
Planning Module	●					
Expansion Capability	●	○	○			
Material, Ornament	○					●

Figure 4 (continued).

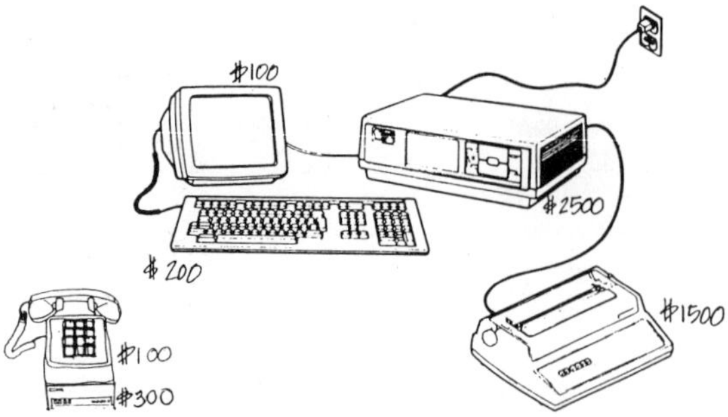

WITH ITS ASSOCIATED QUALITY ASSURANCE TESTS:
component communication; keyboard response/connection; screen contrast, horizontal hold; disk drive noise, time response; printer clarity; clean power, surge control; telephone/modem compatability; etc...

The PC workstation, as discussed in the trade (approximately $5000).

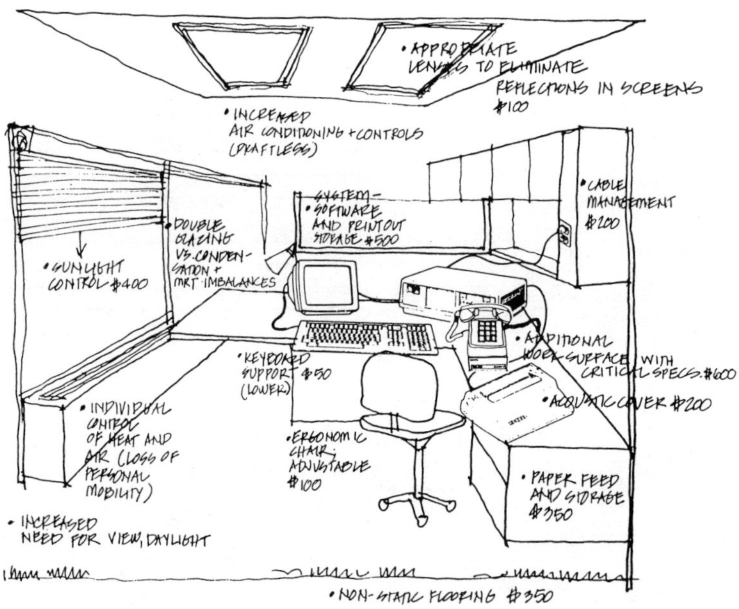

WITH ITS ASSOCIATED QUALITY ASSURANCE TESTS:
OVERALL SOUND ABSORPTION AND TRANSMISSION; OVERALL LIGHT LEVELS AND CONTRAST; OVERALL THERMAL QUALITY, MEAN RADIANT BALANCE + DRAFTLESSNESS; OVERALL SPATIAL/ERGONOMIC QUALITY INCLUDING DIMENSIONAL ADEQUACY, SAFETY, APPEARANCE; OVERALL AIR QUALITY AND RELATIVE HUMIDITY; LONG TERM DURABILITY GIVEN WEAR ASSOCIATED WITH THE TECHNOLOGIES.

The PC workstation that must be discussed for ensuring productive (and humane) work environments (add $3000).

Figure 5

Examples of Integrated Decision-Making Critical to Performance

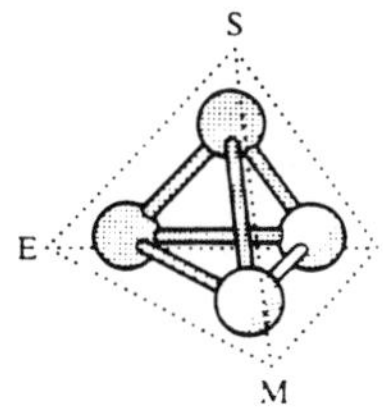

	Integrate Mechanical and Interior Systems for:
Spatial	• service module for present use and flexibility • space volume and ceiling height, functional identity • individual space layout, compartmentalization & flexibility • suitable servicing to individual space and aggregated space • expansion and change potential • individual control/space management
Thermal	• balanced mean radiant temperatures • air distribution, heating and cooling effectiveness, comfort • humidity impact on interior components • volume to be heated, stratification • mechanical efficiency from source to task
Air Quality	• fresh air distribution effectiveness • flushing vs. outgassing • protection from radiant pollution • individual control, compartmentalization vs. pollution migration
Acoustical	• background noise coordinated with appropriate room absorption • minimization of mechanical noise disturbance • minimization of vibration disturbance • effective acoustical compartmentalization around mechanical components
Visual	• adequate lighting distribution, balance • mechanical components for lighting dispersion, reflection • integration for space/function identity • lighting efficiency from source to task (lens) • flexibility potential • individual control, compartmentalization
Building Integrity	• humidity and temperature compartmentalization for arti-fact/machine/human "comfort" • staining and dirt buildup from mechanical on interior • vibration disturbance/destruction • mechanical accessibility, maintenance and repair • fire safety

Figure 6

these systems, their generic component choices and performance (structural engineering, architecture, interior design, and mechanical engineering). In recent years, the complexity of the building's services has yielded other specializations, separating HVAC mechanical systems from: telecommunications and lighting systems; plumbing and fire safety systems; and vertical transportation systems (figure 4) (Hartkopf, Loftness, Mill, 1985).

SET of TECHNOLOGICAL CHOICES	a. Microwave	b. Satellite (dishes)	c. Copper & Coaxial Wire Network	d. Fiber Optic Network	e. 2-Way Radio	f. Multiplexors	a. Copper Twisted Pairs	b. Flat Wire - Non Twisted Cable	c. Coaxial Cable	d. Fiber Optic Cables	e. Universal Wiring	f. Wireless - Radio Frequencies	g. Multiplexors, Concentrators	h. Termination Hardware, Ground Cables	i. Network Architecture: LAN (ring, bus, star..)	j. Wire Management - Hardware & Software	k. ISDN, Smart Wiring	a. Mainframes and Super Mainframes	b. Minis and Super Minis	c. Micros and Super (enhanced) Micros	d. Controllers	e. Architecture	a. Inputers (data, voice, video/graphics..)	b. Processors (read/process, compute, file/store..)	c. Outputers (tapes/disks, printers, plotters..)	d. Services & Service Architecture (e-mail..)	a. Information Technology Management Center	b. Energy/Facilities Management Center	c. Personnel/Technology Training Center
SETTINGS	EXTERNAL SIGNAL PROPAGATION						INTERNAL SIGNAL PROPAGATION											COMPUTERS and NETWORKING					PERIPHERALS and NETWORKING				COMMAND CENTERS		
A. PHYSICAL SETTING																													
1. STRUCTURE	I	C	I	I	I													S	C		C								
2. ENCLOSURE																													
a. Exterior Wall/Window	I	I			I													S	C	C	C		I		I				
b. Roof - Position, Configuration	S	S			C		C		C	C					C			S				I				C	C	C	C
c. Basement - Position, Configuration			S	S			C		C	C				S	C							C					C	C	C
3. BUILDING GEOMETRY																													
a. Exposure and Orientation	S	S			C													I	C	C	C		I		C	C			
b. Vertical Distribution:	S	S	S	S	S									S									S	S	S	S	I	I	I
b1. Floor Plates (Leasing Depths)	I	I	I	I	I		S		S	S	C	C			C	C		S	S		S	S							
b2. Location of Core	C	C	C	C	C	C	S		S	S	C				C	C	C	C	C		C	C							
b3. Core Configuration							S		S	S	C				C	C													
c. Horizontal Distribution:	S	S	S	S	S																		S	S	S	S	I	I	I
c1. Floor to Floor Heights	I	I	I	I	I		S	I	S	S	C				C	C		S	I		I	C							
c2. Structural/Mechanical Space	S	S	S	S	S		S		S	S	C				C	C		S	I		I	C							
d. Space - Quantity &Adjacencies	S	S	S	S	S							C			C	C	C	I	I	I	I	C	S	S	S	S	C	C	C
4. MAJOR SERVICES																													
a. Power (grounding)	C	C				S	C	C	C	C	C			S	C	C		S	S	C	S		S	S	S	S	S	S	S
b. Heating	C	C				C														I									
c. Ventilation	C	C				C												I	I	I	I								
d. Air Conditioning	C	C				C												S	C	I	C			C	S	C	C	C	C
e. Lighting (static)	C	C	C	C	C	C	I	C	I	I	I								I	I	I		C		C	S			
f. Fire	C	C	C	C	C	C	C	C	C	C	C				I	C		C	I		I						S	S	S
5. INTERIOR ELEMENTS																													
a. Ceiling Systems	I	I	C	C			C		C	C					I	I	I	I				I			C	C			
b. Partitioning Systems			C	C			C		C	C	I				I	I	I	I				I			C	C			
c. Flooring Systems			C	C			C	I	C	C	C				I	I	I	C				I			C	C	I	I	I
d. Acoustical Systems																		C		I			C	C	C	C	I	I	I
e. Workstation/ Furniture			C	C			C	I	C	C	C				I	I				C		I	C		C	C	I	I	I
B. ENVIRONMENTAL SETTING																													
1. SAFETY AND SECURITY																													
a. Physical Safety and Security																		C	I		I		I	I	I	C	C	C	C
b. Data Integrity and Security	C	C	C	C		I	C	C	C	C	C	C				I		C	I		I			C		C	C	C	C
2. THERMAL QUALITY																													
a. Air Temperature						I												C	I		I			I	I		C	C	C
b. Mean Radiant Temperature																		I	I	I	I			I	I		C	C	C
c. Humidity/ Static						I												I					I	I	I		C	C	C
d. Air Speed																								I	I				
e. Occupancy Factors and Controls																								I	I				
3. AIR QUALITY																													
a. Fresh Air																									I				
b. Fresh Air Distribution																									I				
c. Restriction of Mass Pollution						I												I	I		I		C	C	C		C	C	C
d. Restriction of Energy Pollution							I	I	I	I	I	I	I	I	I	I	I	C	I		I						C	C	C
e. Occupancy Factors and Controls																													
4. ACOUSTIC QUALITY																													
a. Sound Source Intensity																		C					C	C	S	C	C	C	C
b. Background Noise																							C		S	C			
c. Sound Path - Noise Isolation																		C	I	I	I		C	C	S	C	C	C	C
d. Sound Path - Sound Distribution																							C		C	C			C
e. Occupancy Factors and Controls																							C		C	C			
5. VISUAL QUALITY																													
a. Ambient Light Levels																							C		C	C			
b. Task Light Levels																							C		C	C			
c. Contrast, Brightness Ratios																							C		C	C			
d. Color Rendition																							C		C	C			
e. View/Visual Information and Relief																									C	C			
f. Occupancy Factors and Controls																													
6. SPATIAL QUALITY																													
a. Organizational Layout	C	C	C	C	C	C								I	I		C	I	I	I									
b. Workgroup Layout	C	C	C	C	C									I	I		C	I	I	I			C		C		I	I	I
c. Workstation Layout	C	C	C	C	C									I	I				I				C		C				
7. BUILDING INTEGRITY (durability, maintainability..)	I	I	I	I	I	I	I	I	I	I	I	I	I	I	I	I	I	I									S	S	S
C. ORGANIZATIONAL SETTING																I							I	I	I	I			
LEGEND	S: Show Stopper Conditions										C: Critical Conditions										I: Important Conditions								

Figure 7

A Rationale for Evaluating the Integrated Setting

Successes and failures in building environments are not the result of individual systems or components for the most part, but are the result of the effectiveness of those components within their integrated setting. An acoustic ceiling system may have excellent NRC ratings, but be rendered ineffective by the extensive introduction of flush lens light fixtures, by plastered ceiling areas for decorative effect, or by fresh paint applied in routine maintenance. The cost and success of an office automation effort, for example, lies not with the desktop hardware but with the resulting modifications of the other building systems that determine the office environment (see figure 5). Figure 6 illustrates examples of the importance of effectively integrating two of the five system groups defined – mechanical (HVAC, PLEC) and interior systems (ceiling, floor, wall, furnishings) – for each of the performance areas.

It is important to understand that the performance success of any one of these systems and its components is dependent on its integration with the other systems and its interface with the building's occupancy. The performance of a system cannot be studied in isolation from other systems or from the building occupants (and their sensitivities). Consequently, the building evaluation framework must demonstrate expertise in all of the system areas, their generic choices, and their history of performance in the integrated setting.

Exemplifying Office Automation Studies in the Integrated Setting

Whether the focus of the building evaluation is a particular performance area such as visual quality, or a particular system study such as office automation, the building evaluation must cover each of the systems in the integrated setting. Yet, the task of defining the subset components in each of the building's systems, and the generic design alternatives, is not a trivial one. For example, the common focus of office automation studies is the impact of the peripherals that sit on the worker's desk. Although some studies may broaden to look at mainframes and even cabling architecture, few of them look at the full range of office automation technologies that exist in the "intelligent" office (y axis, figure 7) (Building Research Advisory Board, 1988). However, any field study of the impact of office automation technologies must be studied in relation to the other building systems forming the physical and environmental setting. It is only in this realistic "morass" of integrated systems that one finds the critical conditions for success and failure in the occupied setting (table, figure 7).

The Necessity of Team Evaluations and Team Decision-Making

Building a full list of component and assembly choices in structural, enclosure, interior, mechanical servicing, and occupancy systems is not a trivial task. Understanding the generic choices in each system area, and their field performance, is even more difficult. Finally, identifying the successes and failures in integration between systems and their subsets creates a condition in which team building evaluations become a critical direction. The construct of the team, its expertise in more than one area, and its ability to work in an interactive, debating way should be an area of major emphasis for building evaluation today.

Developing the field evaluation team is a strategic effort. The team must be small enough for interactive field studies and communication, and yet large enough to provide expertise in each of the six performance areas (spatial, thermal, acoustic, visual, and air

Levels of evaluation measurement

1. Plan/archive analysis
2. Expert walkthrough
3. Occupancy and use analysis
4. Simple instrumentation
5. Expert instrumentation
   assign $/time/
   level of confidence

What looked at:
1. Documentation
2. Component—component interface
3. Occupant—component interface
4. Occupant—occupant interface
   % of building
   % of occupancy

Records kept:
1. Checklists
2. Counts
3. Annotated plans
4. Photos/videos
5. Plots
6. Tables
   total $/time/
   level of confidence

Levels of evaluation assessment

1. Expert/informed judgment
2. Pattern recognition
3. Simple algorithms (scalar, curve fit)
4. Statistical assessment
5. Complex algorithms
6. Expert systems
7. Mock-up sensory assessment
   assign $/time/
   level of confidence

Thresholds compared to:
1. Codes/standards
2. Guidelines
3. Project brief
4. Norms
5. Research results

Recommendation type:
1. Specific retrofit
2. Organizational change
3. Maintenance operation change
4. General retrofit
5. Project delivery system change
6. Codes and standards change
7. Data base development
8. Further testing

Figure 8

quality, as well as long-term integrity). Simultaneously, the team must demonstrate expertise in each of the major building systems, with up-to-date knowledge about generic alternatives, subsystem components, and their performance in integrated settings. Indeed, the field evaluation team must have a working knowledge of how buildings and components perform in occupied settings, and of the decision-making process whereby those settings were delivered. The team must also have experience in various levels of evaluation measurement and assessment, in order to manage reliable recommendations with evaluation simplicity. Finally, the team must be capable of working in an interactive manner. It is necessary to establish a workable team decision-making process for the exchange of ideas and for arguing through the ramifications of decisions on the effective design and integration of building subsystems for each performance quality.

Thus, field building evaluations must embrace the range of performance areas and their limits of acceptability within the full setting of integrated systems, including the building occupancy. Although emphasis may be placed on one performance area, or one building system, the field methods must cover each, such that recommendations for action do not lead to other failures. Expert team field evaluations, effectively covering the range of systems and performance areas, for measurement, assessment and recommendation, are one of the most promising directions for building evaluation.

FRAMEWORK FOR EVALUATION 3: COMBINING LEVELS OF MEASUREMENT AND ASSESSMENT

Given the complexity of two framework conditions described, increasing the differentiation between the levels of evaluation measurement and assessment is strongly needed (Leuder, 1986; Architectural and Building Science Directorate, 1982-86). Clearly differentiating the possible levels of measurement and assessment will promote development and deliberate documentation of techniques, and offer a range of choices in: percentage of building covered, cost, time, repeatability, level of expertise needed for measurement, level of expertise needed for assessment, depth of evaluation and assurance of results.

First, a distinction must be made between levels of evaluation measurement (data collection) and levels of evaluation assessment (data interpretation or diagnosis), leading to various levels of recommendations. With further development of these performance measurements and assessment procedures, the field evaluation of buildings can be organized into a strategic package, in response to client needs and resources, as well as available equipment and expertise. In each evaluation package, subjective and objective evaluations are married to enhance the objective understanding of occupant comfort, satisfaction, and productivity, and to clarify the subjective responses of occupants to the quality of their built environment.

Levels of Evaluation Measurement

If one assumes that both subjective and objective field evaluation procedures aim at measuring the quality of a built environment, there are at least five levels of evaluation measurement that can be developed: plan/archive analysis, expert walkthrough analysis, occupancy and use analysis (Rubin, 1980), simple instrumentation, and complex instrumentation (Mill, 1981) (figure 8). Each level of evaluation measurement will feed into records, independently or combined with other levels and other performance qualities. Checklists, counts, annotated plans, photos, videos, plots (curves, histograms), and tables may be carried away from the field measurements. The time taken to create these records will greatly simplify the evaluation assessment. For example, CO_2 measurements could be plotted on a scalar chart already indicating the standards and guidelines of acceptable limits against which the data is to be assessed.

There are certainly relationships between the five levels of evaluation measurement, such as: sequencing (one clearly depending on another), filtering capability (in revealing the scale of the problem of giving identical results a high percentage of the time), and modification capability (results of one allowing quantifiable enhancement of another). It is up to the field evaluation team to develop an effective and strategic field measurement package for the problem at hand.

Levels of Evaluation Assessment

In the past, data assessment procedures to interpret the meaning of data collected (diagnosis) were inseparably tied to the measurement procedures. It is becoming increasingly clear, however, that to ensure reliable performance evaluations and recommendations within budgetary and time constraints, measurement procedures should be developed independently from assessment procedures. The range of assessment procedures might include: Expert/Informed Judgement, Pattern Recognition, Simple Algorithms, Statistical Assessment, Complex Algorithms, Expert Systems, and Mock Up/Sensory Assessment (a statistical form of expert judgement) (figure 8).

Depending on the scale or complexity of the building issue or problem under evaluation, a focused complex instrumentation might be paired with a quick expert judgement, or a widespread plan/archive measurement might be paired with a complex algorithm assessment (such as the energy simulation program DOE2). The independence of the measurement and assessment procedures enables the field evaluation to be tailored for appropriate building coverage, depth of search, repeatability of evaluation, and reliability or confidence in the results, given the constraints of time or money.

1. Three-Dimensional Cable Network (Vertical and Horizontal)
2. Multiple Zone HVAC Network
3. Increased Shared Facilities
4. Improved Lighting/Daylighting Control
5. Improved Noise and Pollution Isolation
6. New, Redesigned Workstations (and Workgroups, with Alternative Workplaces)
7. Individualized Environmental Controls: Light, Heat, Air, Noise, Enclosure
8. Effective Interior/Envelope Interface
9. More Social Settings, Visual Diversity
10. Building Management Trio with CAFM: Facilities Manager, Technology Manager, Personnel Manager

Figure 9. Ten major design changes in the high-tech office.

Recommendations: The Output of Measurement and Assessment Packages

Written evaluations constitute the final step in building evaluation measurements and assessments. There are many types of recommendations that may be offered to the client, but they have major cost and time implications, and may be increasingly pervasive in impact (figure 8). Recommendations for specific building retrofits may be the most common output of field evaluations, but recommendations for widespread retrofits or code and standards modifications may be the most significant. The evaluation team must ascertain the types of recommendations expected, which in turn will drive the selection of measurement and assessment procedures. To have the highest level of confidence, with repeatable and reliable results, it is often necessary to utilize a full menu of field measurement procedures, with extensive building coverage, matched by assessment procedures with the most rigorous algorithms, against the toughest thresholds.

Strategic Packaging of Measurement and Assessment Procedures

Given a measurement and assessment "menu," with a clearly stated level of recommendation, the challenge of completing cost-effective field evaluations of occupied buildings lies in the packaging of appropriate measurement procedures with assessment procedures for responding to client issues, demands, and resources. A professional building evaluation team might develop a filter strategy for uncovering conditions in various performance areas, with increasing levels of time and resource investment only where merited. Again, even with addressing a particular issue or problem the strategic package must still incorporate a full set of performance criteria and the full set of building systems for ensuring recommendations that will positively improve the working environment, without side effects.

FRAMEWORK RESULTS: DEVELOPING NEW MAXIMS FOR DESIGN AND RETROFIT IN THE INTELLIGENT OFFICE

The three conditions discussed as a necessary framework for field evaluations of buildings (packaging levels of evaluation measurement and assessment of total building performance in the integrated setting) have been put into practice. The authors have studied modern office, museum, and courthouse settings with various assembled teams of building performance and building system experts, to uncover performance successes, failures, and calls for action in an effort to promote better environments for building occupants. The results of these studies have led to specific retrofit measures as well as to general maxims. In relation to the "intelligent" office, our building studies have begun the formulation of a series of design maxims that would greatly resolve and enhance the performance of the electronic office (figure 9) (Building Research Advisory Board, 1988).

Many decision-makers in the building sector, owners and architects, envision the creation of a high-tech building as the introduction of the appropriate portable "baggage" into a traditionally well designed office. A few more boxes here, a few more outlets and wires there, and one office adapted for housing the mainframe, printers, and supplies – presto, an instant high-tech office. Today's owners and architects, for the most part, have not been able to grasp the significance of the changes demanded in designing high-tech spaces: changes in the design process, changes in the physical setting, and changes in the delivery and operation of the building. If one could imagine designing a hospital today for the old black bag doctor to bring in his "tools" and then patching in all that high-tech equipment as if it were baggage dropped in each room, you would be imagining the process by which we design high-tech offices.

Instead, one must look at the actual physical and environmental setting changes that must occur in the high-tech office: a three-dimensional cable network (horizontal and vertical cable bundles) so massive and so much in demand that it must be treated like a structural system in conception, an associated HVAC network flexibly subdivided into dozens of zones (instead of five or six) with modularized vertical and horizontal duct plans, a drastically increased amount of square footage for shared facilities (well beyond the coffee pot and copy machine) such as mainframes, patch panels, uninterrupted power supplies, printing rooms, teleconferencing rooms on a building-wide and floor-by-floor level, and a major rethinking of workstation design to incorporate cabling, adequate work surface and storage, adjustability, as well as improved environmental contact and environmental controls for light, air, acoustics. Without tremendous knowledge and innovation, the old equipment, the old telephone closets, the old envelop controls, the old methods of subdividing spaces, and the old furnishings just won't do.

FRAMEWORK RESULTS: INTRODUCING FIELD EVALUATIONS THROUGHOUT THE BUILDING DELIVERY PROCESS

The results of our field evaluations of occupied buildings, however, go beyond maxims for the design of spaces, to a rethinking of the process of design. At least three steps must be reconsidered: predesign programming, team decision-making in design, and field diagnostics in the construction, commissioning, and operation of buildings – if we are to get any closer to assuring quality in the built environment. Each of the steps involves the development of field evaluation methods.

Public Works Canada has instituted a well-defined project delivery system (PDS) (Public Works Canada, 1981) outlining the deliverables and criteria for 10 stages in building production: 1) identification of needs and opportunities, 2) option analysis and selection, 3) project definition, 4) design, 5) working documents, 6) contracting, 7) construction, 8) commissioning, 9) occupancy, and 10) evaluation. The most serious concerns remaining in this division of responsibility for building production are the lack

Stage 1 Identification of Needs and Opportunities

- Establish clients' total performance needs and concerns
- Test existing facilities against these needs
- Test alternative facilities against these needs
- Establish programming team and transdisciplinary method

Stage 2 Option Analysis and Selection

- Gather performance standards and test packages
- Complete transdisciplinary building visits
- Complete site/environmental impact testing

Stage 3 Project Definition

- Write and test performance specifications for function/use
- Establish building delivery process stress factors in each performance area
- Establish performance checks for each design stage
- Establish design team and transdisciplinary method

Stage 4 Design

- Establish advocacy approach to concept development (to reduce stress factors in each performance area)
- Test concept against performance standards
- Establish hierarchical and advocacy approach to preliminary design
- Test preliminary design against performance standards
- Establish hierarchical and advocacy approach to final design
- Test final design against performance standards

Stage 5 Working Drawings

- Establish detail/specification team and transdisciplinary methods
- Test working drawings against performance standards

Stage 6 Contracting

- Set up construction performance tests before tender
- Test performance of various contractors' buildings

Stage 7 Construction

- Test performance quality against project brief regularly
- Supervise known stress factors for performance
- Ensure as-builts are completed
- Install performance "tattletales"

Stage 8 Commissioning

- Avoid early occupancy. Debug base building
- Establish maintenance, operation, user's manuals
- Establish one-year training/turnover design and construction to maintenance and operation
- Balance building to occupancy/functions to meet project brief performance specifications
- Test performance given occupancies/functions

Stage 9 Maintenance, Operation and Use

- Monitor preventative maintenance
- Establish "belt-tightening" operations programs and fine-tuning
- Initiate staged mini-test packages to avoid problems and failures

Stage 10 Evaluation

- Identify transdisciplinary nature of issues/problems
- Develop cost/benefit transdisciplinary test "menu" package
- Develop transdisciplinary recommendations — no side effects
- Feedback into building delivery process
- Institute post-failure evaluation

Figure 10

of teeth (testable limits of acceptable performance quality) and the lack of continuous responsibility of any one player throughout the process.

Performance failures can result from decisions made at any stage in the building delivery process, decisions that reduce the alternatives for each succeeding stage, such that the final product has less chance for success. The concept might be titled "stress factors" or the introduction of decisions in one stage of the building delivery process that significantly narrows the range of possible decisions in successive stages – towards the delivery of spatial, thermal, air, acoustic, and visual quality, as well as long-term building integrity. In fact, a major percentage of the building performance failures that occur in buildings result from "stress factors" that began at the architectural conception of the project. The inability of the range of players in the building delivery process to contribute in any way to the early conceptual ideas, results in decisions that make it increasingly difficult for each succeeding player to ensure performance. The lack of adequate front-end knowledge-building, and the heavy reliance on the architect as the integrated knowledge base from which conceptual ideas must develop, is beginning to take its toll. The performance of modern buildings is becoming worse and worse for both building occupants and their owners.

Consequently, it is critical that the building delivery process begins to reflect more informed front-end decision-making and a structured continuum of accountability. Field evaluation can enhance the building delivery process in at least a dozen areas (figure 10) (Leuder, 1986), with marginal increase in front-end time and cost compared to the massive increase in down-the-line time and cost to the building. Not only is field evaluation critical in the early studies of similar building types (for building performance, given the integrated systems), it is needed as a communication tool in the conceptual development stage by the project team (not the architect alone) (Loftness, Hartkopf, 1988). Field evaluation will also be needed in the working drawing and specification stage and most critically in the construction, occupancy, and one-year commissioning stages.

One key addition to this list of front-end activities in the delivery of a multi-million dollar building, is the addition of a single line project manager with extensive building construction and facilities management experience. At minimal relative cost, this project manager will represent the client's best interest and manage the stages of the building delivery process, with the assurance of continuity in the building project, of accountability, and of effective multidisciplinary (team) communication. The vested interests of the architect, the lack of building management expertise, and the weak knowledge of building failures (or successes) make them as a profession, inappropriate to carry this position of single line authority throughout the building delivery process. The critical contribution of the architect to the design team is key, however, to the building's overall quality – its spatial, functional, and formal performance. Does this chapter end or begin with this reality?

REFERENCES

Architectural and Building Science Directorate, Public Works Canada, working document: Total Building Performance, Hartkopf, Loftness, Mill 1982-1986.

Blanchere, G., 1972, The Notion of Performance in Building: Building Requirements, *Proceedings from Performance Concept in Buildings 1972*, National Bureau of Standards, NBS 361.

Building Research Advisory Board, National Research Council, National Academy of Sciences, 1985, *Building Diagnostics: A Conceptual Framework.*

Building Research Advisory Board, National Research Council, National Academy of Sciences, 1988, *Electronically Enhanced Office Buildings.*

Centre Internationale de Batîment (CIB), 1982, Publication 64, Working with the Performance Approach in Building.

Hartkopf, V., Loftness, V., Mill, P., 1985, Integration for Performance, Chapter 5, *The Building Systems Integration Handbook,* R. Rush, J. Wiley (Eds.) AIA.

Hartkopf, V. Loftness, V., Mill, P., 1983, The Concept of Total Building Performance and Building Diagnostics, ASTM E6.24, *Conference Proceedings 1983,* Bal Harbor, Florida.

International Standards Organization (ISO), 1972, Performance Standards in Buildings, ISO/TC 59.

Loftness, V., and Hartkopf, V., 1988, *Architecture for Art's Sake: Vantage Point Evaluation of Museum Performance for Future Design,* National Endowment for the Arts Grant Publication, Carnegie-Mellon University.

Mill, P., 1981, International Building Thermographic Training Course, Public Works Canada.

National Bureau of Standards, 1972, *Proceedings from The Performance Concept in Buildings 1972,* NBS 361.

R. Lueder (Ed.), 1986, *The Ergonomic Payoff: Designing the Electronic Office,* Nichols Publishing Company, New York.

Rubin, A., 1980, *Building for People: Behavioral Research Approaches and Directions,* Dept. of Commerce, National Bureau of Standards SPS474.

Public Works Canada, 1981, The Project Delivery System.

CHAPTER 13

TOWARDS A POE PARADIGM

John Zeisel

Building Diagnostics, Inc.
Boston, Massachusetts

THE CANADIAN HOSPITAL POE PROGRAM: A SOURCE FOR AGREEMENT

In order for post-occupancy evaluation (POE) to become an accepted and standard part of research and design practice, POE professionals must reach agreement on what POEs comprise. This chapter describes six areas for agreement, employing the Hospital POE Program of the Canadian Department of Health and Welfare (BDI, 1988) to demonstrate these six components in practice. The Hospital POE Program:

1. Establishes a standardized POE methodology including sequencing, methods, and data gathering;
2. Institutionalizes POE in a large and diverse Federal-Provincial health care bureaucracy;
3. Creates a POE database to aid in the development of design standards and guidelines, to improve planning and design decision making, and to set capital funding priorities for new construction and renovation; and
4. Shares the POE approach explicitly and in detail with both the client group and their consultants who will carry out future POEs, and with the rest of the POE professional community.

Agreement on the Term "POE"

Millions are spent annually on post-occupancy evaluation in Canada, New Zealand, Australia, England, France, Japan, South Africa, and the USA, among other countries. For over a decade POE studies have been carried out by in-house professionals and outside consultants for large private corporations and for government agencies. Internationally, such organizations include the New Zealand Department of Public Works, Health and Welfare Canada, Public Works Canada, and Australia's Public Works Department. In the U.S. POEs and POE programs have been carried out by the General Services Administration, the Army, the Postal Service, the California Department of Corrections, the Massachusetts Division of Capital Planning and Operations, and the Veterans Administration, among others.

Many professionals consider post-occupancy evaluation to be a fundamentally sound approach to learning from buildings that employs tested planning concepts, theory and environment-behavior methods; but they think the term "post-occupancy" is a misnomer. Almost everyone who uses the term agrees that POEs are not evaluations of buildings after they have been occupied – post-occupancy – but rather are evaluations of buildings-in-use (Vischer, 1989). Nevertheless, most of us have accepted the acronym POE as an acceptable one for a very common sense approach to learning about buildings, and how to run them, from how buildings actually work. And we are all happier when the acronym POE is used in place of the phrase post-occupancy evaluation.

Such agreement among researchers is truly remarkable in a field that after twenty years cannot agree on its own name! It has been called: Environment-Behavior, or "E-B" (Zeisel, 1984); Environment & Behavior, or "E&B"; Environmental Psychology (Proshansky et al, 1970); and also Environmental Design Research, or "EDR" (Sanoff and Cohn, 1970).

Two forces – the growing market for applied E-B research to solve some real building problem, and the call E-B professionals hear to develop some methodological, even scientific, rigor – seem to be driving us to agree on the term POE to describe applying research methods systematically to assess how a building in use works, and why.

The Need for Agreement

Environment-behavior researchers, while unified on the term POE, carry out POEs in a less unified way: we use limited state-of-the-art knowledge shared piecemeal at EDRA (Environmental Design Research Association) and IAPS (International Association for the Study of People and their Physical Surroundings) conferences through papers and workshops, and through a few books. We insist on forging ahead with POEs, with little apparent concern for the generalizability of their methods and findings.

Vanity and shortsightedness stand in the way of accepting and building on the professional affinity that beckons. Vanity leads researchers, designers and others to believe that the wheel each of us has invented is better than the wheel the others have invented. We don't want to admit that our particular mix of multiple-methods, or our particular approach to increasingly in-depth building assessments, is probably not much better than the next guy's. And it probably isn't. Vanity leads us to believe that the market for POEs wants unique approaches to problem solving, when in reality it wants standardized and proven approaches well executed.

Shortsightedness leads us to mistake each other for competitors, when we are colleagues. As a group we compete with each other for scarce government and private industry funds to carry out POEs. Although I am as interested as the next person in the state of the art of POE research, and know what is happening in the field in Massachusetts, I feel ignorant of such developments in Holland, Germany, France, England and other parts of Europe, Japan, and even a stone's throw away, in New York City, Buffalo, Atlanta, Los Angeles, and Albuquerque.

Our competition is not each other. It is the professional groups who believe that creating buildings is an end in itself, that intuitive building evaluation leading to personal insight is sufficient to complete the learning cycle, and that it makes little difference what methods are used to assess buildings because you can't really learn that much, anyway.

To combat this viewpoint, there is a crying need among POE professionals for more rigorously shared investigative approaches, methods, methodologies, instruments and even scales. Market researchers stormed the Western world by promoting to public and private sector decision makers a rigorous approach for studying consumers.

Rigorous market research approaches were rooted theoretically and methodologically in university departments of sociology, psychology, human relations, and business. The academic base of market research generated intellectually shared knowledge about such topics as focused interviewing, sampling, questionnaire construction, consumer panels, to name just a few. Market research professionals shared their work in detail through schools and research institutes, such as Columbia University's Bureau of Applied Social Research, in order to create a rigorous and generalizable field. POE professionals who want the same credibility, must similarly share their work.

Six Areas for Agreement

In both literature and practice six elements and principles of post-occupancy evaluation studies emerge as fundamental to most POEs. These principles may well serve as a basis for the agreement among professionals that is sorely needed. The six elements are:

1. Building purpose
2. Critical performance criteria
3. Established methods
4. Optional depth of investigation
5. Pre-designed data management
6. Pre-established database structure

1. *Building purpose:* POEs are frequently carried out to compare building-in-use performance to expectations framed before the building was constructed. These expectations have included values. For example, prisons have been evaluated against the value of humane treatment. Expectations have also been framed in terms of the building program – how the designers saw the building being used. For example, a programmatic evaluation might compare actual lighting levels in the building-in-use to those specified in the program. And POEs have compared buildings to abstract goals for which similar buildings have been built.

Employing building purpose as a comparative benchmark for POEs has different implications from values, the building program and goals (Zeisel, 1986). Building purpose implies ownership. For the public or private "owners" of most if not all buildings on which POEs are carried out, their buildings exist to fulfill several purposes. In most cases, these tend to be organizational purposes such as supporting the health care the organization is set up to provide, presenting a new corporate image, increasing organizational effectiveness, or creating greater communication among employees.

In addition to implying some form of "ownership," taking building purpose as the driving force behind POE investigations implies that buildings are tools to achieve a set of ends. From this follows a more important derivative principle; namely, that buildings can be separated from their purposes. Such separation enables analysis of the building and its structure, analysis of the purposes and their role in the organization, and constructive debate and comparison of the two to each other.

2. *Critical performance:* POEs often assume that because all buildings must meet certain standards, the performance of all buildings is comparable. In a very broad sense, this premise is true: all buildings must stand up, enclose some behavior or objects, and have a certain life cycle. However, while there are some "general" performance criteria for buildings, investigation into general performance criteria for a building is unlikely

to yield robust data and information useful in design decision making about that building type.

"Critical" performance criteria on the other hand can be distinguished from general performance criteria in that they reflect only those building characteristics without which the building's particular purposes cannot be fulfilled – critical performance criteria can be imputed from building purposes. Because building types differ by purpose, the critical performance criteria used in POEs of different building types differs. Taking critical performance criteria as the basis for POEs implies building type differences.

The term "critical" also implies that not all performance criteria will be the subject of each POE. Rather only those that are most important in meeting the building's purposes are selected.

3. *Established methods:* Through lengthy trial and error in field studies, and through methodological analysis, post-occupancy evaluators have developed a well established set of methods for use in their evaluation. This multi-method methodology includes building walkthroughs, interviews, questionnaires, unobtrusive measures, behavior observation, and secondary analysis, to name the major components. The methods have been tested in comparative studies for their reliability and validity. If methods be the measuring stick, POEs are part of a building research and design paradigm that has come of age.

The methods POE researchers use regularly are no longer idiosyncratic to the investigator or the country in which the POE is being carried out. The methods are no longer individualized; they are shared nationally and internationally. This implies that lively debate within the field needs to take place about details of method in order to improve them. But outside the field, if professionals compete to determine who best applies methods and interprets results, rather than who invents the best methods, the field of POE is served best.

POE methods and their results are well enough developed to be put to the same test as scientific approaches in other more established fields. In established sciences, methods, findings and theories are tested and improved through attempts at "falsification," the presentation of alternative explanatory hypothesis to determine if those methods, findings and theories can stand up to rigorous attacks. (Popper, 1972)

4. *Optional depth of study:* Just as building design and construction are carried out through a series of increasingly detailed drawings and decisions; POEs are carried out with the option for increasingly detailed investigation, data gathering and analysis. Just as doctors need to carry out a series of increasingly detailed and focused tests to diagnose a complex illness, POE investigators have found it both impractical and nearly impossible to totally evaluate a complex building in a single shot.

The most useful approach to POEs appears to be a diagnostic one which begins with a broad brush look at the building to determine if there are any major problems or qualities. If there are, and if the users of the research find it practical to look further at the building, the investigator weeds out less significant and focuses on more significant building elements. The investigator selects which problems to focus on, depending on the resources available, the investigator's research interests, the client's practical problems, and the potential usefulness of the results.

Optional depth and focus of study as part of a diagnostic POE approach takes away from POEs the burden of having to be systemically exhaustive; enabling POEs to be applied as practical problem solving tools. Theory becomes the handmaiden to practicality, rather than an end in itself.

5. *Pre-designed data management:* Data for POE analysis collected haphazardly with no clear set of organizing hypotheses are less powerful decision making tools than those selectively collected with a clearly organized analytic framework developed beforehand. Exuberant researchers learned this lesson in the early days of questionnaire research in sociology. They mistakenly felt they had important findings when they found statistically significant relationships between answers to the many questions they logically – but with no predesigned set of hypotheses – included in their field questionnaires.

Successful POE investigators employ their understanding of building performance and user reactions to construct models of user responses to environments as the basis for constructing questionnaires and survey instruments. For example, Vischer's building-in-use assessment system questionnaire, described in her Environmental Quality in Offices (1989), is based on a tested model of the interrelated reactions building users have to the set of basic indoor ambient quality characteristics in office buildings. Because there are structured hypotheses in her standardized data management and data analysis methodology, the results are transferable between buildings.

Pre-designed data management systems have also been developed for hospitals and courthouses (Building Diagnostics Inc, 1988a and 1988b). As such statistical data handling systems become standardized, and shared between investigators, the POE field increasingly is able to develop sound theoretical approaches.

6. *Pre-established database structure:* No matter what methods are used or data are gathered, if the data are not organized to be useful for building improvement, they remain data; they do not become information. Here the term database describes a system for organizing data so that they are accessible to decision makers, so that groups of data that are related can be accessed together, and so that the format of the data corresponds to the format of the decisions that need to be made.

Just as data gathered to test hypotheses are more powerful than data gathered haphazardly, databases that are pre-established and pre-structured are more efficient and powerful for making decisions.

Database design implies that the data are being gathered for a purpose, and that behind that purpose are users of the data. This in turn implies that decision makers who want to take the most informed decisions, employ data gathering methods, data, and databases for displaying the data that are most reliable and valid because they are interconnected by a common theoretical and practical thread.

In sum, when shared by POE investigators, the following themes make POEs a powerful force in environmental design. First, building purpose as a vehicle for organizing POE data gathering and analysis, and from which critical performance criteria are derived for particular building types. Second, tested methods that enable all POEs to be compared to one another and thus can be used to build a sound and firm shared body of knowledge. Third, structured hypotheses that connect and serve to structure data gathering methods, statistical data analysis and databases that result from POEs.

The principles underlying these six elements might be considered the beginning of a POE paradigm.

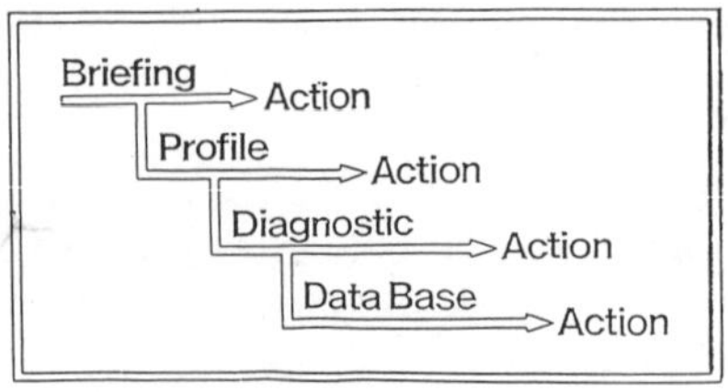

Figure 1

THE CANADIAN HOSPITAL POE PROGRAM: AN OVERVIEW

This six-point approach to POEs has been developed in a specific context – the Canadian health care system. A national POE methodology was developed for joint use by the Federal Health and Welfare Department and the Provincial Health Authorities. POEs of eight hospitals were carried out during the methodology development phase, and three hospital POEs were carried out during the POE Program's first full year.

Phasing POEs

Each Hospital POE carried out as part of the program is phased to focus increasingly POE efforts on items of design and facility operations that are significant for future design improvement and to screen out unimportant ones. Phasing the Hospital POE enables the POE team to approach these complex facilities and organizations a bit at a time, peeling off layer by layer until the most salient lessons are learned. The phased POE process helps the team make sense of what otherwise might be an overwhelming task. The following diagram represents this approach:

Phased Post-Occupancy Evaluation: The Four Phases

The Canadian Hospital POE process comprises four phases. In the first, or "briefing" phase, the POE team prepares a report describing the hospital in general terms using data gathered without going on site. In the "profile" and "diagnostic" phases, the POE team goes on-site to determine successes and problems, to collect and analyze data, and to suggest recommendations for action. These two sequential site visits enable the team to collect some data, reflect on it, and then focus on only the most important data in the second hospital visit. Finally, the "database" phase provides the opportunity to compare the findings from various POEs and draw more generalizable conclusions.

- The Briefing
- The Profile
- The Diagnostic
- The Database System

The Briefing: Briefing activities begin when the Province identifies the hospital at which the POE will be carried out. Members are selected for the POE team, or a POE consultant is hired, with a project manager in charge of organizing and coordinating the POE activities. Data for preparing a Briefing Report, collected from the hospital by phone and mail, include a chronological history, staffing, reference information, and site and floor plans that describe the organizational and physical aspects of the hospital. Team members review the Briefing Report and the POE schedule in a formal briefing session before visiting the hospital.

The Profile: The Profile visit to the hospital results in a written Profile Report that identifies the design successes and problems of individual departments and specific spaces based on hospital staff responses recorded in standardized interviews and questionnaires. Before meeting with a group of key hospital staff members in a focus interview the POE team members take an orientation walkthrough tour of the entire hospital. At the focus interview the POE team members explain the POE purpose and interview methods and, through group discussion, build an overall impression of the topics of staff concern. Team members then carry out interviews in the individual departments.

Two to three days at the hospital allow the team members to collect data through interviews, questionnaires, photographs, and measurements which are then analyzed for the Profile Report. The statistical results of the questionnaire analysis for the hospital are compared to national data norms stored in the Database system, and then are added to the Database to be included in the national norms. The Profile Report describes in-depth substudies to be carried out at the hospital in the Diagnostic phase of the POE.

The Diagnostic: In the Diagnostic phase the POE team, generally a POE project manager and several experts, returns to the hospital on the follow-up diagnostic visit, to carry out the more detailed substudies. Team composition may differ, depending on the skills required for the particular substudies. During the substudies team members examine design problems and successes using several combinations of research methods for each study. These substudies may be specific to a department or subsystem of the hospital, concern a hospital-wide system or issue, or be generic and generalizable to all hospitals.

During a three-day Diagnostic visit ten to fifteen substudies are carried out in time segments varying from one hour to six hours. Following data analysis, recommendations for further action are written on the basis of results from each substudy, with some substudies yielding more than one recommendation. These recommendations may include correcting the problem described, conducting further study, or modifying policies, design guidelines and standards. Put into a standardized Action Recommendation format, the recommendations are compiled in a Diagnostic Report and added to the national database system.

The Database System: As Diagnostic recommendations are generated in POEs throughout the country, they are added to the national database, which also holds questionnaire responses from the profile studies. The database system provides users in the Provinces and Territories rapid access to recommendations from across the country and thus to the most up to date hospital design and POE experience. For each recommendation, key words entered into the system enable users to call up recommendation topics relevant to their hospital design tasks. This database, overseen by a POE Analyst, is used to update design guidelines and standards so they keep up with state of the art knowledge.

The following diagram illustrates how the Canadian Hospital POE process moves through the steps for gathering data, leads to a system for managing data, and feeds into future decisions.

Canadian Hospital POE Program Methodology: Multiple POE Methods

During Profile and Diagnostic site visits, investigators in the Canadian Hospital POE program select from the following list of methods in carrying out each investigation and substudy on specific topics.

1. Interviews
2. Questionnaire Results
3. Behavior Observation Studies
   - Behavior Log
   - Behavior Mapping
   - Behavior Tracking
4. Signs Systems Study
5. Records Analysis
6. Instrument Measurements
7. Space & Plan Measurements
8. Policy & Procedures Analysis
9. Physical Conditions Studies
   - Damage, Wear & Tear Study
   - Interior Finishes & Materials Study
   - Furnishing & Equipment Study
10. Building Systems Studies
    - HVAC Study
    - Lighting & Electrical Systems Study
    - Communications System Study
    - Plumbing & Gas Distribution Study
    - Building Enclosure Systems Study
11. Photographic Documentation

This list, or one essentially equivalent, could easily serve as an agreed-upon POE methods menu for all POEs, with each individual POE study or substudy employing a subset of these methods. As more and more such studies are carried out principles will emerge as to which methods are best suited to solve particular POE problems.

PERFORMANCE CRITERIA FOR HOSPITALS

The entire Hospital Post-Occupancy Evaluation is tied together by hospital facility critical performance criteria. These criteria underlie the observational POE methods, the interviews and questions, and their analysis.

The Hospital POE criteria are generated by and derived from the purposes for which hospital buildings are built. Each part of the hospital has one or more purposes that together help the entire facility fulfill its overall purpose. For example, the overall purpose of a hospital is to provide health services. Food service, medical records, housekeeping, emergency, surgery, inpatient departments, and even the design of parking areas, all contribute to the hospital's overall purpose by meeting their own individual purposes. Two purposes of hospital parking areas that support the overall purpose of hospital health services are to accommodate all of the traffic to and from the hospital with minimal confusion, and to make sure that people in cars who have emergencies can park and get to their destination quickly. When these purposes are met,

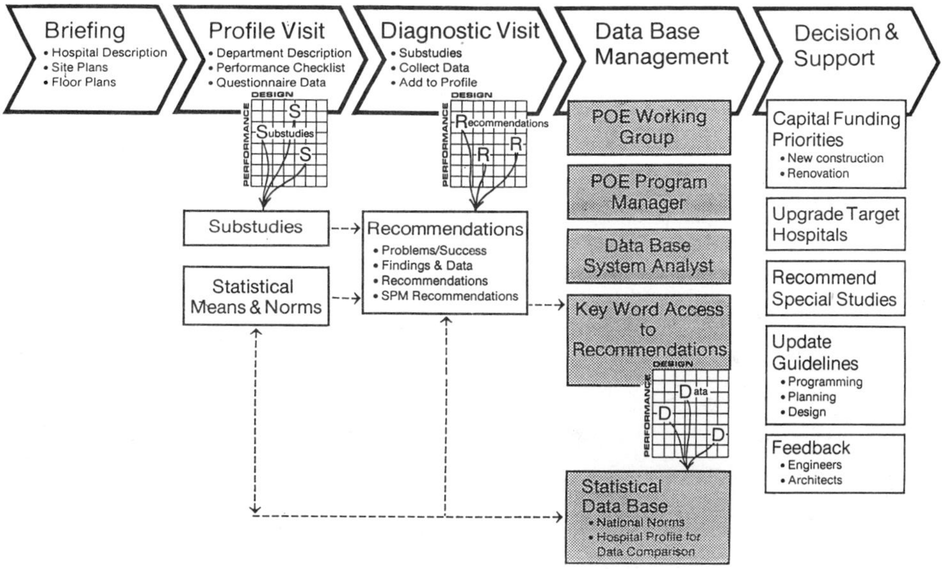

Figure 2. Hospital POE methodology.

together with meeting the purposes of other areas and departments, the hospital's overall purpose of providing quality health care is met.

The Hospital POE Program is based on the construct that design of the physical environment contributes to meeting a building's purposes by addressing its "Performance Criteria", the effects the building's physical environment must have to support its specific purposes. A building's purposes are met by both "General" and "Critical" Performance Criteria. For a hospital, the following "General Performance Criteria" apply (listed alphabetically, not in order of importance).

- Accessibility
- Air Quality/Ventilation
- Cleaning Ease
- Comfort
- Convenience
- Efficiency
- Function
- Health
- Hygiene
- Information Clarity
- Privacy
- Safety
- Security
- Size Adequacy
- Stress Reduction
- Surveillance
- Treatment Support

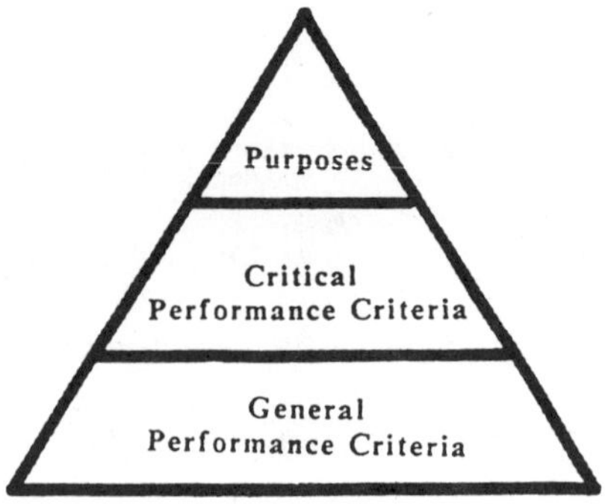

Figure 3

Critical Performance Criteria

Performance criteria that are particularly fine-tuned to building type differences are more critical to a building's meeting its purposes than are general performance criteria. For example, in order for an operating suite to support carrying out safe, hygienic, effective and timely surgical procedures it must at least: be located conveniently to inpatient departments; have adequate and appropriate systems support in treatment areas; provide a way to control infection from people coming from outside the surgery; be easily cleaned and sterilized; provide nurses with easy surveillance over recovering patients; include a place where staff members can relieve stress after intense concentration on operations; and include lighting and surface materials that reduce anxiety and stress for patients. These are some of the "critical performance criteria" for a hospital operating suite.

Figure 3 illustrates the hierarchical relationship between purposes, critical performance criteria, and general performance criteria.

Critical performance criteria distinguish hospital environments from one another just as they distinguish different building types. While general hospital performance criteria are important for every location within a hospital, some criteria are more critically applied to certain areas than others. For example, air quality is more critical in the operating suite because of infection control and in the laboratories because of staff health and safety than in the cafeteria or even inpatient departments. Security is more critical in psychiatric units and in the emergency department than in labor/delivery or the cafeteria. For this reason, general performance criteria which are relevant throughout every hospital department, are included and studied in a POE as "critical" only in those departments where they directly influence the department's purposes.

In each Hospital POE, the critical performance criteria guide and drive data gathering and analysis so that they lead to useful recommendations and actions. The list of what is particularly important to look at and to understand (i.e. Critical Performance Criteria) enables investigators to focus quickly and efficiently on those design elements which have the main impact on how the facility fulfills its purposes. For this reason, Critical Performance Criteria are built into all data gathering methods in the Canadian Hospital POE Methodology: interview schedules, questionnaires, inventories, observation guides, and so on.

To build the bridge between performance criteria and design of the physical environment, either directly or via standards and guidelines, one more set of linkages

Department Or Area	Critical Performance	Physical Element	Characteristics	Measurement Criteria
▪ *Parking*	▪ *Adequate for required number of cars*	▪ *Parking lot*	▪ *Size* ▪ *Layout*	*This can be a formula, a specific measurement (SF, inches) or reference to a guideline or standard*
	▪ *Adequate separation of visitors, emergency, staff and doctors*	▪ *Parking lot*	▪ *Layout*	
		▪ *Entrance barriers to parking lots*	▪ *Type* ▪ *Flexibility* ▪ *Location* ▪ *Size* ▪ *Colour*	

Figure 4

must be made. These are the linkages between performance criteria, physical elements, characteristics of elements, and criteria for assessing if the design meets the performance criteria. Taking our simplified example of hospital parking areas, the following table demonstrates these connections:

In the Canadian Hospital POE Program, Critical Performance Criteria for Hospitals are divided into five categories representing groupings of physical spaces that have similar purposes within the whole hospital. These categories are useful in design and design review, in that they are referred to in this form in guidelines and standards, and as a group are considered by hospital programmers and designers to be a relevant hospital planning typology.

1. Common Spaces
2. Administration
3. Support Services
4. Diagnostic and Treatment
5. Inpatient Departments

The Database System – Purpose

As part of the national POE Program, Hospital POEs are being carried out by the Provinces on hospitals across Canada. The reports and recommendations for each hospital facility are used first by the particular hospital that is the subject of the study. Comparison with the results from other similar hospitals are both useful to the particular hospital and transferable to hospitals generally. The Hospital POE Database System provides the ability to access POE results in the context of results from hospitals across the country.

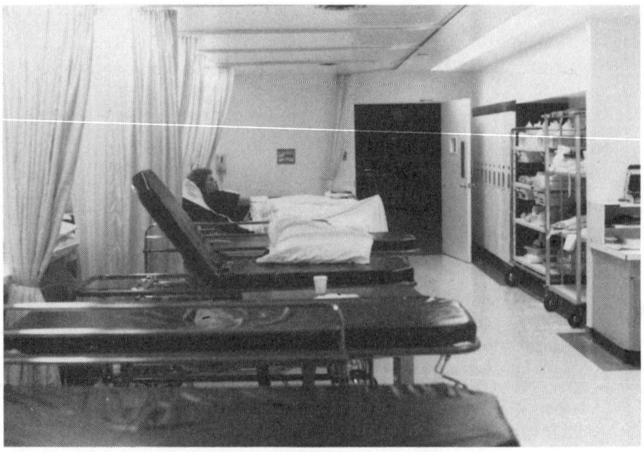

Figure 5

The database comprises questionnaire data, hospital improvement recommendations, and summaries of major POE findings. The purpose of storing POE data in a centralized database is to facilitate retrieval of data to solve specific problems and the degree to which the design and management of the database are responsive to users, determines the usefulness of the POE Program to decision-makers. The database does not contain the raw action recommendations from Diagnostic POEs, but these form the backbone of the database and are available when requested.

For example, a hospital planner seeking information on day surgery design could access POE findings in the database on day surgeries in existing hospitals around the country. Similarly, an administrator needing information on teletracer systems to help decide whether or not to install one in his/her hospital could access the teletracer recommendations that resulted from various POE studies. A third use of the database is for designers to access hospital design guidelines and standards in conjunction with the POE findings that describe how hospitals built to those guidelines actually work; some space standards, for example, may work better than others.

SUMMARY

Post-Occupancy Evaluations can be carried out simply and easily by designers, administrators and evaluation researchers trained in using standardized procedures and approaches. The methods employed in POEs, their sequencing and the database for analyzing and accessing data do not have to be reinvented each time a study is carried out. In fact, the less innovation there is in this area, the more reliable and interesting findings from POEs can be.

Hospital POEs have been employed in Canada, for example, to develop design criteria for new and rapidly changing treatment areas. One such changing treatment area in many hospitals is the Day Surgery, where presently more than half of all hospital surgical patients are admitted and cared for. This figure is up from under ten percent of all surgery only a few years ago. Most Day Surgeries we examined in our POEs have been designed like Emergency Rooms, with small bed areas, little privacy and no space for visitors. But patients in Day Surgery are very different from Emergency Room patients. Their visits are planned. They spend almost no time in waiting rooms, but a

Figure 6

lot of time waiting to go into surgery, and they spend a lot of time after surgery and recovery being observed by nurses and seen by a doctor before being released. They are often accompanied by a relative or friend who leaves the hospital only during the surgery itself, if at all, but wants to be with the patient the rest of the time. All these findings have important implications for space planning in Day Surgery departments.

Day Surgery, then, must incorporate the privacy required of a short term inpatient area where patients are examined by a doctor and where confidential discussions take place, as well as the openness of an observation room where staff can efficiently keep an eye on recovering patients. In addition there is a need for companions to be able to wait with patients and to calm them before surgery, but not to disturb other patients recovering or waiting alone. Finally, most treatments after anesthesia require that patients eat something to test their stomachs before leaving, requiring a nourishment center in addition to changing rooms, toilets and secure lockers.

POEs, through cooperation and agreement on the basics of POE methodology, can easily uncover situations like these and develop useful findings and recommendations.

In sum, the Canadian Hospital POE Program is useful to Canadian hospital facility planners, designers and policy makers, but just as important it provides a way for POE researchers to share approaches and compare results. By agreeing on reliable and valid methods and methodologies, professionals can concentrate on reaching agreement on purposes of different building types, critical performance criteria for building types and their sub-units, the appropriate application of multiple methods, POE phases, statistical data handling, and database design. With such agreement, the next steps of testing methods and refining approaches can take place still more effectively, and from that a POE paradigm can be built.

REFERENCES

Building Diagnostics, Inc., 1988a, *Hospital Post-Occupancy Evaluation: Methodology and Prototypes*, Ottawa, Health Facilities Design, Health and Welfare Canada.

Building Diagnostics, Inc., 1988b, *Post-Occupancy Evaluation: Bristol County Courthouse*, Boston, Office of Programming, Division of Capital Planning and Operations, Commonwealth of Massachusetts.

Popper, K. R., 1972, *Objective Knowledge: An Evolutionary Approach*, London, Oxford University Press.

Proshansky, H. M., Ittelson, W., and Rivlin, L. (Eds.), 1970, *Environmental Psychology: Man and His Physical Setting*, New York, Holt, Rinehart & Winston,.

Sanoff, H. and Cohn, S. (Eds.), 1970, *EDRA 1: Proceedings of the First Environmental Design Research Association Conference*, New York, John Wiley & Sons.

Vischer, Jacqueline C., 1989, *Environmental Quality in Offices*, New York, Van Nostrand Reinhold.

Zeisel, J., Building Purpose: The Key to Measuring Building Effectiveness, M. E. Dolden and W. Robertson, Jr. (Eds.), 1986, *The Impact of the Work Environment on Productivity*, Washington D.C.: Architectural Research Centers Consortium, Inc..

Zeisel, J., 1984, *Inquiry by Design: Tools for Environment-Behavior Research*, New York, Cambridge University Press.

CHAPTER 14

EVALUATING THE BUILT ENVIRONMENT FROM THE USERS' POINT OF VIEW: AN ATTITUDINAL MODEL OF RESIDENTIAL SATISFACTION

Guido Francescato

Department of Housing and Design
University of Maryland, College Park

and

Sue Weidemann and James R. Anderson

School of Architecture
University of Illinois, Urbana-Champaign

INTRODUCTION

The literature in the evaluation of built environments has had a tendency to be long on techniques and on examples of applications but short on theory. This is not surprising, as evaluation is a practical activity concerned mainly with the performance of existing environments in use. Its aim is primarily that of providing information which can be applied to improve unsatisfactory environments. Within this limited scope, theory would appear at first glance to be of little relevance, just as it is not important for the driver of an automobile to be conversant with the theory of internal combustion engines or that of aerodynamics in order to drive safely and successfully.

So, for the evaluator interested simply in applying certain techniques in order to gather information there is little reason to dwell on theoretical considerations. However, theory, conceptual frameworks, paradigmatic assumptions, and general philosophical perspectives take on fundamental importance for evaluators interested in using information to plan and design new satisfactory environments, and for those concerned with developing appropriate techniques, understanding their uses and limitations, and assessing their utility for the purposes they are claimed to serve. In any field, advances are impossible in a vacuum of theory. The evaluation of the built environment is no exception. As Canter and Kenny (1982) have noted, the value of any empirical work is ultimately a function of the theoretical formulations on which it is based:

> "Unless there is an understanding of the role that the physical environment plays in people's lives it is extremely difficult to know which aspects of that environment to measure and how to argue for the significance of any relationships which are found between the environment and human actions or experience." (p. 147)

We are not suggesting that environmental evaluation has no conceptual foundations. Certain theoretical assumptions do underlie environmental assessments and more specifically the processes of Post-Occupancy Evaluation (POE). These assumptions have to do with the interaction of the activities of planners, designers, and others responsible for making the environment with those of researchers intent on evaluating the resulting environment. For example, Zeisel (1981) has developed a view of the design process in which "image formation" (the invention of environmental configurations, or forms) is periodically compared to "test information" (the knowledge derived from empirical research). Thus, the results of POEs of existing environments are seen as guiding the design process toward the goal of an "acceptable response," that is, one which will be largely free of objectionable qualities.

Zeisel's formulation and other similar frameworks are undoubtedly useful vantage points from which to consider the potential applications of POEs to the professional practice of planners and designers. But before embarking on such applications, it would be desirable to gain a clearer understanding of the evaluation process itself. To this end, the theory of environmental evaluation needs to be sharpened: efforts should be made to describe the nature of the process, to examine a variety of approaches and the assumptions that underlie them, and to develop conceptual and empirical frameworks that account for the relationships among the various aspects, factors, or characteristics of the phenomena being evaluated.

This chapter is intended as a contribution to the theory of the evaluation process. It focuses on the concept of users' satisfaction and relies primarily on research conducted by the authors and others on a specific environmental type: that of multifamily housing. But the perspective presented here is equally applicable to other types of environments. Indeed, the construct of satisfaction has been widely used in non-residential settings, such as work environments (e.g. Sundstrum, 1987). In the first section of the chapter, residents' satisfaction is defined as an evaluation criterion. The reasons for its utilization are outlined. In the second, the utility of models is discussed. The third section presents the evolution of a number of models of residential satisfaction. A model of satisfaction based on attitude theory is described in some detail and its implications are examined in the fourth section. In the fifth, a comprehensive model of the relationships among environment, satisfaction, and behavior is proposed. Finally, a number of issues in satisfaction research and in practical applications to planning and design are addressed.

RESIDENTS' SATISFACTION

To evaluate is to assess performance with respect to a criterion. But in order to be assessed, performance must first be measured. This already presents a problem, as there are always multiple aspects of performance, some of which will be relevant to the intended assessment and some of which will not. For example, the performance of an automobile may be measured in terms of maximum speed and acceleration, comfort, or fuel economy, among many other factors. But some of these factors will be more relevant than others as a function of the type of assessment sought. Fuel economy and comfort will probably not be important in the case of a race car, while racing speed will not be a relevant factor in a normal passenger vehicle. The use of the object to be evaluated determines to a large degree the aspects of performance that need to be measured.

POE stresses the value of assessing environments in use, rather than prior to occupancy – as is frequent in architectural criticism – precisely for this reason. Indeed, one of the applications of POE has been the comparison between the use that the designer intended for an environment and that to which it was put by its users. But is the designation of the use of an environment sufficient to determine the relative importance of the aspects of performance to be measured? This is a more difficult question because it involves the value attributed to certain aspects of an environment, as well as to the environment as a whole, by a person or social group. Hence the need for a criterion that will express the value attributed to the whole, and to which the values attributed to each aspect of that whole may be related. Residential satisfaction is such a criterion.

The value attributed to the whole will vary across people and environments (as well as over time). Therefore, the relationship between a number of aspects of performance and a criterion can be examined by the use of appropriate statistical analyses in order to discover which of these aspects, taken together, predict the criterion (that is, which aspects significantly account for the variation in the criterion). In addition, the relative importance of each aspect in predicting the criterion can also be determined by these same means.

The choice of residential satisfaction as a criterion is governed by a number of considerations that have been fully discussed in prior publications (e.g., Francescato, Weidemann, and Anderson, 1987). Here it will be sufficient to note that residents' satisfaction stresses the point of view of the inhabitants themselves. The assumption behind this emphasis is that many problems in the built environment are in fact the result of neglecting the users' point of view.

The nature and meaning of the concept of satisfaction have been frequently discussed in the literature and a number of definitions have been offered. For instance, Canter and Rees (1982) interpret residential satisfaction as a reflection of "the degree to which [the inhabitants] feel [that their housing] is helping them to achieve their goals" (p. 185). The attitudinal model of satisfaction presented in this chapter suggests that more comprehensive interpretations are possible.

In operational terms, the authors have defined residential satisfaction simply as an index of the answers to the following questions:

- How satisfied are you with living here?
- How long do you want to live in this housing development?
- If you move again, would you like to live in another place like this?
- Would you recommend this place to one of your friends if they were looking for a place to live?

There are two reasons for choosing this index. The first is purely technical: the reliability of the criterion can be increased by using the index rather than the single question "How satisfied are you with living here?". The second is conceptual, reflecting the hypothesis that satisfaction is a multifaceted construct in which cognitive, affective, and conative variables coexist. (More about this later in this chapter.) In this context, it is important to stress that results from studies using other operational definitions cannot be directly compared with those of research that uses the above index as a criterion.

The criterion of resident satisfaction is not without its limitations. These have been addressed in detail elsewhere (Francescato et al., 1987). It is sufficient to note here that these limitations must be kept in mind when interpreting the results of satisfaction research but do not invalidate the utility of the criterion.

UTILITY OF MODELS

Moore, Tuttle, and Howell (1985) suggest a scheme in which theoretical orientations, frameworks, and models are seen as leading to the formulation of explanatory

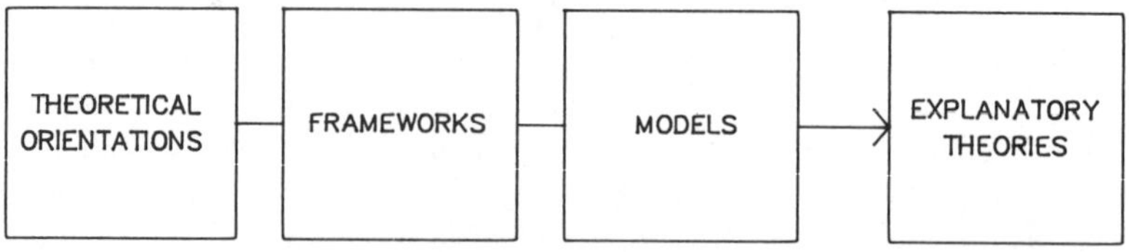

Figure 1. Dimensions along which theories can be characterized.

theories (figure 1). They stress the need for explicitness and for models and explanatory theories that can be tested.

At present, it is premature to propose an explanatory theory of POE. Among other unresolved questions, the uncertainties surrounding the feed-forward process (that process by which POE findings are seen as helpful in creating a body of knowledge applicable to solving future design problems) loom too large to permit the construction of a full-fledged explanatory theory. (There is a link between the weaknesses of evaluation theory and the uncertainties in feed forward. Research results–useful as they may be for diagnostic and intervention purposes–are not knowledge per se. Knowledge implies generalizability and an understanding not only of the specific problem at hand, but also of the underlying mechanisms and processes that operate in environmental/ behavioral transactions.)

But it is possible to develop models that can make explicit the theoretical orientations and the assumptions that underlie a research approach. This explicitness is essential in the process of interpretation to which the results of any study must be submitted. Interpretation confers meaning to data, so that they can make a contribution to knowledge and serve as a guide for action. But interpretation is impossible in the absence of some kind of conceptual formulation.

Models can also illuminate the potential linkage to work done in other fields, either directly, by describing congruences with existing models in different areas of study, or indirectly, by making clear what is the domain of concern. For example, the model presented in this chapter identifies a linkage to attitude theory, proposes a representation of the relationships among environment, satisfaction, and behavior, and makes it possible to compare this perspective with other approaches.

Finally, models provide a structured means with which research utilizing different approaches or focusing on specific concepts, sets of factors, or groups of variables can be classified. Classification of research is not only desirable to gain a clearer understanding of the strengths and weaknesses of each study, but also to identify overlaps or their absence.

EVOLUTION OF SATISFACTION MODELS

Weidemann and Anderson (1985) have provided a comprehensive discussion of conceptual frameworks of residential satisfaction, which, among other things, clarifies a number of operational definitions used by different authors. Interested readers are referred to that work, but a summary of the evolution of residential satisfaction models is presented here as a basis for discussing the attitude-based model proposed later on in this chapter.

Two somewhat different sets of models evolved from research by Marans and Rodgers (1975), Marans (1976), and Marans and Spreckelmeyer (1981), on the one hand, and from studies by Francescato, Weidemann, Anderson, and Chenoweth (1979), on the

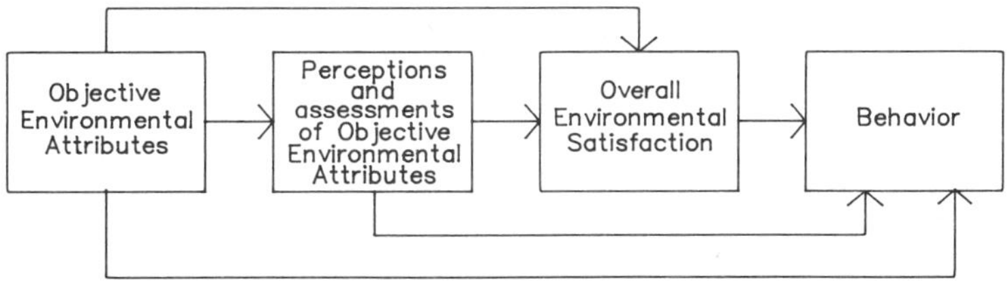

Figure 2. Conceptual model of residential satisfaction.

other. The basic model resulting from the former stream of work is a conceptual formulation and is shown in figure 2. The models ensuing from the latter stream of research were derived empirically. They represent relationships among factors found to be significantly related to the residential satisfaction index described earlier. These empirically derived models are shown in figures 3 and 4.

The models in figures 3 and 4 are strictly limited to depicting relationships among perceptions. No attempt is made to introduce or hypothesize any relationship with behavior. The model of figure 2, on the other hand, advances the hypothesis that behavior is directly affected by perceptions and assessments of objective environmental attributes, by the attributes themselves, and by overall environmental satisfaction. But the literature in social psychology has increasingly recognized the role of intentionality, strongly suggesting that cognitive and contextual aspects are not the only predictors of behavior. This does not mean, of course, that such aspects can be ignored, but rather that their effect on behavior is likely to be significantly mediated by intentions. For example, MacIntyre (1981) states that:

> "We cannot... characterize behavior independently of intentions, and we cannot characterize intentions independently of the settings which make those intentions intelligible both to agents themselves and to others. (p. 192) And Weidemann and Anderson (1985) note that one of the best-documented and most discussed issues in the literature dealing with attitudes is the finding that attitudes often do not directly predict behavior. Fishbein and Ajzen (1975) use an intervening variable, which they refer to as behavioral intentions, to diminish this attitude-behavior discrepancy. (p.157)"

Weidemann and Anderson, therefore, proposed a model in which they attempted to integrate all the above formulations by including aspects such as person characteristics, attitudinal and affective variables, and intentions to behave with respect to the environment (see figure 5). This model represents an important advance with respect to prior ones, but it also introduces two unresolved issues.

First, while the Marans-Spreckelmeyer model proposed definite causal paths, as indicated by the direction of the arrows, the Weidemann-Anderson model identifies only one causal link (between objective attributes and perceptions about them). All other aspects are interconnected, but the hierarchy of the aspects and the direction of their effects on each other is not made explicit.

Second, while the "age-old trilogy" of cognitive, affective, and conative variables that people display in response to any social object (Ajzen and Fishbein, 1980, p. 19) is identified in the model, satisfaction – as an independent criterion – is not. Indeed, Weidemann and Anderson propose an interpretation of satisfaction in purely affective terms:

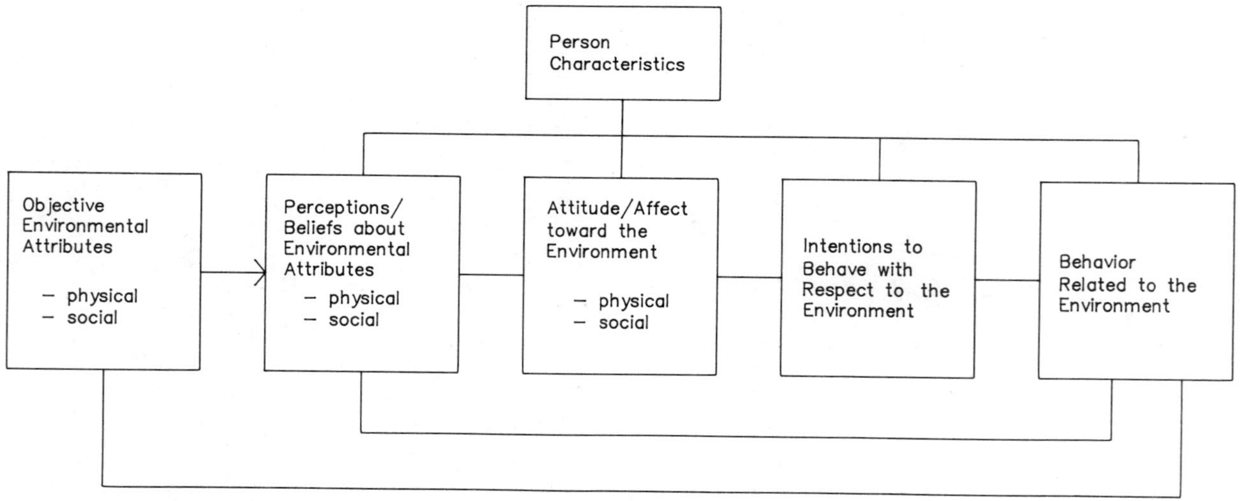

Figure 3. Empirical model of residential satisfaction, depicting the results of principal component analysis. Factors related to satisfaction and significant relationships among them are shown.

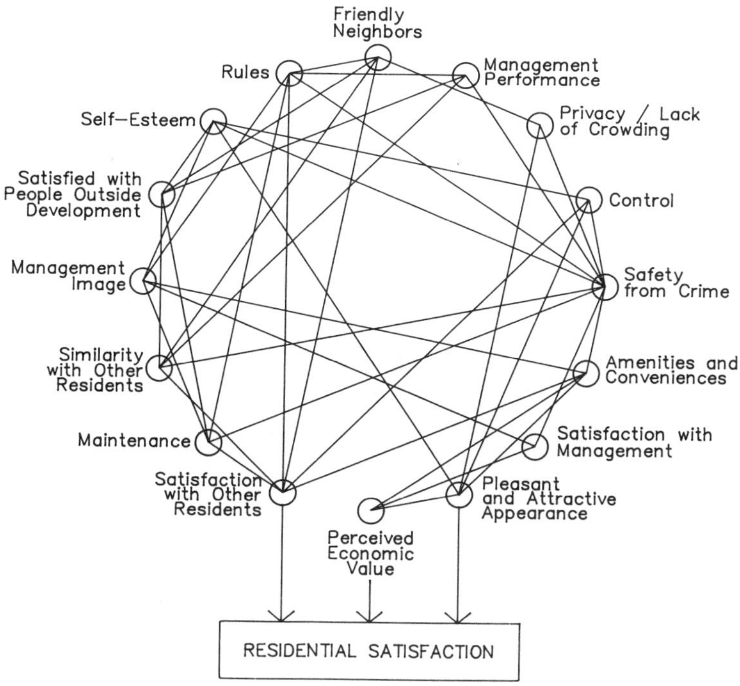

Figure 4. Empirical path model of residential satisfaction. R-square values are shown in nodes, path coefficients at links.

> "It is in this way that we conceive of residents' satisfaction with where they live. It is the emotional response to the dwelling, the positive or negative feeling that the occupants have for where they live. As such, it is a global representation of the affective response of people to the social-physical environment in which they live." (p. 56)

This is in line with the definition of global appraisals offered by Zajonc (1980), who called them "affective reactions" and viewed them as somewhat independent of cognitive processes (Russel and Snodgrass, 1987). On the other hand, some hold that assessments of psychological objects are cognitive events, or "evaluative cognitions" (Mandler, 1982, 1984). Still others doubt that "terms such as emotion, cognition or affect are defined in a precise enough way to allow us to decide whether affective appraisals are really cognitions or really emotions" (Russel and Snodgrass, 1987, p. 249).

In this chapter, we propose a means to reconcile these mutually contradictory notions, by harking back to a classical construct in social psychology that defines global evaluations of psychological objects as attitudes. In reviewing the development of this construct, Ajzen and Fishbein (1980) mention that by the end of the 1950s attitudes were already considered by most psychologists "as complex systems comprising the person's beliefs about the object, his feelings toward the object, and his action tendencies with respect to the object" (p. 19), that is, systems made up of cognitions, affects, and conations.

In the ensuing years, a number of studies focused on attitude and its components. Rosenberg (1956) proposed an expectancy-value model of attitude in which evaluations

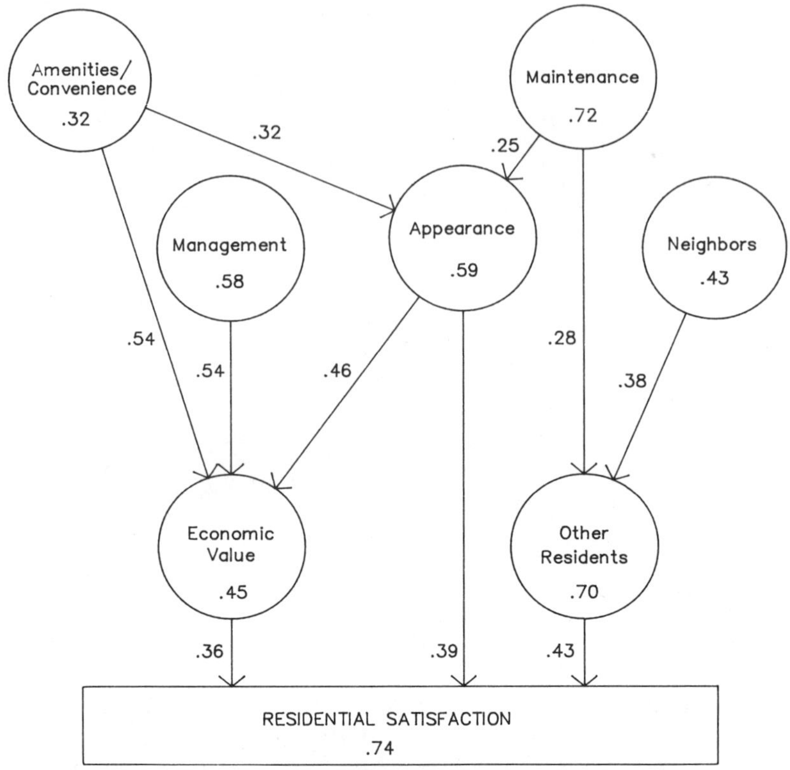

Figure 5. Integrated conceptual model of residential satisfaction.

were seen as strongly dependent upon people's expectations or beliefs that the evaluated object furthered or hindered the attainment of their goals. (Cf. the definition of satisfaction by Canter and Rees, 1982, mentioned earlier.) Triandis (1964) and Fishbein (1964) showed that affective and conative variables are highly interrelated in evaluative processes involving attitudes. D.T. Campbell (1947), Bettelheim and Janowitz (1950), and Ostrom (1969) provided further evidence that measures of cognition, affect, and conation all contribute to explain attitudes.

Ajzen and Fishbein (1980) developed their Theory of Reasoned Action, in which they suggest that the link between attitudes and behavior is mediated by intentionality with respect to a specific behavior. They also postulate an important distinction: that between attitudes toward "targets" (people, institutions, objects) and attitudes toward behavior. Further, they determined that while attitudes toward targets only indirectly affect behavior, attitudes toward behavior, when combined with measures of subjective norms and behavioral intentions, do indeed predict behavior. (More on this in the next section.)

Ajzen and Fishbein (1980) acknowledged that "it is possible to discern, in the last few years, a trend to return to the unidimensional definition of attitude as evaluation or affect with respect to a psychological object" (p. 26). But they pointed out that "there is still a widespread consensus that, in addition to affect, attitudes also contain cognitive and conative components." (p. 26). It is on the basis of this consensus that we adopt the view they concisely expressed in this statement:

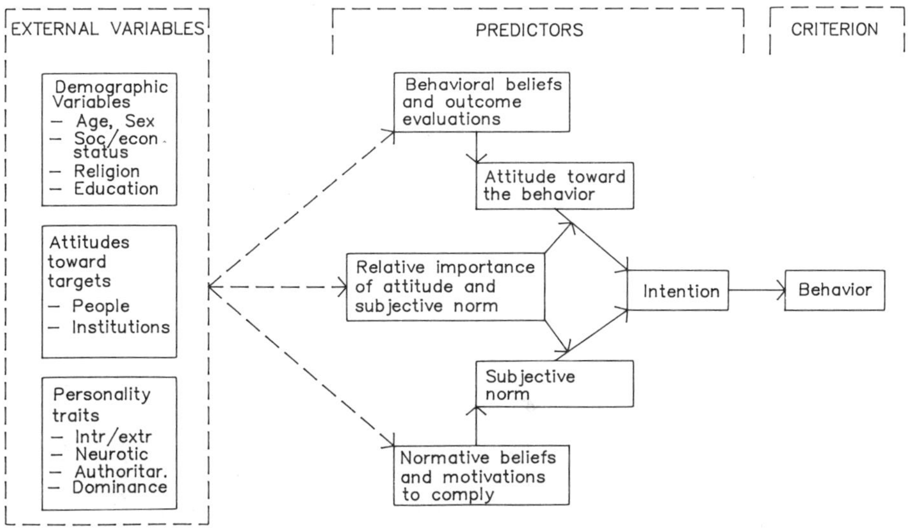

Figure 6. Attitude model of residential satisfaction.

> "We restrict the term "attitude" to a person's evaluation of any psychological object and we draw a clear distinction between beliefs, attitudes, intentions, and behaviors." (pp. 26, 27)

Within this line of thinking, the construct of residential satisfaction can be conceived as a complex, multidimensional, global appraisal combining cognitive, affective, and conative facets, thus fulfilling the criteria for defining it as an attitude. This is the point of departure for the construction of the comprehensive model of satisfaction presented in the next section. It is worth noting that this view of the construct is consistent with the nature of the satisfaction index described earlier, as the questions in the index span the entire spectrum of cognition, affect, and conation. Indeed, having chosen this index as a criterion, it would be conceptually inappropriate to regard satisfaction as a unidimensional concept.

AN ATTITUDINAL MODEL OF SATISFACTION

Central to the model proposed here is a further distinction elaborated by Ajzen and Fishbein (1980): that between prediction and understanding. In this sense, variables that have the power of strengthening prediction are considered direct "determinants" (i.e., predictors) of the criterion (e.g. of behavior). Their relationships to the criterion are defined as "stable". On the other hand, variables which do not improve the accuracy of prediction are considered "external." External variables are still worth measuring and including in models, because they still exhibit some relations to the criterion. Consequently, they are useful when attempting to gain a better understanding of the phenomenon under study, but their effect on the criterion tends to be indirect and to have volatile, or low predictive strength.

It follows that the specific universe of variables to be included in a study is a function of that study's objective. That is, is the investigation aimed merely at identify-

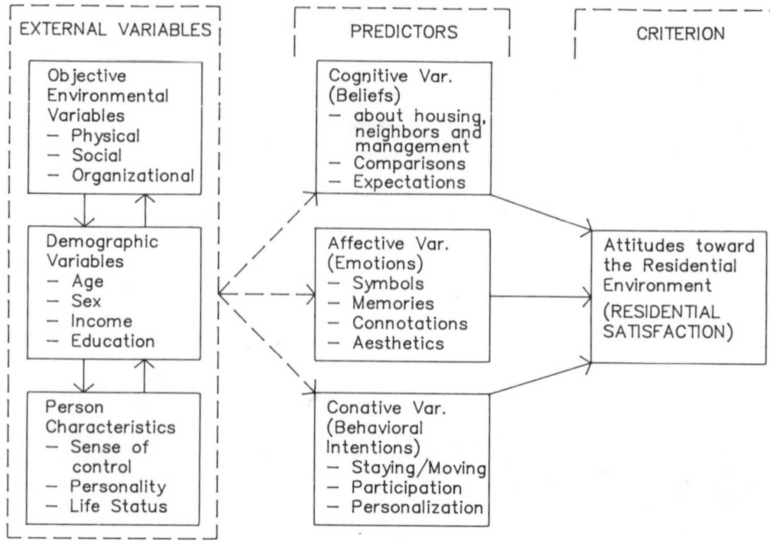

Figure 7. Attitude-behavior model.

ing factors which will accurately predict a criterion, or is it aimed at explaining the relationships and mechanisms underlying the system being studied?

For instance, a number of POEs have been carried out in order to determine which housing modernization strategies were more likely to increase the inhabitants' satisfaction with their housing environment (e.g., Anderson and Weidemann, 1979; Edwards, Kaha, and Anderson, 1985; Selby, Westover, Anderson, and Weidemann, 1987). To the extent that these studies were aimed primarily at predicting the likely success of specific interventions, it was not necessary that they should include external variables. But mere prediction of preferred interventions was not necessarily the only purpose of these investigations. It was also hoped that the results would augment the researchers' understanding of the phenomena under study – that is, they would yield knowledge generalizable to similar situations elsewhere. With these considerations in mind, it was indeed advisable to measure and consider external variables as well.

With the distinction between external and predictor variables clearly established, it is possible to construct a conceptual model of residential satisfaction based on attitude theory (figure 6). In such a model, the external variables consist of the objective characteristics of the physical, social, and organizational environments; the demographic variables; and the personal characteristics of the respondents.

Items in the physical environment may include such objective measures as the overall size and the density of the housing complex, the number of rooms per dwelling and per inhabitant, the layout of both buildings and dwellings, the presence or absence of amenities, the number and location of parking spaces, the level of maintenance of buildings and grounds, the level of noise in public and private spaces, and behavioral traces such as the presence or absence of personalization, the degree of care of private spaces, and the extent to which the inhabitants may have modified the original building and/or dwellings.

Demographic variables may consist of the usual items of age, sex, income, and educational attainment, as well as the length of time the respondent has lived in the community and in the current complex or dwelling, the degree of choice among

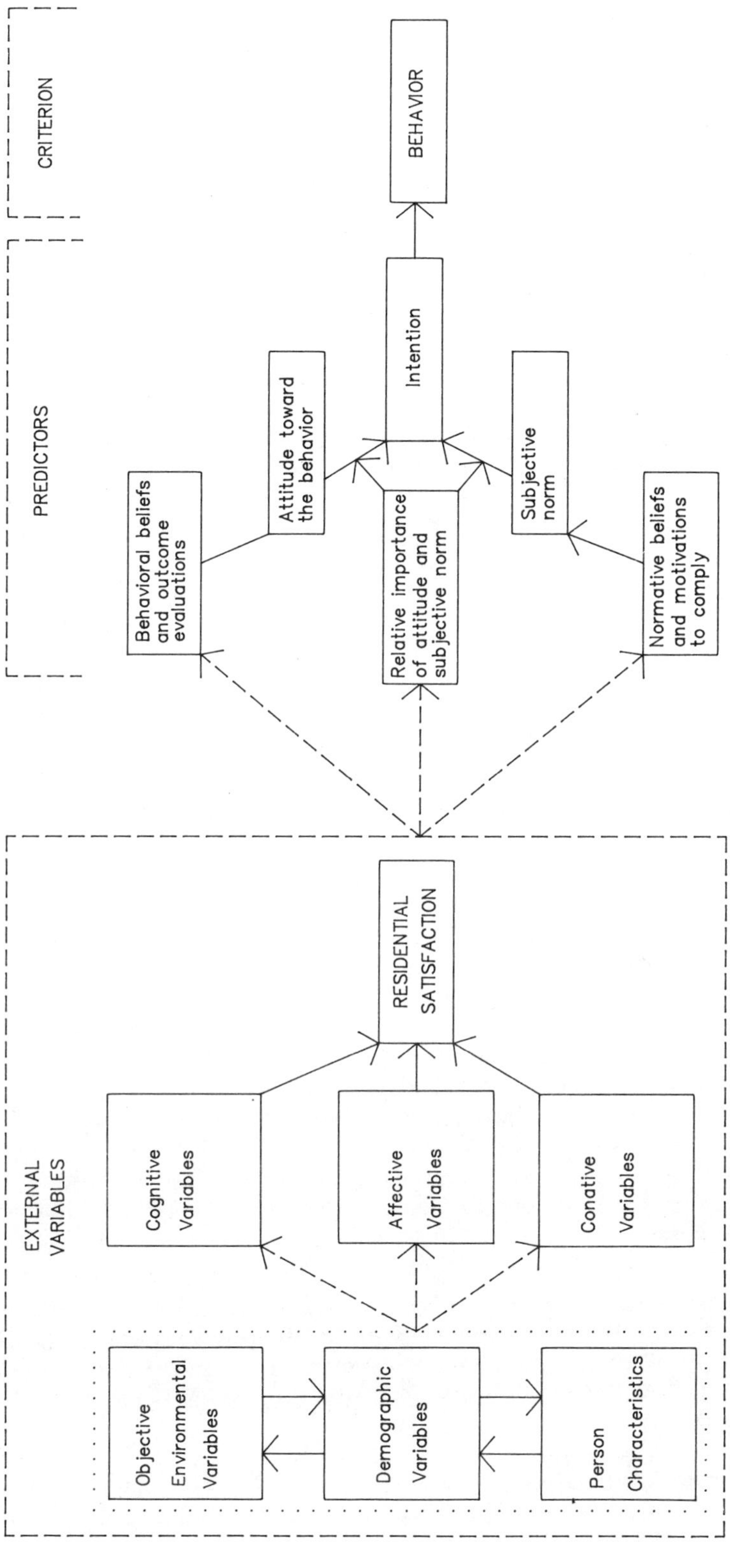

Figure 8. Comprehensive model of the environment-satisfaction-behavior relationship.

alternative living quarters available to the respondent, the length of time the respondent lived in the prior home, and the size of the town in which the respondent grew up.

Among the items that refer to the respondent's personal characteristics, various measures of personality traits, the degree of control over one's life, and life-cycle measures may be included.

The predictor variables entail cognitive aspects (beliefs), affective variables (emotions), and conative aspects (behavioral intentions). Beliefs may be about the physical environment of the community, the housing complex, and the dwelling; about one's neighbors and proper norms of behavior among neighbors; and about management policies, rules, and practices. These beliefs tend to be strongly affected by comparisons with past experiences and expectations about the future.

Emotions or feelings may encompass symbolic values attached to one's home, including again comparisons to prior living experiences, especially those of childhood; memories that are evoked by comparisons with the current home; sense of protection; aesthetic feelings; and connotations elicited by physical and social characteristics of the current residence; as well as hopes, expectations, and aspirations.

Behavioral intentions may involve desire for staying or moving, interest in and willingness to participate in tenant organizations and activities, wish to personalize and modify one's dwelling, propensity to take an active role in maintenance, security, and safety, or in negative activities such as vandalism, and inclinations to supervise the activities of one's children and teenagers.

Inspection of the pattern of relationships displayed in the model of figure 6 begins to suggest explanations for certain research findings that have been occasionally puzzling. For example, there has been a long tradition of belief among planners and designers that improving certain objective physical environmental conditions (such as structural soundness, sanitation standards, and thermal comfort) would lead to more satisfying living environments. Experts discussing "housing quality" continue to emphasize such objective criteria (e.g., CIB, 1988). But research has often shown weak correlations between such improvements and satisfaction (e.g. U.N. Economic Commission for Europe, 1973; Francescato et al., 1979).

Conversely, objectively less desirable conditions were found to correlate with high levels of satisfaction under certain conditions. In a specific example, G. I. Bill students housed in temporary barracks while attending college after their return from World War II were highly satisfied with their residential environment in spite of its poor objective quality (Schorr, 1963). In another study, residents living in multifamily high-rise developments, but aspiring to move to single-family detached homes, were more satisfied the more they believed they would be able to move to the environment of their choice in the future (Michelson, 1977).

The model proposed here accounts for these findings by clearly identifying the mediating role of beliefs, emotions, and behavioral intentions between objective variables and satisfaction.

In addition, the model suggests a high degree of interaction among objective environmental variables and other external variables such as demographic and personal characteristics. This would lead us to expect significant differences in overall satisfaction within the same objective environmental conditions as a result of differences in age, sex, income, education, personality, and life status, but not a direct predictive effect of any such variables on residential satisfaction. Results of empirical research corroborate this expectation. For example, objective aspects such as the location of a housing development, its density, the type of site layout, and whether the housing was high-rise or low-rise, were not significant predictors of satisfaction in a national study of publicly assisted housing. But a number of socio-demographic characteristics including income, education, and lifestyles were nevertheless related to differences in satisfaction levels and in satisfaction with a number of aspects that, in turn, were predictors of satisfaction (Francescato et al., 1979).

The model also addresses an important issue in evaluation research: that of the collective predictive strength of the variables that comprise the universe of a study. This strength is expressed as the percentage of variance in the criterion (in our case residents' satisfaction) that is accounted for by the significant predictors as a group. The greater this percentage, the greater the likelihood that the measured variables, taken together, will exhibit a strong relationship with the criterion. (This issue is highly relevant to a very practical question in POE, that is, what can one do to improve users' satisfaction with an existing environment? If certain aspects of the physical, social, and organizational environment have a cause-effect relationship with the criterion, they can be modified through a variety of interventions in order to obtain higher residential satisfaction levels.)

Because one can never measure everything that can possibly affect a criterion, it is extremely useful to identify as economical as possible a universe of variables; that is, to maximize the predictive strength of the variables while minimizing their number. Statistical procedures are useful in determining which variables are weak predictors and can therefore be removed from the universe. But data analysis cannot be used to discover which variables that are not a part of the universe should be added in order to increase the overall predictive strength. Comparisons between the conceptual model suggested here and empirically derived models such as those of figures 3 and 4 offer a means for identifying aspects of the total universe of variables that are likely to increase total predictive strength. For example, a comparison of figures 4 and 6 leads to the hypothesis that the inclusion of certain affective variables, especially those having to do with symbolic meaning of home, with connotations of specific aspects of the physical and social environment, and with memories of one's childhood home, would increase overall predictive strength.

Space does not permit an exhaustive review of all the research issues that can be addressed by further examination of the attitude model of residential satisfaction. Here we have presented only a few examples of such issues among those which have been frequently discussed by both researchers and professionals interested in POE. But there is an important area of concern that deserves special attention. The next section addresses this concern.

SATISFACTION AND BEHAVIOR

A number of researchers have been interested in examining potential links between satisfaction and behavior. They have reasoned that improvements in unsatisfactory environments should result in changes not only in the users' satisfaction with the environment, but also in changes – one would hope for the better – in the social behavior of environmental users. For instance, O. Newman (1973, 1976) devoted a great deal of attention to design features that might deter crime. Researchers at BOSTI (1981) examined the relation between space planning and interior design of office environments and employees' productivity. In its most extreme form, this view has been known as "environmental (or physical) determinism," that is the belief in a direct link between design and behavior. This extreme view is naive, and has become discredited among the more thoughtful designers and researchers (e.g. Franck, 1984). Nevertheless, it often continues to underlie the expectations of POE evaluators and their clients.

Does the attitude model presented here provide a means for illuminating the nature and pattern of the relationships between environmental design, satisfaction, and behavior? In order to answer this question, it is necessary to restate Ajzen and Fishbein's (1980) distinction between attitudes toward a target and attitudes toward a behavior. Residential satisfaction, as should have become clear from the preceding discussion, must be viewed as an attitude toward a target, that is toward a system composed of people, institutions, and physical objects representing the event "living here," not as an attitude toward a behavior. Consequently, before attempting to hypothesize the links between satisfaction and behavior, it will be necessary to examine the Ajzen and Fishbein (1980) model, in which the pattern of relationships among aspects affecting behavior is displayed (figure 7).

It is apparent that this model is structurally similar to the attitude model of satisfaction of figure 6. But there are significant differences between the two models in the aspects hypothesized as predicting the criterion. A new aspect, intentionality, is placed in a critical predictive position with respect to behavior. In turn, the intention to engage in a particular behavior is seen as directly affected by the attitudes toward that behavior, the subjective norm, and the relative importance attributed to attitude and subjective norm by an individual or social group. Behavioral and normative beliefs are seen as underlying the attitude and subjective norm. Demographic variables, personality traits, and attitudes toward targets are considered to be external variables, the relationship of which to the criterion is not stable, and is, in any event, mediated by the predictors.

Of particular importance in interpreting this model is the insistence of its authors that attitudes, intentions, and subjective norms refer to a specific behavior. So, if in a study of housing we are interested in predicting a particular housing-related behavior, such as vandalism or moving, it is the attitude and subjective norm toward, and the intention to engage in vandalism or moving that will have a direct effect on the behavior under study, not the attitude toward housing. This is clearly shown, for instance, in a study that measured intentions toward the physical environment of a business district, intentions toward using the district, and actual usage (Qualkinbush and Anderson, 1978). Thus, improving the residents' satisfaction with their living environment may have an indirect effect on their moving behavior or on their vandalism, but it certainly does not predict either outcome.

At this point, we need but to take a simple step – the combination of the models of figures 6 and 7 – to obtain a comprehensive conceptual representation of the relationships between the two. Such a representation is shown in figure 8.

Structurally, this model is equivalent to the attitude-behavior model of Ajzen and Fishbein shown in figure 7. Consequently, there is one important change with respect to the attitude model of satisfaction displayed in figure 6: residential satisfaction, as an attitude toward a target, can no longer be included among the predictors, because it is only the attitude toward the behavior that will predict a specific behavior. But there is also a difference with respect to the model of figure 7: unlike the Ajzen-Fishbein model, ours includes objective environmental characteristics among the external variables, on the strength of MacIntyre's (1981) already cited view that intentions cannot be characterized independently from the settings that make them intelligible.

When viewed as aspects of behavior (rather than aspects of satisfaction), environmental settings, demographic and person characteristics, beliefs, emotions, behavioral intentions toward the target, and attitudes toward the target (residential satisfaction) must all be considered as external variables. As such, their relationship to a specific behavior (e.g. moving or vandalism) is not "stable," that is, it has low and volatile predictive strength. As cautioned earlier, one should not conclude that there are no relationships between environment, satisfaction, and behavior. What the model shows is that such relationships are indirect and mediated by the predictors of specific social behavior. Ajzen and Fishbein (1980) refer to these links as "possible explanations for observed relations between external variables and behavior" (p. 84), indicating in this manner that while external variables do not predict behavior, they must be considered as parts of the universe of variables that are of concern if understanding of the attitude-behavior connection is sought.

SUMMARY AND CONCLUSIONS

In this chapter, we have developed a conceptual framework aimed at offering a clearer and more robust explanation of satisfaction as an environmental evaluation criterion. After briefly reviewing the evolution of the satisfaction construct, the utility of conceptual models, and prior models that had been proposed to explicate the relationships between environment, satisfaction, and behavior, we have presented a model of satisfaction based on attitude theory, and in particular on Ajzen and Fishbein's Theory of Reasoned Action (1980). In this model, satisfaction is seen as an attitude toward a target, that is, toward a specific physical, social, and organizational environment.

Viewing satisfaction in this perspective allows a clarification of the link between satisfaction and behavior. Within this optic, we have developed a comprehensive model of the environment-satisfaction-behavior relationship. The comprehensive model shows how satisfaction changes from a criterion to an external variable when viewed as an aspect of behavior. In turn, this permits us to account for the relationships between environment, environmental satisfaction, and behavior, which have been often hypothesized and, in a few cases, observed. At the same time, it offers an explanation for the non-predictive character of those relationships.

There are two important aspects of the models presented here that have not been discussed: time and environmental scale. There are no agreed-upon conceptualizations of the time and scale aspects of evaluations, though there has been considerable discussion of these issues in the literature (e.g. Altman and Rogoff, 1987, for time; Canter, 1983, for residential scale). The scope of this chapter prevents us from examining these aspects in any detail. But we should still point out that time affects evaluations as it affects other human enterprises. Thus, it is important to understand that the results of any assessment are but a snapshot taken at a specific moment. Whenever possible, the appropriate mode for environmental evaluation should be that of monitoring. Environmental scale also needs to be taken into account. In the residential environment, for instance, one can identify the scales of room, dwelling, building, site, neighborhood, community, city, and region. These different levels of scale entail different levels of environmental experience; hence the relationships among these levels should be a part of a further expanded model of satisfaction.

What are the implications of the theoretical frameworks proposed here for environmental evaluation research? Perhaps first and most important is the need for empirical testing of the conceptual models. This, in turn, suggests that more careful attention be given to the selection of variables measured in environmental evaluations: we have already alluded, for instance, to the need for inclusion of more comprehensive and refined measures of environmental affect and behavioral intentions. If the attitude model of satisfaction is correct, the percentage of variance in satisfaction accounted for should be significantly increased by the inclusion of such measures. The role of external variables (especially that of objective environmental descriptors) and their interrelationship should also be empirically tested.

Second, studies should be conducted that explicitly address the comprehensive environment-satisfaction-behavior relationship. In the residential area, this would seem an especially fertile territory for inquiry in regard to such behavior as modifying and personalizing dwellings, maintaining private and semi-private yards, monitoring development and neighborhood security, interacting with neighbors, controlling children's and teenagers' behavior, engaging in or discouraging vandalism, moving, and participating in tenant organizations. Perhaps surprisingly, except for a small number of studies that have attempted to examine some connections among environmental, perceptual, attitudinal, and behavioral factors (e.g. Michelson, 1977; Newman and Duncan, 1979), there has been little comprehensive research of this kind.

Third, it should be clear from our discussion of the satisfaction- behavior relationship that environmental interventions on the physical, social, and organizational aspects of satisfaction (the demographic and personality characteristics being usually not modifiable), while likely to increase satisfaction, are unlikely to affect behavior in a direct fashion, and even less likely to affect outcomes (for instance productivity in work environments or learning in educational ones). This suggests that high levels of users' satisfaction with the environment should be pursued on their merit (that is for their social desirability), not to elicit behavior that the interveners deem desirable.

Satisfaction, as we have seen, can be a powerful construct with which to perform assessments sensitive to the users' point of view. But it does not follow that it should constitute the only basis for evaluation. In other words, satisfaction is a necessary but not sufficient criterion. Ideally, environmental evaluations should also include other criteria, such as economic, ecological, technological, and functional soundness. This is, of course, nothing but common sense. But it is surprising how often advocacy of the satisfaction criterion has been taken to mean that one can achieve comprehensive evaluations without taking into account other obviously important aspects of the total environmental performance.

Environments are systems with multiple "customers," hence with multiple objectives (Francescato et al., 1987). Therefore, they must be evaluated using multiple criteria. In deciding whether or not a particular environment is satisfactory, one must ask: satisfactory for whom? In the case of housing, it is not only the inhabitants who must be satisfied, but the planners and designers, with their professional ideologies and interests, the developers, who must be able to make a profit, the governmental bodies that regulate and perhaps provide financial assistance, with their political ideologies and constraints, and so on.

In engaging in comprehensive POE, one must still define a universe of criteria and, within that universe, a ranking or weighing of the contribution that each single-criterion evaluation should make to a comprehensive assessment. However, defining and ranking a universe of criteria cannot be done through empirical research. It is still an activity that requires judgment in balancing the often contradictory objectives of the various customers against each other.

FIGURE CREDITS

1. From Moore, Tuttle, and Howell, 1985 (see below). Reprinted with publisher's permission.

2. From Marans and Spreckelmeyer, 1981 (see below). Reprinted with authors' permission.
3. From Francescato, Weidemann, Anderson, and Chenoweth, 1979 (see below). Reprinted with authors' permission.
4. From Francescato, Weidemann, and Anderson, 1987 (see below). Reprinted with publisher's permission.
5. From Weidemann and Anderson, 1985 (see below). Reprinted with publisher's permission.
7. Adapted from Azjen and Fishbein, 1980 (see below). Reprinted with publisher's permission.

REFERENCES

Ajzen, I. and Fishbein, M., 1980, *Understanding Attitudes and Predicting Behavior*, Prentice-Hall, Inc., Englewood Cliffs, New Jersey.

Bettelheim G. and Janowitz, M., 1950, *Dynamics of Prejudice: A psychological and sociological study of veterans*, Harper and Row, Pub., Englewood Cliffs, New Jersey.

BOSTI (Buffalo Organization for Social and Technical Innovation,) 1981, *The impact of office environment on productivity and quality of working life: Comprehensive findings*, Author, Buffalo, New York.

CIB (International Council for Building Research Studies and Documentation,) 1988, *CIB 89: Quality for building users throughout the world, Call for papers*, Centre Scientifique et Technique du Batiment, Paris.

Campbell, D. T., 1947, *The generality of a social attitude*. Unpublished doctoral dissertation, University of California, Berkeley.

Canter, D. and Kenny, C., 1982, Approaches to environmental evaluation: An introduction, *Int. Review of Appl. Psych.*, 31:145.

Canter, D. and Rees, K., 1982, A multivariate model of housing satisfaction, *Int. Review of Appl. Psych.*, 31:185.

Edwards, S., Kaha, A., and Anderson, J. R., 1985, *Planning for change: Resident participation in Westwood family housing improvements*, Housing Research and Development Program, University of Illinois, Urbana-Champaign.

Fishbein, M., 1964, The relationship of the behavioral differential to other attitude instruments, *American Psychology*, 19:540.

Fishbein, M. and Ajzen, I., 1975, *Belief, attitude, intention and behavior: An introduction to theory and research*, Addison-Wesley, Reading, Massachusetts.

Francescato, G., Weidemann, S., Anderson, J. R., and Chenoweth, R., 1979, *Residents' satisfaction in HUD-assisted housing: design and management factors*, U. S. Department of Housing and Urban Development, Washington, D. C.

Francescato, G., Weidemann, S., and Anderson, J. R., 1987, Residential satisfaction: Its uses and limitations in housing research, in: *Housing and neighborhoods: theoretical and empirical contributions*, W. van Vliet, H. Choldin, W. Michelson, and D. Popenoe (Eds.), Greenwood Press, New York.

Franck, K., 1984, Exorcising the ghost of physical determinism, *Environment and Behavior*, 10:411.

MacIntyre, A., 1981, *After virtue*, University of Notre Dame Press, Indiana.

Mandler, G., 1982, The structure of value: Accounting for taste, in: *Affect and cognition*, Clark, M. S. and S. T. Fisk (Eds.), Erlbaum, Hillsdale, New Jersey.

Mandler, G., 1984, *Mind and body: Psychology of emotion and stress*, Norton, New York.

Marans, R., 1976, Perceived quality of residential environments: Some methodological issues, in: Perceiving Environmental Quality: *Research and Applications*, Craiks, K. H. and E. H. Zube (Eds.), New York, Plenum Publishing Corp.

Marans, R. and Rodgers, W., 1975, Toward an understanding of community satisfaction, in: Metropolitan America in Contemporary *Perspectives*, Hawley, A. and V. Rock (Eds.), New York, Halsted Press.

Marans, R. and Spreckelmeyer, K. F., 1981, *Evaluating Built Environments: A Behavioral Approach*, Institute for Social Research and the Architectural Research Laboratory, University of Michigan, Ann Arbor.

Michelson, W., 1977, *Environmental Choice, Human Behavior, and Residential Satisfaction*, New York, Oxford University Press.

Moore, G. T., Tuttle, D. P., and Howell, S. C., 1985, *Environmental design research directions: Process and prospects*, New York, Praeger.

Newman, O., 1973, *Architectural design for crime prevention*, Washington, D.C., U. S. Government Printing Office.

Newman, O., 1976, *Design guidelines for creating defensible space*, Washington, D.C., U. S. Government Printing Office.

Ostrom, T. M., 1969, The relationship between the affective, behavioral, and cognitive components of attitude, *Journal of Experimental Social Psychology*, 5:12.

Qualkinbush, S. and Anderson, J. R., 1978, A study of intentions to use downtown Urbana, Illinois, in: *Priorities for environmental design research*, Weidemann, S. and J. R. Anderson (Eds.), Washington, D.C., Environmental Design Research Association.

Rosenberg, M. J., 1956, Cognitive structure and attitudinal affect, *Journal of Abnormal and Social Psychology*, 53:367.

Russel, J. A. and Snodgrass, J., 1987, Emotion and environment, in: *Handbook of environmental psychology*, Stokols, D. and I. Altman (Eds.), New York, John Wiley & Sons.

Schorr, A, 1963, *Slums and social insecurity*, U. S. Department of Health Education, and Welfare, Social Security Administration, Washington, D.C.

Selby, R. I., Westover, T., Anderson, J. R., and Weidemann, S., 1987, *Resident satisfaction: A means to better housing*, Housing Research and Development Program, University of Illinois, Urbana-Champaign.

Sundstrum, E., Work environments: Offices and factories, in: *Handbook of environmental psychology*, Stokols, D. and I. Altman (Eds.), New York, John Wiley & Sons.

Triandis, H. C. 1964, Exploratory factor analysis of the behavioral components of attitude, *Journal of Abnormal and Social Psychology*, 68:420.

U.N. Economic Commission for Europe, 1973, *Housing requirements and demand: Current methods of assessment and problems of estimation*, Geneva, United Nations.

Weidemann, S. and Anderson, J. R., 1985, A conceptual framework for residential satisfaction, in: *Home environments*, Altman, I. and C. Werner (Eds.), New York, Plenum Publishing Corp.

Weidemann, S., Anderson, J. R., Butterfield, D. I., and O'Donnell, Residents' perception of satisfaction and safety, *Environment and Behavior*, 14:695.

Zajonc, R. B., 1980, Feeling and thinking: Preferences need no inferences, *American Psychology*, 35:151.

Zeisel, J. 1981, *Inquiry by design: Tools for environment-behavior research*, Monterey, California, Brooks/Cole Publishing Company.

CHAPTER 15

ADVANCES IN POE METHODS: AN OVERVIEW

Robert B. Bechtel

Department of Psychology
University of Arizona
Tucson, Arizona

ABSTRACT

Design research should now be differentiated into Pre-Design Research (PDR) and Post-Occupancy Evaluations (POEs). While most methods overlap, the different purposes of these two types of research influence selection of methods and analyses and modify methods to suit each. Simulation is the only method peculiar just to PDR and is rapidly advancing to a micro stage through electronic advances. The self-report controversy continues for all methods that use self-report forms of answers. The newest field, the transactional approach, also stirs controversy. Industry and government clients are preferring expert and focused groups and these may be in danger of crowding out the more tried methods of social science.

PRE-DESIGN RESEARCH

When Bechtel and Srivastava (1978) reviewed POEs done on housing, it became evident that POEs constitute the bulk of work done in the entire field of E&B research. They discovered over 1,300 published studies on housing alone and many more have been done since then. Furthermore, the survey did not include POEs done on other buildings, such as schools, hospitals, and so on.

Another type of research has been confused with POEs; it is better called Pre-Design Research, or PDR. The purpose of this kind of research is to collect new information so that a new entity, such as a building, park, or city, can be designed. While POEs focus on design as it exists in a standing building, PDR has a broader range of information to collect in order to pursue the many hypotheses of a future design or to go beyond hypotheses to collect information on the broadest scope of user needs. PDRs are information-gathering operations, rather than the evaluating or testing operations of a POE.

FURTHER DISTINCTIONS

In POEs the design decisions have already been made. In PDR the information is needed in order to make those decisions. POEs are retrospective; PDR is prospective.

While most of the methods used in E&B research will overlap, at least one method, simulation, is almost exclusively used for PDR.

The statistics used to analyze POE and PDR data greatly influence the methods chosen. Because PDR is broad-gauge and inclusive, the statistics tend to be more those of association, while the statistics of the POE, because they are testing and looking for significance of differences, tend to be tests of significance. The PDR uses more correlations, factor analyses, and similar tests; the POE uses more t-tests, ANOVAs, and other tests of significant differences.

A POE evaluates the behavioral response of a building's occupants to the building, as they experience it. As such, it is a performance measure, because it measures whether the building performs as intended. A deeper significance is applied to POEs, however, as over time they become an instrument of quality control. This means that, in much the same way that quality control in manufacturing influences the design and construction of products, the POE becomes the measure of quality in building design. The Australian government (see Pegrum and Bycroft, chapter 17) has adopted a POE system of quality control for its buildings. The Building Research Institute in Sweden has had POEs as part of its evaluation system for more than a decade.

PDR, by contrast, measures the needs and requirements of people who will occupy a future building, and the data becomes input for the programming process of design. This is where confusion arises, since POE information on past designs is also used in programming. In fact, any good designer will borrow from both POEs and PDR as part of programming. Also, PDR is in many ways a performance measure, but only a suggested performance measure for the future, which must await a POE for confirmation as a tested performance standard.

In another ideal sense PDR is a pre-measure while POE is a post-measure for that experiment which is a building. Unfortunately, this ideal situation is seldom realized. The most obvious reason is the design fee process, which is more favorable to PDR than it is to POE. Seldom does a client feel the need for evaluating a building after construction if funds have already been stretched to make a building as good as possible.

Two sources of additional conflict also arise from this circumstance. The first is the understandable reluctance of a client to evaluate a considerable investment when chances are it could have been done better. The second is an ethical and scientific question of whether the people performing the PDR should be the same as those who do the POE. This is both an ethical and scientific issue because if the same people do both they may influence the POE outcome favorably. The PDR performers will always be suspect if they obtain favorable results doing a POE. The pressures toward favorable results are often very subtle, and therefore, it is best to avoid suspicion by having an entirely new team do the post-occupancy evaluation.

In my own experience, however, I sometimes find that the team doing the POE does not know enough about the events of programming to do an adequate job. This is the PDR-POE dilemma: how can the PDR team do its job well enough so that the information helps programming in such a way that the programming decisions are clear enough for the POE? The only answer is meticulous documentation of procedures and decisions.

PDR TECHNICAL ADVANCES

As mentioned previously, simulation is the one method that seems to be used almost exclusively for PDR. An excellent example of innovation in this method was the

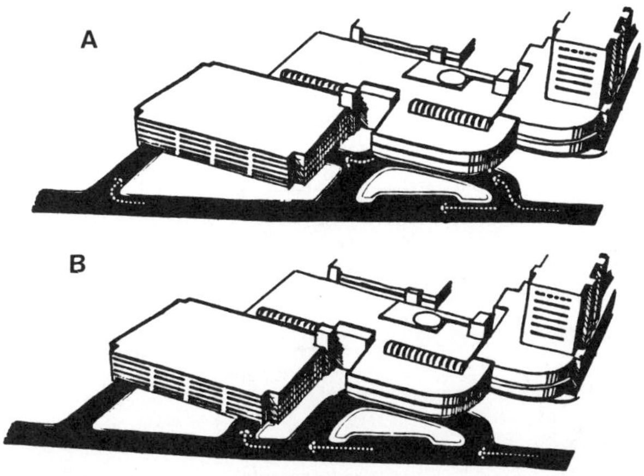

Figure 1. A is with the entrance, B without.

Reizenstein Carpman, Grant, and Simmons (1985) study. The problem addressed was whether a hospital should place a parking entrance near a patient drop-off point, as a convenience for the people bringing patients. A problem could develop because other people coming to the hospital might see the entrance and choose it rather than the regular parking entrance, causing traffic congestion. The study simulated automobile trips by moving a fish-eye camera first around a model with the new entrance and then around a model without the entrance (see figure 1). Subjects were asked which entrance they would pick as the video paused at decision points. Even though signs were clearly visible in both models, subjects chose the new entrance enough times to produce a near certainty of congestion. The new entrance was not built.

This study took only a few weeks and only a little over two hundred dollars to complete. It provided a clear and efficient PDR method for making an important decision and it provided an example of how modern electronic methods can make PDR information more available at this micro level. No longer must an entire building be contemplated before PDR is justified: it can now be applied to smaller units, such as entrances.

Having said that simulation is used almost exclusively for PDR, it is worth noting that there is one post-event (though not POE per se) use for simulation: photographing models via videotape to reconstruct accidents (Severson, 1988). This is a useful device for court cases as evidence for either the defense or prosecution.

CONTROVERSIES OVER POE AND PDR METHODS

Focus Groups

A method that has been around for some time and has been borrowed from marketing research is the use of "focus groups." This method is similar to the charette and earlier participatory design methods from the sixties (see Daish and Kernohan,

1986). The only reason for including this, as a recent advance, is that it has been rediscovered and is becoming increasingly popular.

Focus groups are collections of people brought together to answer questions about a building that exists or about a proposed design. Hence, these groups can be used for either POE or PDR. Usually the group is chosen for its expertise in some area, such as maintenance or management, but it could also be a group of housing residents. Usually these groups are non-randomly collected so there are problems of representativeness (see Marans in Bechtel, Marans and Michelson, 1987). The controversial aspect of focus groups comes from their use without sampling procedures in government and industry.

The bulk of POE work has shifted from academic settings, which peaked in 1973 (Bechtel and Srivastava, 1978), to government agencies and private industry. More and more government agencies are requiring PDR and POE for their buildings (e.g., Vischer, 1985; Picasso, 1985). In this regard, the work of the General Services Administration (GSA), the Veterans' Administration, the Air Force, and the US Army Corps of Engineers has already been cited (see National Academy Press Report, 1987). The US Postal Service (Kantrowitz et al., 1986) is a recent addition to the list of government agencies.

Sweden, with its Building Research Institute, and New Zealand (Daish, 1980), have also experimented with institutionalizing POEs in the building process; Sweden regularly incorporates POE findings in code reviews.

Eichinger (1985) reports how the VA uses expert panels – selected from personnel who are veterans of previous design changes – who are brought to the site of a new hospital and given a set of pre-tested questionnaires to evaluate significant new designs. The same method is reported by Taylor et al. (1987) as the mainstay for evaluation of scenic outdoor environments.

Daish (1980) reported a method that has yet to be tested in the US, the "most knowledgeable person" technique. This is a variant of the expert panel. In this case the most knowledgeable person in a given area, e.g., Maintenance, is chosen by an expert panel as the most knowledgeable person in their field. This method has yet to be compared with more traditional methods, including the expert panel; its largest recommendation is the saving of time and money. Pegrum and Bycroft (see chapter 17) also use this method in Australia.

Standardized Questionnaires

A search of PDR and POE literature shows that most of the work continues to be eclectic, using more than one method as a matter of course. An exception to this observation is the new and revised set of social climate scales developed by Rudolph Moos and his colleagues at Stanford University (Moos and Trickett, 1986; Moos and Moos, 1986; Moos and Spinard, 1986). Usually the scales are composed of 90 or so items and are divided into six to ten subscales which measure different environmental qualities. They have been used in a variety of settings, such as classrooms, homes, offices, nursing homes, and mental hospitals. Although the main purpose of these scales is the global assessment of social climate, they have been used to point out issues relating to job stress, environmental support, and other factors.

Moos has always had a strong environmental interest (Moos, 1985), but the social climate scales have not been very widely used in either PDR or POEs. The reasons for this neglect are not apparent. Pegrum and Bycroft (see chapter 17) use the satisfaction scales developed by David Center.

Transactional Research

The transactional model of research is producing a new method for both POE and PDR work (Oxley et al, 1986) but it is not without controversy (Kaplan, 1987; Altman et al, 1987). Essentially, the transactional approach seeks to recapture the holistic quality of the environment by collecting data through extensive observation and interview follow-ups based on that observation. It seeks comprehensiveness while recognizing the unique qualities of every setting. This method is somewhat reminiscent of Barker's (1968) behavior setting survey, but is less quantified and less arduous. Critics (Kaplan, 1987) state that it does not provide the traditional controls found in hypothesis-testing research. Proponents of the work (Altman et al, 1987) assert that traditional methods leave too much information out of the data collection and do not permit a truly contextual picture of behavior. Bechtel (1988) raises the question whether the transactional methods provide anything better than does ecological psychology.

So far, too few studies have been done to evaluate the usefulness of the transactional model (for it is more a model of method use than a new method), but usually these more comprehensive methods of data collection have proven useful to PDR and POEs – especially PDR, where the emphasis is on more global behavior.

The Subjective Report Controversy

Over time some instruments and methods of data collection have gained favor. Chief among these have been subjective self-report measures like the semantic differential (see Bechtel, 1975; 1987). The controversy over self-report measures is now entering its thirteenth year. It continues to be discovered by new researchers and uncomfortably avoided by some older ones. Daniel and Ittelson (1981) have pointed out that self-report measures can often confound the verbal response with (what should be) an environmental response. In their classic study, Starr and Danford (1979) showed that the words "lawyer's office" produced the same semantic profile of responses from subjects as did the actual environment of a lawyer's office. Thus, there was no difference in self-reports between those who were looking at words that labelled an environment and those who were actually responding to a specific environment labelled by those words. If self-report measures like the semantic differential are to be used for measuring attributes of the physical environment, there must be some assurance that it is the physical environment that is being measured and not a verbal label of that environment. In short, there can be no assurance that it is the environment that is being measured, unless the instrument used has been shown to discriminate between the verbal and the distal world and between generically similar environments. This is the requirement for discriminant validity first raised by Campbell and Fiske in 1956 (see also Fiske, 1982, and the exegesis in Bechtel, Marans and Michelson, 1987).

Much misunderstanding still exists concerning this controversy. Some still feel it demands that no self-report measures should ever be used to measure environmental attributes. On the contrary, this would only eliminate one class of possible confoundings. The fact is, all measures which do not discriminate between verbal and nonverbal responses should be pre-tested for discriminant power before being used to measure any environmental attribute. This certainly does not mean that one can never use semantic differentials or adjective check lists, but it may mean that certain adjectives which do not discriminate between verbal and environmental stimuli should never be used.

The Scenic Beauty Estimate, or SBE (Daniel and Boster, 1976; Brown and Daniel, 1984) continues to be used on outdoor POEs; it avoids this confusion by using a nonverbal numerical scale, and also pre-tests for discriminant validity. The SBE produces mathematical modeling for management decisions in parks and forests.

Generic Studies

Marans (see chapter 18) introduces the intriguing concept of a generic POE: that is, one done with such systematic sampling and measurement technique that it can be applied as a standard for other buildings. While Marans limits his first attempt to measuring lighting in the workplace, the potential exists for extending it to many other dimensions. In addition, the data are stored on a disk with a software component, LISREL, which provides one potential for universal use by designers and also provides the capacity for adding new data as more studies are completed. The use also greatly helps keeping POE results as an archive.

In a similar vein, Ventre (see chapter 19) describes a portable multiple measurement cart that simultaneously records luminance and illuminance, acoustics, air and radiant temperatures, relative humidity, and air movement as well as CO_2 and CO_2 ratios. Ventre's method is generic because it randomly samples spaces within a large building, thus being less generic than Marans' wider sampling but much more extensive than previous POEs, which were usually limited to smaller buildings.

The Marans and Ventre studies demonstrate the potential for statistical sampling as being capable of extending POEs to a virtually limitless application spanning buildings and geographical limits. Ventre's measurement list makes practical the physical measurements in POEs which have been much neglected in the past. One could hope, not unrealistically, for a combination of both these methods in the future.

CONCLUSIONS

It may yet be too early to call it a trend, but time and budgetary demands of industry are creating pressures against the use of the more costly and time-consuming methods which are the stock in trade of social scientists. There seems to be a preference on the part of many clients in these agencies and industries for the quicker and more prestigious expert panel. The expert panel enables the client to rub elbows with the best experts in a given field. This has a tendency to boost the client's ego and to produce unchallenged statements from the experts. A relatively anonymous social scientist has some difficulty sounding expert among the luminaries and challenging the elite.

Yet the validity of this method must be challenged. Can it be demonstrated that experts can provide better and more economic information than more conventional methods for PDR and POEs? Perhaps there are circumstances where one could be used more effectively than the other, or where both need to be used simultaneously. At present there is no basis for making judgments about the viability of traditional social science over the panel-of-experts approach.

The only solution to this problem is a carefully crafted experiment comparing and contrasting these methods with standard social science techniques. Does the expert panel produce more useful information than a survey of user needs? Does the expert panel cost compare favorably with such a survey? The larger issue is whether the social scientists will have the courage to gather the resources and make such a test. If not, I see the pressures toward more frequent use of ad-hoc groups forcing out the use of the only methods known so far to test validity and reliability of results.

Meanwhile, sampling procedures and electronic techniques are advancing the applicability of POE methods into ever-wider areas, and the principle of the POE as a quality control of building design is moving toward worldwide acceptance.

REFERENCES

Altman, I., Werner, C., Oxley, D., and Haggard, L., 1987, "Christmas Street" as an example of transactionally oriented research, *Environment and Behavior*, 19:501-524.

Barker, R., 1968, *Ecological Psychology*, Stanford University Press.

Bechtel, R., 1975, The semantic differential and other paper and pencil tests, in: W. Michelson (Ed.), *Behavioral Research Methods in Environmental Design*, Dowden, Hutchinson and Ross, 41-78.

Bechtel, R., 1988, Back to the future again: a perspective on ecological psychology, in: Hoogdalem, Prak, Van de Voordt and Wegen (Eds.) *Looking Back to the Future*, Vol. II, 87-91, Proceedings of IAPS 10.

Bechtel, R., Marans, R., and Michelson, E. (Eds.), 1987, *Methods in Environmental and Behavioral Research*, Van Nostrand Reinhold.

Bechtel, R., and Srivastava, R., 1978, *Post-Occupancy Evaluation in Housing*, Dept. of Housing and Urban Development.

Bechtel, R., 1985, Symposium: Successful use of POEs in building delivery systems, in: S. Klein, R. Wener, and S. Lehman (Eds.), *Environmental Change/Social Change*, Proceedings of EDRA 16, 340-341.

Brown, T. and Daniel, T., 1984, *Modeling Forest Scenic Beauty: Concepts and Applications to Ponderosa Pine, Rocky Mountain Forest and Range Experiment Station.*

Campbell, D. and Fiske, D., 1956, Convergent and discriminant validation by the multi-trait, multi-method matrix, *Psychological Bulletin*, 56:81-105.

Daish, J., Gray, J., and Kernohan, D., 1980, *Post-Occupancy Evaluation of Government Buildings*, Victoria University of Wellington.

Daish, J. and Kernohan, D., 1985, Putting the POE to work: A case study in which POE is combined with participatory programming, in: S. Klein, R. Wener and S. Lehman, *Environmental Change/Social Change*, Proceedings of EDRA 16, 275-286.

Danford, S. and Willems, E., 1975, Subjective responses to architectural displays: A question of validity, *Environment and Behavior*, 7:486-516.

Daniel, T. and Boster, R., 1976, "Measuring Landscape Aesthetics: The Scenic Beauty Method," U.S. Dept. of Agriculture Forest Research Paper RM-167, Rocky Mountain Forest and Range Experiment Station.

Daniel, T. and Ittelson, W., Conditions for environmental research: Reactions to Ward and Russell, *Journal of Experimental Psychology*: General, Vol. 110, 153-157.

Eichinger, C., 1985, In search of innovation: Examining the POE process in the VA, In Symposium: Successful uses of POEs in building delivery systems, in: S. Klein, R. Wener and S. Lehman (Eds.), *Environmental Change*, Proceedings of EDRA 16, 340-341.

Fiske, D., 1982, Convergent-discriminant validation in measurements and research strategies, in: D. Brinberg and L. Kidder (Eds.), *Forms of Validity in Research*, 77-92.

Icofano, D., 1985, Participatory Decision Making: Some directions for evaluating theory and practice, in: S. Klein, R. Wener and S. Lehman, *Environmental Change/Social Change,* Proceedings of EDRA 16, 257-266.

Kantrowitz, M., Farbstein, J., and Schermo, B., 1985, The Image of Post Office Building: First Findings, in: J. Wineman, R. Barnes, and C. Zimring (Eds.), *The Costs of Not Knowing...,* Proceedings of EDRA 16, 259-265.

Kaplan, R., 1987, Validity in environment/behavior research, *Environment and Behavior,* 19:495-500.

Moos, R., 1985, Creating healthy human contexts: Environmental and individual strategies, in: J. Rosen and L. Solomon (Eds.), *Prevention in Health Psychology,* University Press of New England.

Moos, R. and Moos, B., 1986, *Family Environment Scale Manual,* 2nd Ed., Social Ecology Laboratory.

Moos, R. and Trickett, J., 1986, *Classroom Environmental Scale Manual,* 2nd Ed., Consulting Psychology Press.

Oxley, D., Haggard, L., Werner, C., and Altman, I., 1986, Transactional qualities of neighborhood social networks: A study of "Christmas Street," in: *Environment and Behavior,* 18:640-677.

Picasso, G., 1985, Corporate assessment of real estate: An exploratory study, In symposium: Successful use of POEs in building delivery systems, in: S. Klein, R. Wener and S. Lehman (Eds.), *Environmental Change/Social Change,* Proceedings of EDRA 16, 342.

Reizenstein Carpman, J., Grant, M., and Simmons, D., 1985, Hospital Design and Way Finding: A Video Simulation Study, *Environment and Behavior,* 17:296-314.

Reizenstein Carpman, J., Grant, M., and Simmons, D., 1986, *Design that Cares,* American Hospital Association.

Starr, N. and Dansford, S., 1979, The invalidity of subjective ratings of the physical environment, in: W. Rogers and W. Itellson (Eds.), *New Directions in Environmental Design Research,* Proceedings of EDRA 9.

Taylor, J., Zube, E., and Sell, J., 1987, Landscape Assessment and perception research methods, in: R. Bechtel, R. Marans and W. Michelson (Eds.), *Methods in Environmental and Behavioral Research,* Van Nostrand Reinhold, 361-393.

Severson, E., 1988, Courtroom productions return juries to the scene, *The Arizona Daily Star,* February 4, page 1D.

Vischer, J., 1985, Public Works Canada experience with POEs, In symposium: Successful use of POEs in building systems, in: S. Klein, R. Wener and S. Lehman (Eds.), *Environmental Change/Social Change,* Proceedings of EDRA 16, 342.

CHAPTER 16

A HOSPITAL EVALUATION: THE PROBLEM-SEEKING METHOD

Steven Parshall

CRS Sirrine, Inc.
Houston, Texas

INTRODUCTION

Of all the steps involved in realizing a health care building program, evaluating the final product is an important but often overlooked step. The primary gauge that is used for measuring success is user and client satisfaction. Evaluation is part of an ongoing process that looks back, but its benefits are in the future, to both the health care administrator and the architect.

There are many reasons why an evaluation might be undertaken. Following are some common intentions:

- *To justify actions and expenditures.* Large building programs, especially those undertaken by an institution like a hospital, may have a great demand for accountability. This may be especially true if public funds are involved. (See figure 1.)
- *To measure design quality (conformance to requirements).* A facility represents policies, actions and expenditures by management. The quality level of the end product should reflect these. Ultimately, quality is conformance to the client's expectations for that end result. (See figure 2.)
- *To fine-tune a facility.* With such a sophisticated and specific building type as a medical facility, some fine-tuning may be necessary to ensure that it functions as effectively as possible. (See figure 3.)
- *To adjust a repetitive program or to prepare for a future building program, renovation or expansion.* A hospital may be built in phases, with additional wings, or freestanding clinics may build several facilities. Either one may demand certain repetitive elements. The results of a POE are most useful when they contribute to the program and design of a subsequent step in accomplishing a master plan. This might involve a new unit or upgrading existing facilities. An evaluation can also help iron out the snags that were encountered the first time around. (See figure 4.)
- *To research man-environment relationships.* A medical facility can be a stressful place for its patients and staff. Helpful information may be

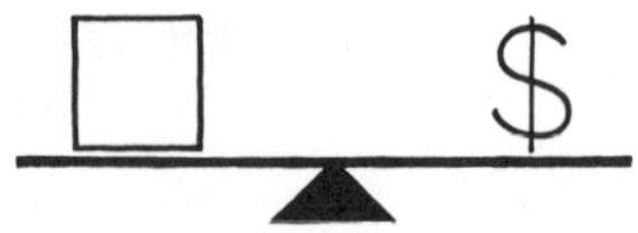

Figure 1

gained that can then be applied in improving other facilities, or the performance of the evaluated facility. (See figure 5.)

- *To test the application of new ideas.* Innovation involves risks, but new ideas are necessary to make advancements. These should be tested before further application is made. (See figure 6.)
- *To educate past and future participants.* This is helpful to both the client and architect. New personnel arriving at any point in the program could benefit greatly from lessons learned through an evaluation, and by participating in the evaluation process. (See figure 7.)

While there are many approaches to conducting an evaluation study, this chapter describes a comprehensive yet practical method that is feasible within the scope, budgets, and time commitments of a majority of CRSS clients. An important consideration in the development of this method is economy of effort while achieving the greatest value for our clients.

METHOD

The most common approach is to evaluate the performance of a facility once it is occupied – a post-occupancy evaluation, or POE. This way the evaluation team can consider responses from facility users, who are the best information source. After solving the shakedown problems, and after the novelty has worn off, the first major performance evaluation should take place between six months and two years after occupancy.

The evaluation team should represent different backgrounds and points of view. The collection of a diverse group contributes to a more objective evaluation. The size of the team will vary between three and seven, with one person having the responsibility of team leader. A team might encompass the following roles:

1. Administrator
2. Facilities Manager
3. User Representative
4. Programmer
5. Designer
6. Project Manager

Different data gathering techniques may be used by the evaluation team. Observation requires touring the facility. Interviews, either formal or informal, allow a deeper investigation. Surveys reach a greater number of people than interviews, and can be more comprehensive and specific. Hospital staff, physicians and patients are the prime source of information through interviews or surveys.

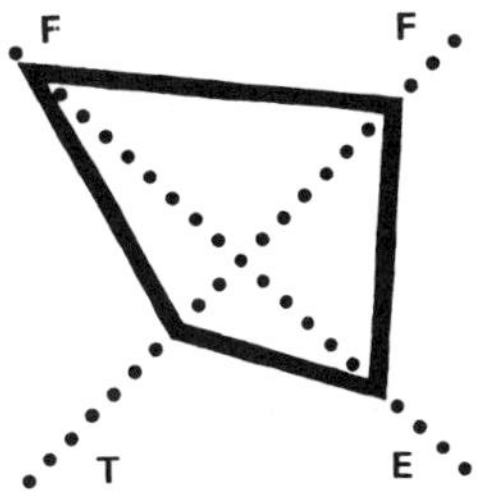

Figure 2

The first step in most building projects is programming, or accumulating and analyzing data from the client before design begins. This data is gathered from interviews with the client/users and compiled into a statement of requirements. A direction for the building design is the result of programming, based on goals and concepts the client wishes to achieve, and the design then proceeds on that basis. A record of the initial programming requirements and design intentions is essential for a rigorous evaluation of the result.

FIVE STEPS AND FOUR CONSIDERATIONS

Hospital facilities are very specific building types. They must fulfill certain demands that are unique, and the design must respond to those requirements. However, hospital facilities are easily evaluated when a method such as this one is applied. It is pragmatic: it is comprehensive, yet simplified enough for practical application.

The process has five steps:

1. Establish the PURPOSE;
2. Collect and analyze QUANTITATIVE information;
3. Identify and examine QUALITATIVE information;
4. Make an ASSESSMENT;
5. State the LESSONS LEARNED;

and four major considerations:

FUNCTION, FORM, ECONOMY, and TIME.

Evaluating, like programming, involves an organized process of inquiry, which is comprehensive in content. Organization of an evaluation (feedback) should correspond to the framework used for programming (feedforward). The similarity of organization, content, and format will increase the usefulness of the POE results for programming and design in the future.

1. *Purpose*

It is essential that everyone involved has a clear understanding of why the evaluation is being undertaken. Successful evaluation depends on cooperation. All participants – owners, users, architect – should agree on what they hope to gain.

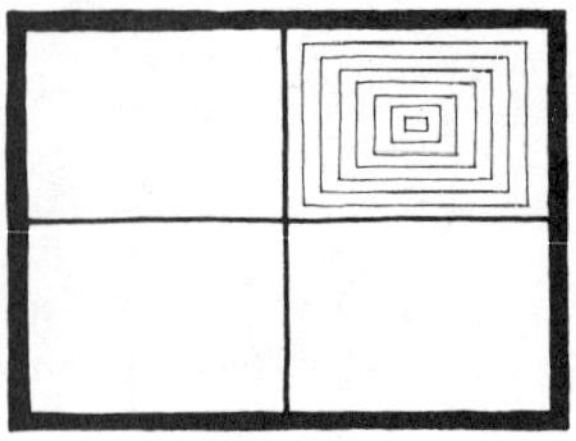

Figure 3

The purpose will also affect the way in which the evaluation is done. An expenditure justification might differ from a man/environment relationship evaluation. Users or participants on the team may be different. More than one purpose may be served during an evaluation, however.

2. *Quantitative Description*

The second step, preparing a quantitative description, includes collecting factual data on the building as designed; for example, the floor plans. Analyzing parameters provides a basis for comparing this facility with similar ones.

Functional Adequacy: A measure of the amount of area per the facility's primary unit of capacity. Example: gross area per hospital bed for inpatient facilities, or patient visits for ambulatory facilities. The analysis might also compare the capacity for planned procedures with the actual operations performed.

Space Adequacy: Net assignable areas include the sum of all spaces required to serve functional requirements. Unassigned areas consist of all other spaces in the building, specifically circulation areas, mechanical areas, public toilets, janitor closets, unassigned storage, walls and partitions. Gross area of a building is the sum of the net assigned area and the unassignable area. The ratio of net assignable area to the unassigned area measures the building efficiency.

Construction Quality: The unit cost associated with the quality level of the building measured as the building cost per gross square feet. It is also helpful to note unique building systems which minimized cost, and unusual constraints, such as codes or site location which increased costs.

Technical Adequacy: The cost of fixed and special equipment (such as renal dialysis or laser surgery devices), measured as a percentage of the building cost, though it is also possible to represent it as a unit cost.

Energy Performance: A measure of the amount of energy per gross square foot consumed for the standard operation of a building.

User Satisfaction: Obtaining some form of a reading on how satisfied users are with the facility.

3. *Qualitative Description*

A qualitative description includes examining the client's goals for the facility, the programmatic and design concepts for achieving those goals, and the statements representing design problems that the designer intended to solve. This step also

Figure 4

includes identifying changes that have taken place since occupancy, and unresolved issues.

Goals convey the client's stated intentions. Sometimes clients express great aspirations that are not fully achievable in the end, due to budgetary or operational constraints.

Concepts are ideas for realizing goals. Programming concepts represent abstract relationships and functional arrangements. Design concepts are physical responses that provide a unifying theme to the building solution.

Problem statements represent a recognition of the critical project conditions, and a direction for the design effort.

Changes since occupancy are indicators of new requirements, or inadequacies. Changes are actions taken to alleviate undesirable conditions.

Issues are unsettled and controversial decisions that are in dispute. They are posed by the occupants or owner of the facility during the evaluation, or raised by the evaluation team.

4. *Assessment*

The evaluation criteria are standard questions that reflect important values. (See "Example of Question Set for Facilities Evaluations," below.) The evaluation team should review the question set to understand the meaning of the criteria. Each evaluator forms a subjective response as to the degree of excellence attained by the facility.

Quality is a value judgment that varies with every individual. It is subjective. Nonetheless, quantification is useful.

First, rating provides a mechanism for identifying the differences in perception of a building by the various evaluators. Better understanding is possible when the evaluation team discusses these differences.

Second, rating provides an explicit pattern of how the parts contribute to the whole assessment. Clearer knowledge of the strengths and weaknesses is possible when the evaluators compare these patterns and discuss them.

5. *Lessons Learned*

Lessons learned are conclusions about strengths or weaknesses. Rarely should an evaluation conclude with more than twelve statements. At a minimum, four statements will cover each of the major considerations: function, form, economy and time.

Function. When evaluating functional performance, refer to the original goals and concepts of the program. The original program provides an immediate focus on the important client decisions that influenced the design.

Form. The evaluation must include aesthetic standards to determine the physical

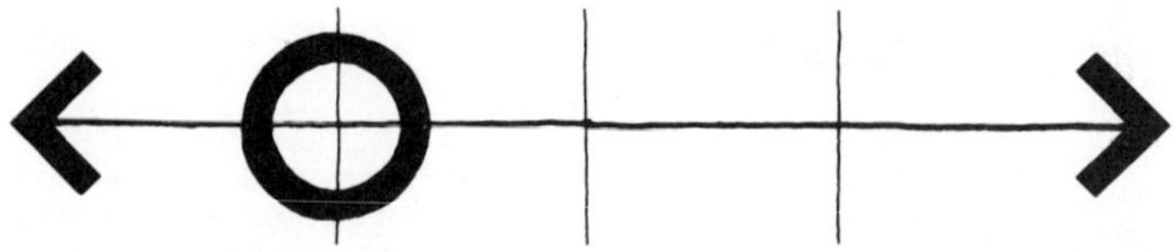

Figure 5

design excellence of the building. This is the most difficult part of the evaluation, since aesthetic standards are ever-changing.

Economy. It is important to consider the original quality goals for the facility – the quality commensurate with the initial budget. It is unrealistic to expect grand quality results if the original budget allowed for no more than an economical level.

Time. Because two or three years may elapse between programming and occupancy, the initial users may be different from those involved in the original planning. A certain amount of user satisfaction, therefore, depends on periodic interior design, or on the degree that partition and utility service changes are possible within the basic structure.

EXAMPLE OF QUESTION SET FOR FACILITIES EVALUATIONS

Instructions

1. Use this form for rating a facility.
2. Review the criteria for each consideration and agree on the meaning for the particular building type being evaluated.
3. Enter score from 1 to 10 for each criterion:

1	Complete Failure	6	Good
2	Critically Bad	7	Very Good
3	Far Below Acceptable	8	Excellent
4	Poor	9	Superior
5	Acceptable	10	Perfect

4. Add the scores and divide by 6 to get an average score for each major consideration.

FUNCTION

		Score
A.	THE OVERALL ORGANIZATIONAL IDEA (the big functional concept)	_____
B.	EFFECTIVE ARRANGEMENT OF SPACES (activities and functional relationships)	_____
C.	WELL-PLANNED CIRCULATION (entry, orientation, flow)	_____
D.	ADEQUATE SPACE ALLOCATIONS/PARKING (net assignable/unassigned area, parking)	_____
E.	RESPONSE TO USER PHYSICAL NEEDS (comfort, safety, convenience)	_____

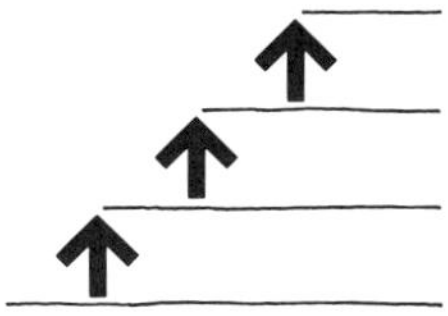

Figure 6

F. RESPONSE TO USER SOCIAL NEEDS ____
(privacy, interaction, sense of community)

SUM TOTAL ____
DIVIDE BY SIX ____6
AVERAGE FUNCTION SCORE ____

FORM

		Score
A.	CREATIVITY AND EXCELLENCE IN DESIGN (imagination, innovation)	____
B.	PERFORMANCE OF BUILDING SYSTEMS (structural, mechanical, electrical, signal)	____
C.	RESPONSE TO SITE CONDITIONS (physical, climatic, aesthetic)	____
D.	PROVISION FOR ENVIRONMENTAL CONTROLS (light, sound, temperature, ventilation)	____
E.	RESPONSE TO USER PSYCHOLOGICAL NEEDS (order, color, variety, views)	____
F.	APPROPRIATE SYMBOLISM (image, character, scale)	____

SUM TOTAL ____
DIVIDE BY SIX ____6
AVERAGE FORM SCORE ____

ECONOMY

		Score
A.	REALISTIC SOLUTION TO A BALANCED BUDGET (initial cost control)	____
B.	RETURN ON INVESTMENT (most for the money)	____
C.	MAXIMUM EFFECT WITH MINIMUM MEANS (elegance, multiple purpose)	____
D.	EFFICIENT PLAN AND SHAPE (unassignable area, volume)	____
E.	EASE OF MAINTENANCE (materials and building systems)	____

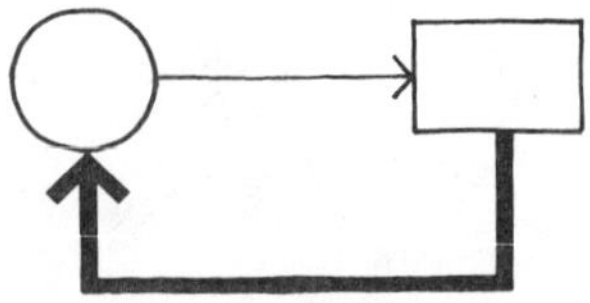

Figure 7

F. COST-EFFECTIVE OPERATIONS (energy efficiency, minimum upkeep) ____

SUM TOTAL	____
DIVIDE BY SIX	____6
<u>AVERAGE ECONOMY SCORE</u>	____

TIME

		Score
A.	CONVERTIBLE SPACES FOR CHANGES IN FUNCTION (dynamic activities, universality)	____
B.	FIXED SPACES FOR SPECIFIC ACTIVITIES (major static activities)	____
C.	PROVISION FOR GROWTH (expandibility, shell space)	____
D.	VITALITY AND VALIDITY OVER TIME (sustaining quality, holding power)	____
E.	HISTORICAL AND CULTURAL VALUES (significance, continuity, familiarity)	____
F.	USE OF MATERIAL AND TECHNOLOGY (expression of the times or advanced systems)	____

SUM TOTAL	____
DIVIDE BY SIX	____6
<u>AVERAGE TIME SCORE</u>	____

(See figure 8.)
1. Use this graph for illustrating the pattern of scores.
2. Enter the average score for each consideration on the graph.
3. Mark the point on the graph representing the score.
4. Connect the points on the graph to form a quadrilateral.

EVALUATION ACTIVITIES

The typical sequence of activities:

1. Initiation – This meeting is to establish the purpose of the evaluation and to identify the background data requirements.
2. Preparation – This task requires background research to prepare the quantitative and qualitative descriptions. It may also entail the user surveys to record their

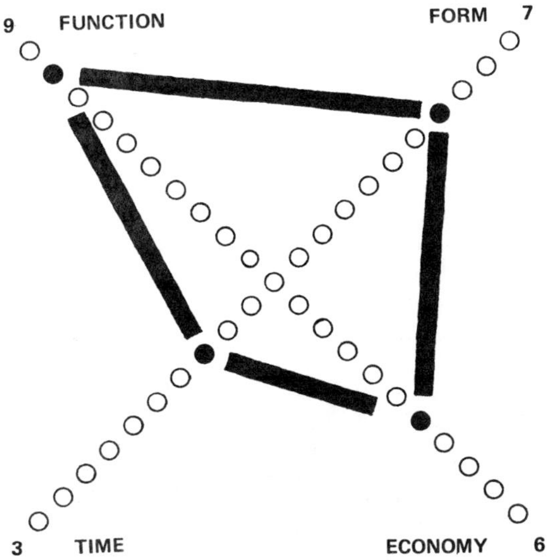

Figure 8

perceptions. These surveys should encompass three major areas: 1) description of the users, 2) their satisfaction with the facility, and 3) evaluation of particular features of the health care facility design.

3. Tour – The evaluation team makes a visual inspection of the facility. During the tour, evaluators may undertake random interviews with users and probe for responses about performance.
4. Discussion – After the tour the evaluators meet to discuss their observations.
5. Assessment – Each evaluator makes a judgement as to the facility's success by assigning a score. The ratings are recorded on a special graph, which illustrates the pattern of each assessment.
6. Summation – The evaluation team reviews the quantitative and qualitative descriptions, along with the assessment ratings, and prepares a statement of the lessons learned.
7. Presentation – The team leader presents the conclusions.
8. Documentation – The team leader prepares a report of the evaluation process.

CASE STUDY

In 1981, University Hospital, the teaching hospital of the University of Washington in Seattle, embarked on a 5-year expansion program. This included the addition of a new inpatient 193-bed tower and outpatient clinic facilities. This phase of the building program was completed in 1986 resulting in a 450 bed capacity for inpatient care. (See photo 1.)

In January 1987, the project team held a post-occupancy evaluation of the new East Wing Addition. This evaluation came nine months after occupancy, allowing time for its occupants to settle in and make adjustments to their space.

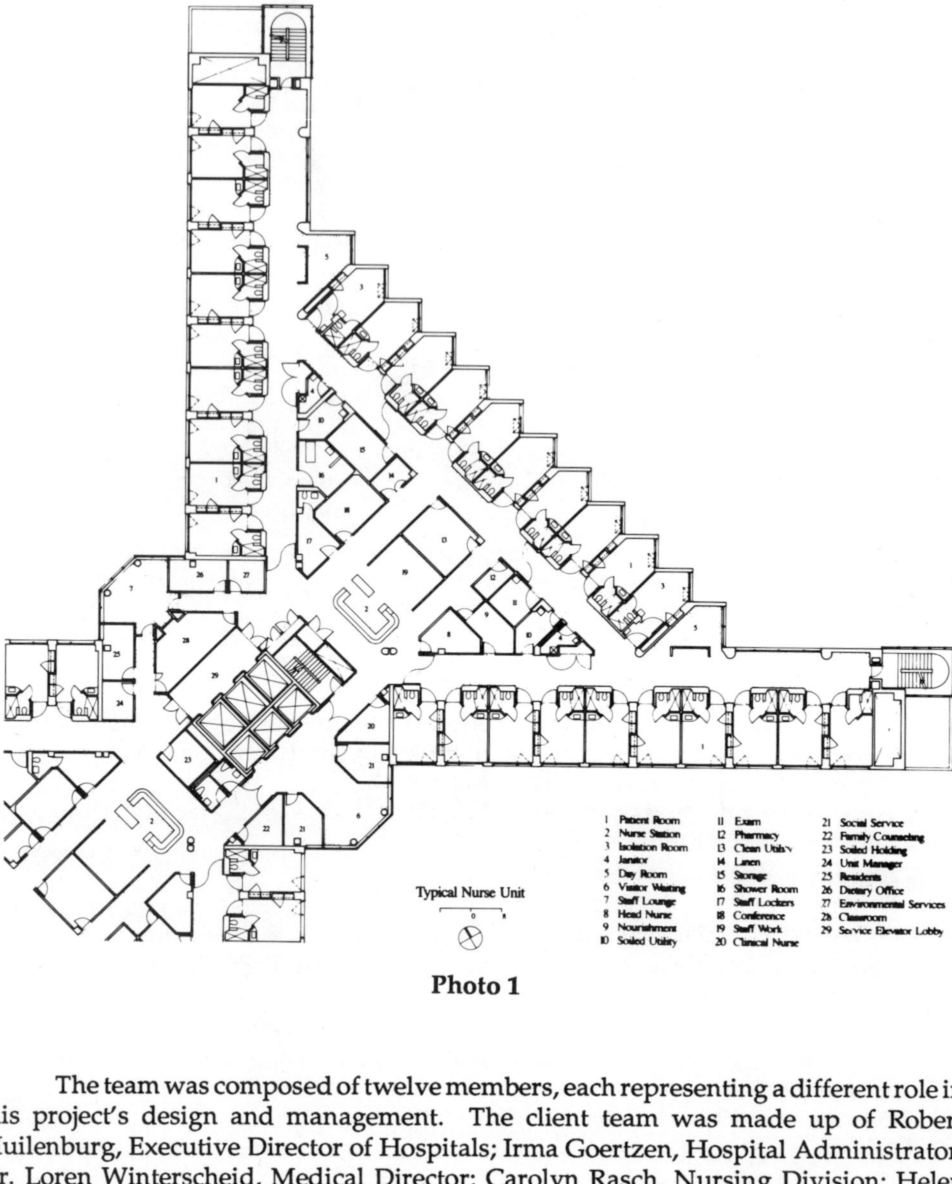

Photo 1

The team was composed of twelve members, each representing a different role in this project's design and management. The client team was made up of Robert Muilenburg, Executive Director of Hospitals; Irma Goertzen, Hospital Administrator; Dr. Loren Winterscheid, Medical Director; Carolyn Rasch, Nursing Division; Helen Shawcroft, Director of Planning; Richard Aradine, Assistant Administrator, Facilities Construction; and Warren Burkland, University Architect, Project Manager.

The consultant team involved three companies: from CRSS, Jay Bauer, Executive Vice President, Steven Parshall, Director of Research, Michael Shirley, Project Architect; from WPFSA, Chris Griffes, Project Architect; from RDA, Robert Douglass, Health Care Consultant.

The reasons for undertaking such an evaluation were laid out beforehand by the team to establish direction for their efforts. The purpose agreed upon was threefold: to provide a basis for updating the master plan; to learn lessons from the new and renovated facilities that might pertain to the programming and design of future renovations or additions; and to fine-tune the recently completed facilities.

Photo 2

The team (see photo 2) held a one-day evaluation meeting following the CRSS Problem-Seeking method of post-occupancy evaluation. It adheres to the five step method that has proven effective with many diverse building types. The fifth and final step in the evaluation is the compilation of lessons learned. These are insights that have been gained through the programming, design, and construction of the facility, and which are brought to light through the evaluation process.

The lessons learned from the University Hospital POE are as follows:

Overall Organizational Idea

The integration of the functional programming and design team under one contract saved one year in the overall project delivery, and resulted in fewer programmatic changes during design and construction.

The participatory design process and hospital management's commitment to "making this project happen," improved occupant buy-in and cooperation during the difficult renovation/move-in period.

Since the team's focus was on the overall building form, and on the individual departmental requirements, more attention was needed on linking the parts within the whole, including hospital-wide systems and functions within public spaces.

Well-Planned Circulation (Orientation)

Although flow within the building has been improved by the "concourse" concept (a major circulation mall on the main level), the multiple levels of entry and remaining unrenovated area require further consideration to unify the overall complex.

The bed tower's triangular shape, along with the lack of orienting features and clear interior lines of sight, suggest that special attention be given to orientation within the space when using this configuration. (See photo 3.)

Photo 3

Creativity and Excellence in Design

This project was approached from a patient outlook which led to a focus on features and amenities for patients and staff. The response to unique qualities of the site and the imaginative use of "captured space" created special people spaces.

The hospital's commitment to quality, investment in a "best in class" image and acknowledging its community responsibility has created a "nice place" that has a special value. The pay-off: making it an attractive place where the medical faculty want to practice medicine.

The continuity of the client/consultant team, which upheld the values established at the project's beginning, allowed for achievement of the initial goals and aspirations.

Building Systems/Environmental Controls

Not fully integrating the engineering functions into the project team led to some unsatisfactory outcomes with building systems and environmental conditions.

Realistic Solution to a Balanced Budget

Favorable building economic conditions and timely action contributed to overall success of the project and captured opportunity for improved overall performance of hospital services. In the end, "no amount of planning could replace dumb luck." The key, however, was the will to proceed in face of adversity, and balancing the budget at each step of the process beginning with programming.

Ease of Maintenance/Cost Effective Operations

Some unique features of the design may lead to higher life cycle costs, but these were trade-offs the hospital was willing to make in exchange for achieving the overall goals. Future projects might establish operating and maintenance criteria at programming to make this a more important consideration during design.

Vitality and Validity Over Time

Implementation of the master plan concepts represents a significant investment. The hospital must maintain these major features by updating the master plan for both external conditions and internal arrangements on an on-going basis. Professional advisor overview on even "small changes" will help maintain these concepts.

Provision for Growth

The site is somewhat constrained and has limited growth options, therefore, future master planning should consider: 1) follow through with the clinic facilities addition as originally planned; 2) looking for opportunities for decentralization of functions to affiliate hospitals, thereby "freeing up" new opportunities at the University Hospital campus.

SUMMARY

This method is practical enough for use in evaluating most health care facilities. The five-step process suits many purposes.

The criteria are comprehensive, encompassing four major considerations: function, form, economy and time.

The method acknowledges user satisfaction, though the final assessment requires the judgements of an evaluation team.

Six months to two years after initial occupancy is the best period for conducting a post-occupancy evaluation.

As an aid to the programming process, however, an appropriate time for conducting an evaluation is prior to initiating a new building program, remodeling, or discontinuing the use of a facility.

Strive for the format and organization of the evaluation results to be compatible with those informational formats used in programming and design.

The procedure allows for quick starts and for timely completion.

CONCLUSION

CRSS programmed its first project in 1951 using the Problem-Seeking method. The first publication about the method appeared in 1969; and in 1987, the AIA Press published the third edition of Problem-Seeking. While earlier editions contained chapters on evaluation techniques, the latest edition added a chapter, for the first time, on "Facilities Evaluation."

Programming has been practiced for over 35 years; however, its wide-scale adoption and routine practice came much later. Evaluating facilities is only recently gaining acceptance as a meaningful practice. More development of tools and techniques will improve its implementation. Yet the drive for wider application of evaluation will come from client demand. When post occupancy evaluations achieve value for both clients and users of facilities, then POE will also experience demand comparable to that of programming.

Hospital projects are highly complex in the diversity of considerations that must be resolved. Often necessary trade-offs are made along the way from inception to final completion. Moreover, the process which may last from three to five years, involves a multi-participant team, comprising both client and architect. An evaluation offers this team an excellent opportunity to visit the implemented plan and reconfirm the priorities and concepts set forth in the program and design.

Has the investment achieved the ideas and ideas that we set out to achieve? An affirmation strengthens the health administrator's plans to follow through with the operational activities necessary to achieve the full value of the asset. The evaluation is also an opportunity to recognize the minor adjustments that will realize the fullest performance of the facility in place.

BIBLIOGRAPHY

Friedman, Arnold (Ed.), 1978, *Environmental Design Evaluation*, New York, Plenum Publishing Corp..

Marans, R. W. and Spreckelmeyer, K. F., 1981, *Evaluating Built Environments: A Behaviorial Approach*, Institute for Social Research, Ann Arbor.

Parshall, S. A. and Pena, W. M., 1982, *Evaluating Facilities: A Practical Approach to Post-Occupancy Evaluation*, CRS Sirrine, Houston.

Pena, W. M., Parshall, S. A., and Kelly, K. A., 1987, *Problem Seeking: An Architectural Programming Primer*, Washington, D.C., AIA Press.

Preiser, W. F. E., Rabinowitz, H. Z. and White, E. T., 1988, *Post Occupancy Evaluation*, New York, Van Nostrand Reinhold.

Post-Occupancy Evaluation of Hospitals, 1975, *AIA Journal*, January 1975.

Post-Occupancy Evaluation Practices in the Building Process – Opportunities for Improvements, National Research Council, Building Research Board, 1987, Washington, D.C.

CHAPTER 17

QUALITY DOWN UNDER: BUILDING EVALUATION IN AUSTRALIA

Roger Pegrum and Peter Bycroft

Construction Group
Department of Administrative Services
Canberra, ACT, Australia

PUBLIC WORKS IN AUSTRALIA

The bicentennial of European settlement in Australia has led to a number of public and private questions about how Australia has performed in the first two hundred years and where it might be heading in the future. The most obvious changes in Australia since 1788 have been environmental, cultural, and political. It is therefore not surprising that architects and environmental designers generally are looking again at their performance, and the the Australian government is leading the charge towards a proper system for environmental assessment.

The office of the Australian Government Architect is part of the Construction Group in the federal Department of Administrative Services. It is now and has always been one of the biggest builders in the country. History and political agendas have now produced a very clear commitment to evaluation of the built environment and the social effects of our work. In the simplest terms, government architecture, like all public service enterprises, is being asked to show an efficient and effective use of public money. In both public and private sectors, there is also a growing public expectation that design policy will produce healthy environments. The challenge of ensuring quality standards in design and performance is now one of the most pressing items on the government agenda. Evaluation of buildings during design and documentation, the monitoring of detail and construction techniques and assessment after completion are now built in to the government delivery process in Australia.

The Australian States and Territories federated into the Commonwealth of Australia in 1901. The Commonwealth government was then responsible mainly for defense, communications, and customs infrastructures, but the role of the Commonwealth and its relationship with the States has changed considerably in the three-quarters of a century since the union. State governments are now responsible for law and order, education, and housing within their own boundaries. The system of local government which divides the States into municipalities and shires also has an active political and physical role, mainly in the area of public health and the monitoring of building standards.

The Department's corporate objectives are to provide for the needs of Australians in the built environment by :

1. advising the Government on policies and practiceds within the Department's area of responsibility - housing, public works and construction industry policy.

2. administering efficient and cost-effective Government housing programs.

3. providing a service for the design, execution, maintenance and (where appropriate) operation of Commonwealth works with particular regard to :

   - quality, time and cost
   - use of the best available resources both internal and external to the Department
   - consideration of all practicable alternatives to satisfy the clients' needs
   - community objectives and allocation of national resources.

4. in consultation with industry, other public authorities and users, promoting efficiency and cost-effectiveness in the housing and construction industries.

5. undertaking and encouraging social, economic and technical research and ensuring the appropriate dissemination of the re sults.

6. striving to meet community objectives by protecting the environment and conserving energy and other scarce resource in Departmental operations.

7. maintaining a working enviroment to attract and retain skilled and motivated staff.

Figure 1

The federal government sits at the top of the pile and is the major provider of buildings for public and institutional purposes. Its regular clients include the Department of Defence, the Attorney-General's Department, the aviation authorities, Commonwealth research organizations, and a large number of banks and semi-government organizations. Formal liaison among federal, State, and local government agencies occurs through the National Public Works Conference (NPWC) and the National Committee on Rationalised Building (NCRB). Cooperation and assistance in technical matters is also provided through participation in the work of the Standards Association of Australia, the Australian Model Uniform Building Code (Building Code of Australia), and on many single-issue groups such as the National Occupational Health and Safety Commission.

The Construction Group of the Department of Administrative Services is the latest form of the Commonwealth government public works authority, which has existed complete but in various guises since 1901. Its corporate objectives are outlined in figure 1. For most of that time federal agencies have been tied to this organization for the acquisition of their facilities. Lately there has been a move to free up the process and allow a more competitive search for the best possible building solution. There are, of course, dangers in this. One thing which makes the work of government architects unique is their commitment to owning and operating the completed projects. This

Findings

1. Reductions in the safety, reliability and quality assurance work force at Marshall and NASA Headquarters have seriously limited capability in those vital functions.

2. Organisational structures at Kennedy and Marshall have placed safety, reliability and quality assurance offices under the supervision of the very organisations and activities whose efforts they are to check.

3. Problem reporting requirements are not concise and fail to get critical information to the proper levels of management.

4. Little or no trend analysis was performed on O-ring erosion and blow-by problems.

5. As the flight rate increased, the Marshall safety, reliability and quality assurance work force was decreasing, which adversely affected mission safety.

6. Five weeks after the 51-L accident, the criticality of the Solid Rocket Motor field joint was still not properly documented in the problem

Figure 2

concern for life-cycle costing is quite different from that of the private sector. However, there may also be advantages, particularly if this competition works to ensure that the performance of government architects throughout the country at least equals that offered by some of our better private practices.

Australian government architects operate in all States and territories and represent the government overseas in the construction of embassies and in foreign aid work. Each mainland State has a regional office which works as a multi-disciplinary unit responsible for the work within its boundaries. At the center of each office are architects, engineers, building economists, landscape designers, and interior designers, handling a construction program with an annual total value well over one billion dollars. In the Central Office in Canberra, the Directors of Architecture and Engineering oversee the professional performance of the whole group and together with principal architects and engineers from the States ensure consistency of performance across the country. Central to this monitoring process is a series of evaluation procedures, of which post-occupancy evaluation is the most recent addition.

The awareness of evaluation as a tool in measuring the effectiveness of government programs has reached all levels of government in Australia. State public sector organizations are now required to prepare, report, and implement rolling five-year plans for evaluation of their programs. In addition, organizations are required to report annually on the findings of evaluations, action taken as a result of the findings, and consequent improvements to organizational performance (Allan, 1987).

The new interest in evaluation is founded primarily on the notion of accountability, and it is this philosophy which dominates government discussion of all programs. The Australian holy trinity, which until recently was cost, time, and quality, has been reworded as quality, cost, and time, with quality first among equals. This small but important shift reflects public opinion and supports overseas developments. Considerable emphasis was given in Australia to the findings of the Presidential Commission on the accident of the space shuttle Challenger. Of the six findings of the Commission, four implied a lack of attention to quality assurance (see figure 2).

Such catastrophic events in other countries have strengthened the commitment of evaluators to broad-based evaluation, despite continuing pressure for quicker and cheaper results. The Australian government has set out guidelines for program evaluation and has required all agencies to develop plans to achieve consistent results. These rules seek the development of "reliable indicators of performance which can be used for ongoing monitoring." (Linard, 1987) These initiatives suggest that government programs be measured for effectiveness every three to five years, a very difficult task if it is to be done in a reliable manner. Government moves nevertheless have had noticeable results in professional circles. A plethora of publications on evaluation is matched by the growth of the Australasian Evaluation Society, whose annual conferences are dominated by papers on assessments of government programs. Concurrent moves to introduce program budgeting are sympathetic to changes in program management techniques and herald major structural changes in public service organization. In the not-too-distant future today's experiments in program evaluation will produce an integrated network of performance indicators by which a program, its component projects, and the initiating government policy will all be assessed as a matter of course.

The delivery of public works in Australia has also been affected by this growing interest in the science of evaluation. Both the National Public Works Conference and the National Committee on Rationalised Building have produced guidelines on how to undertake project evaluation (Leslie, 1986; NPWC, 1986), and the NCRB maintains a working sub-committee on building evaluation. This sub-committee has representatives of three large organizations (the Australian and New South Wales governments, and Telecom Australia) which between them produce more than thirty post-occupancy evaluations each year. In addition, several other major organizations in Australia have developed, or are currently developing, post-occupancy evaluation programs. These include the Victorian Public Works Department, Department of Health and Country Fire Authority, the NSW Department of Housing, the New South Wales and South Australian Departments of Technical and Further Education, and the Western Australian Building Management Authority. The most influential Australian POE program, and the one which is most commonly used or adapted by these organizations, is that undertaken by the Construction Group of the Department of Administrative Services.

QUALITY MANAGEMENT IN AUSTRALIA

In June 1987 the Australian government released the Report of the Committee of Review of Standards, Accreditation and Quality Control and Assurance (known as the Foley report after its chairman Kevin Foley, the former chairman of the Australian Industrial Research and Development Incentives Board). The Foley committee was a multidisciplinary industry-based committee charged by the government to conduct a comprehensive review of the national system of standards and quality control, and to determine the extent to which official and unofficial procedures and attitudes have hindered the development of a more efficient and competitive Australian industry.

Although aimed primarily at export industries, the Foley report highlighted a growing interest at government level in the notion of "quality." In the introduction to the report, Australian Prime Minister Bob Hawke stressed the need to follow the

SYSTEM ELEMENT	Category A		Category B		Category c	
	Clause	Procedure	Clause	Description	Clause	Description
Tender and contract	3.1	A	4.1	B	5.1	C
Design Assurance	3.2.1	A C/R	4.2	N/R	5.2	N/R
Design Verification	3.2.2	A	4.2	N/R	5.2	N/R
Documentation	3.3	A	4.3	B	5.3	N/R
Measurement and Testing Equipment	3.4	A	4.4	B	5.4	C
Procurement	3.5	A	4.5	B	5.5	N/R
Inspection and Test Plan(s)	3.6	A	4.6	B	5.6	C
Incoming Inspection	3.7	A	4.7	B	5.7	C
In-process Inspection	3.8	A	4.8	A*	5.8	C
Final Inspection	3.9	A	4.9	A*	5.9	C
Inspection Status	3.10	A	4.10	A*	5.10	N/R
Identification	3.11.1	A	4.11.1	A*	5.11.1	A*
Traceability	3.11.2	A C/R	4.11.2	A* C/R	5.11.2	A* C/R
Handling and Storsage	3.12	A	4.12	A*	5.12	N/R
Manufacture/ torage	3.13	A	4.13	A* C/R	5.13	N/R
Special Processes	3.14	A	4.14	A*	5.14	N/R
Packaging and Shipping	3.15	A	4.15	A*	5.15	N/R
Records	3.16	A	4.16	A*	5.16	C
Nonconformance	3.17	A	4.17	B	5.17	C
Corrective Action	3.18	A	4.18	A*	5.18	B
Customer-Supplied Products or Services	3.19	A	4.19	A*	5.19	A*
Statistical Techniques	3.20	A	4.20	A*	5.20	N/R
Quality Audits	3.21	A	4.21	B	5.21	N/R

LEGEND FOR PROCEDURE AND DESCRIPTION COLUMNS

A - Category A procedures apply

A* -Category A system elements documented as descriptions rather than procedures (less than Category A)

B -Category B system elements documented as descriptions rather than procedures (less than Category A*)

C -Category C system elements documented as descriptions when specified in the contract (less than Category B)

C/R -To be documented and implemented only when specified in the contract.

N/R -No requirement to be satisfied.

Figure 3

Japanese example of significantly improving the quality of our manufactured goods. A major thrust of the report's recommendations is "designing and building quality into the product at the outset and at each stage through production."

The Foley report was much influenced by Canadian Standard Z299 and by British Standard BS5750, which calls for "the provision of a total system to ensure that the service or product reliably provides a performance that matches its promises" (Baden-Hellard, 1987, p. 35). The new Australian Standard AS2990 "Quality Systems for Engineering and Construction Projects" validates the recommendations of the Foley report. The report and the Standard together highlighted the need for a change of attitude to quality from one based on the inspection of finished products to one which prevents failure by control of the process.

AS2990 identifies twenty-three quality system elements which cover the range of project delivery processes and project complexities (see figure 3). Projects in category A have complexity or high technology characteristics requiring detailed preventative-based quality systems. Categories B and C are progressively less demanding. Many of the quality elements relate to formal inspection and testify procedures during construction. A number are less quantifiable. Design Assurance and Design Verification are two elements which together relate to the commonly applied practice of Value Management.

Post-occupancy evaluations by this office support the argument that extended application of Design Assurance and Design Verification procedures will reduce failings in the finished buildings. Design reviews and evaluation are part of the internal systems already in place, and these have been generally most satisfactory in achieving the desired

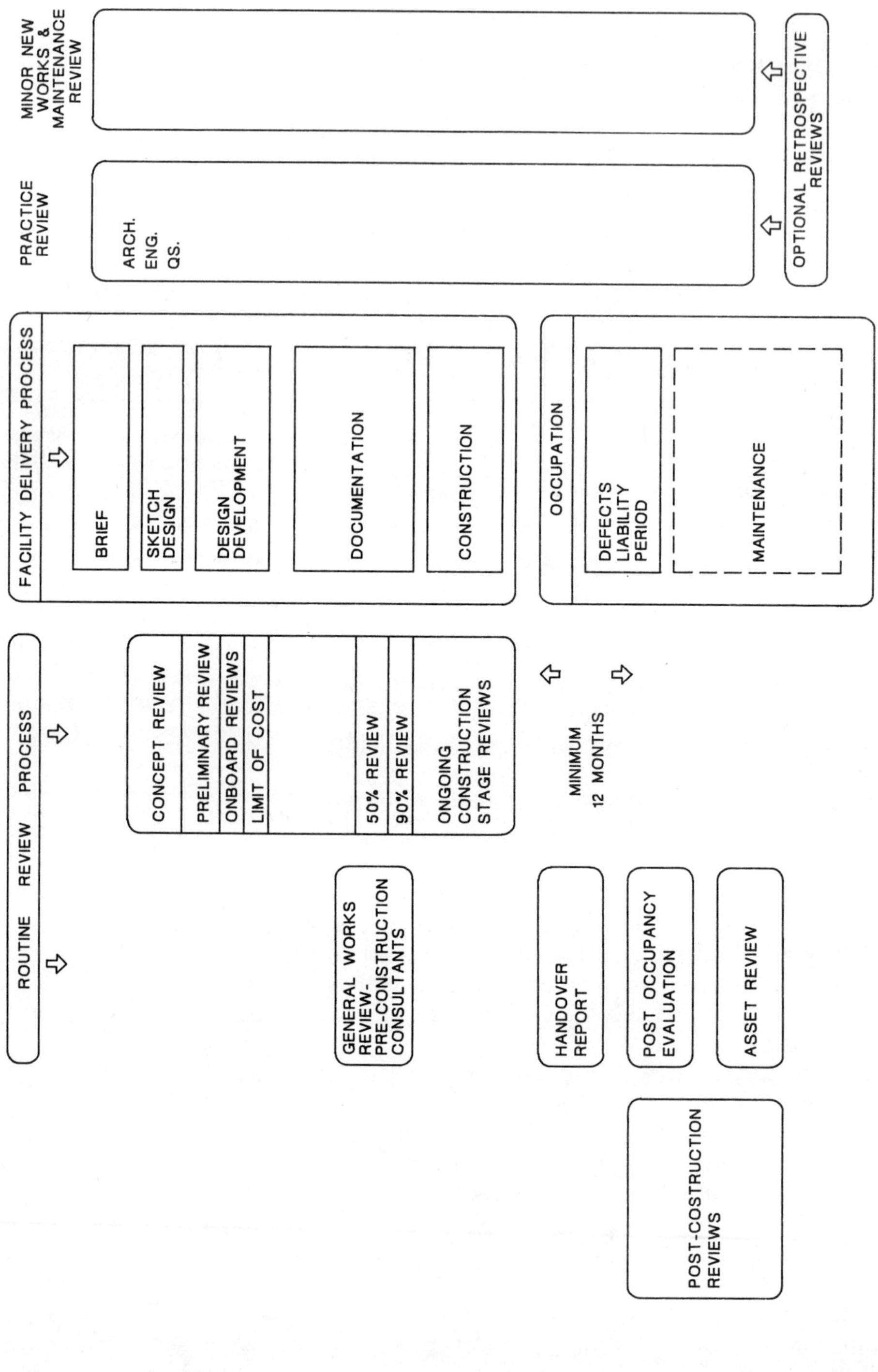
ROUTINE REVIEW PROCESS
FACILITY DELIVERY PROCESS
PRACTICE REVIEW
MINOR NEW WORKS & MAINTENANCE REVIEW
GENERAL WORKS REVIEW-PRE-CONSTRUCTION CONSULTANTS
CONCEPT REVIEW
PRELIMINARY REVIEW
ONBOARD REVIEWS
LIMIT OF COST
50% REVIEW
90% REVIEW
ONGOING CONSTRUCTION STAGE REVIEWS
BRIEF
SKETCH DESIGN
DESIGN DEVELOPMENT
DOCUMENTATION
CONSTRUCTION
ARCH.
ENG.
QS.
OPTIONAL RETROSPECTIVE REVIEWS
POST-COSTRUCTION REVIEWS
HANDOVER REPORT
POST OCCUPANCY EVALUATION
ASSET REVIEW
MINIMUM 12 MONTHS
OCCUPATION
DEFECTS LIABILITY PERIOD
MAINTENANCE

Figure 4

satisfaction ratings in planning conceptual areas. Greater attention is now being given to building services, one area which stands out in post-occupancy evaluation as the source of discontent.

Coincidentally, the growing interest in quality systems at government level has occurred at a time when the construction industry had been developing and implementing related concepts. A key factor in these developments has been the growing emphasis on life-cycle as opposed to capital costing of facilities. Works organization in Australia have increasingly endorsed the need for project design review procedures which take into account the likely operational and maintenance costs throughout the life of a facility (NCRB, 1984). Estimates vary in Australia as to the proportion of the total costs attributable to the cost of occupation. If debt servicing and wages are included in operational costs, initial capital cost can be as low as 10% of total life-cycle cost. If wages are excluded, estimates of the proportion of life-cycle cost attributable to operational cost remain in excess of 50% to 60%.

There is a view, widely held but misleading, that operational and maintenance costs are inversely proportional to capital cost. This view suggests that the more expensive the initial investment, the lower the likely operational costs. But the fact that some designs can achieve a minimization of operational costs emphasizes the importance of well-refined project review procedures. The key procedure capable of achieving solutions which have both low capital and low operational costs is value management. These procedures are well documented and are practiced in most of the industrialized world (see Dell'Isola, 1982; Walters, 1986; Rosenbauer, 1986). Value management is gradually being introduced as a highly effective design management tool in the construction industry of Australia.

Value management techniques are employed at the design stage and involve an intense function-for-costs analysis of design alternatives focusing on likely life-cycle cost. The procedures require an information base developed from operational and maintenance reviews of previously completed facilities. They involve the integration into the design process of data from previous project reviews, asset management reviews, and post-occupancy evaluations. Together these concepts establish a quality management system for the project delivery process. To establish these as an integrated quality system, and thereby close the feedback loop, life-cycle costing needs to be in place, value management needs to be in place, project reviews need to be in place, post-construction reviews need to be in place... and the data categories used throughout this process need to be agreed upon and standardized.

No public works authority in Australia has yet managed to integrate its quality assurance procedures to this extent, but the Construction Group has in place a synoptic framework which will in due course achieve the results sought by the federal government. Capital works and asset management programs are organized on a client basis, with an established suite of reviews which covers all aspects of the client service. Known presently as Operational Procedures, these reviews monitor and evaluate work at the following critical stages:

Program reviews – review the range of works activities to see that they are efficiently and effectively managed to meet the time and cost criteria established with the client agency.

Project reviews – evaluate the progress and quality of individual major and minor new works through briefing, concept design, and developed design prior to contract committal.

Pre-construction consultations – examine the impact of the design on construction activities, post-commissioning operation and maintenance.

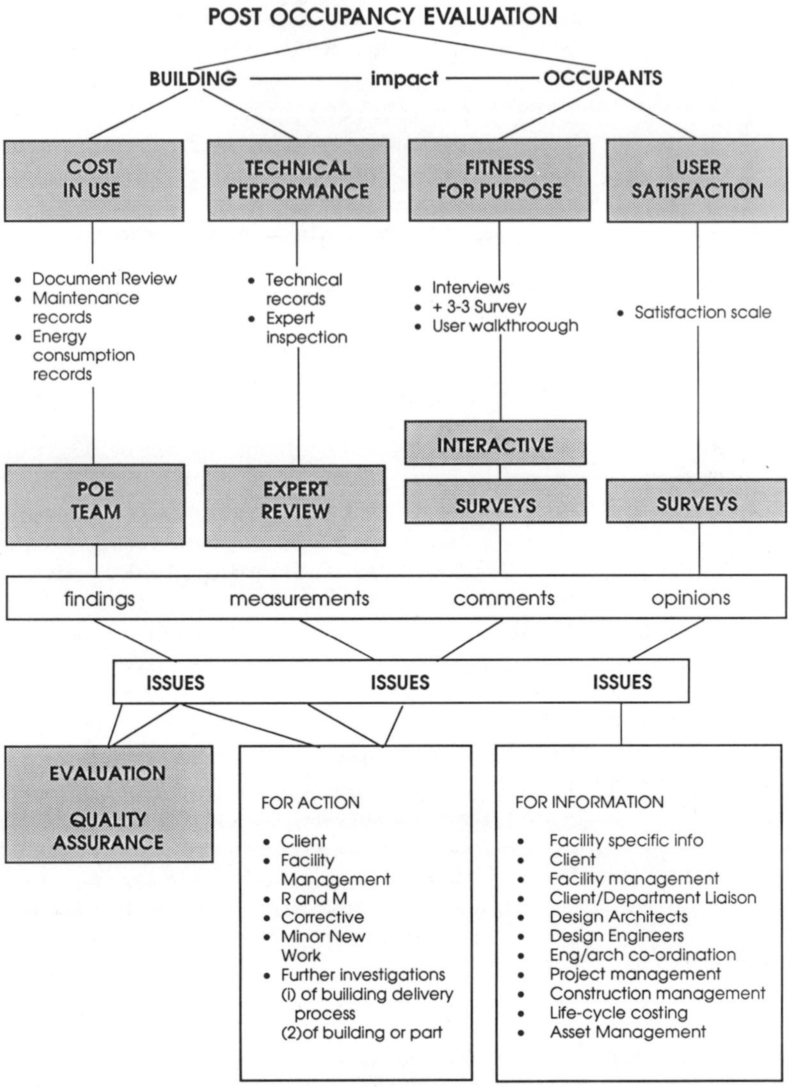

Figure 5

Pre-commissioning and handover consultations – aim to ensure a smooth transition of responsibilities for the proper operation and maintenance of an asset.

Asset reviews – assess the appropriateness of the asset management plan for an individual asset and follow up the assessment of quality over time as measured by post-occupancy evaluation.

Post-occupancy evaluation (POE) – studies the quality of a completed facility and assesses particularly value-for-money to the client with respect to capital and continuing costs in use.

Practice reviews – review professional, administrative, and management procedures and practices in use within the Group with specific reference to architecture, engineering, and quantity surveying standards.

The operation and reporting of each of these procedures is itself constantly under review. The extent of the Construction Group's commitment to evaluation is demonstrated by the continuing use of such reviews, which serve also to allocate a clear role to post-occupancy evaluation. The relationship of POE and the other reviews in the facility delivery process is shown diagrammatically in figure 4.

Post-occupancy evaluations by the Construction Group, although small in number to date, support the argument that extended application of design assurance and design verification procedures can reduce failings in the finished buildings. Design reviews and evaluations have been generally most satisfactory in achieving the desired satisfaction ratings in planning and conceptual areas. Greater attention is now being given to detailing and building services, two areas which stand out in post-occupancy evaluation as sources of discontent to almost all building users.

POST-OCCUPANCY EVALUATION IN AUSTRALIA

The general thrust of the emphasis on program evaluation in Australia has been from the top downwards. Department of Finance guidelines place the onus on senior management in each department to establish an evaluation program and to implement it. The impetus for the Construction Group POE program was generated in this way. There are, however, a number of management frameworks which can lead to the development of post-occupancy evaluations in government and semi-government agencies. In Australia, there are at least four different approaches presently used. These can be described as:

1. *Ad-Hoc*: Where POE is undertaken on an occasional and project-specific basis with little or no corporate management framework. POE in the Victorian Department of Housing and most POE undertaken through academic institutions falls into this category.
2. *Management-Initiated*: Where POE is seen as a performance indicator in a suite of project review techniques directed at improving the delivery of a particular service. The Construction Group has developed its POE program within this framework which is driven by directive from senior management.
3. *Professionally Integrated*: Where POE is seen as part of a general approach to delivering public services based on participatory methods. New South Wales State Public Works has developed a system in their delivery of primary and secondary schools where participations by users in post-occupancy evaluation is but one element in a suite of associations including public and student participation in design, the production of user manuals for each school, and a general concern about public equity in professional decision-making.
4. *Staff-Initiated*: Where, usually because of technological or related changes to the working environment, staff associations or staff trade unions request and participate in post-occupancy evaluation. The POE program in Telecom was initiated in this manner, and it is an excellent example of cooperation between management and staff to ensure the provision of adequate working conditions.

Despite the significantly different organizational objectives for the various POE programs in Australia, it is of interest that there is a general agreement on the methodologies used in these programs. Post-occupancy evaluation has a shorter history in Australia than in Europe or the United States, but there has been constant awareness of what has happened in overseas programs. The draft International Standard on Building Performance is well known in Australia, and the Construction Group has endorsed the four major factors in building performances studies identified in that Standard, i.e., user requirements, spatial requirements, building fabric subsystems, and performance in use.

In the late 1970's, government organizations in Australia and New Zealand began investigating the relevance of POE to their work. The New Zealand Ministry of Works and Development (MWD), working with consultants from the School of Architecture at Victoria University in Wellington, extended the techniques developed by the U.S. Army Corps of Engineers into a standard technique for POE (Daish et al, 1983). Australian developments in POE have drawn heavily on the New Zealand experience, although more recent practice in the Construction Group is derived from the European tradition, specifically from the issues raised in the draft International Standard on Building Performance.

Investigations of various evaluation methods commenced in 1978 at the Australian Experimental Building Station (now the National Building Technology Center) in North Ryde, NSW. Methods developed in New Zealand by the Ministry of Works and Development were modified for Australian use in a series of pilot projects in the early 1980's, and by 1985 some 20 POE studies had been completed to test methodologies appropriate to the government building program. The government procedures were formalized in an Operational Procedure in 1985, and a draft Field Manual and Explanatory Manual were released the following year (Department of Housing and Construction, 1986).

The objectives of the Construction Group POE program have been identified as:

a. Measurement of departmental performance in terms of quality of product (asset) and service to client;
b. Appraisal of facilities in use, and:
c. Assessment of the success or otherwise of facilities provided by the Group.

The nature of POE varies among organizations, but generally the major performance factors recommended in the draft International Standard are followed. Throughout Australia, for instance, the NCRB recommends that the following issues be covered in POE evaluations:

a. Technical performance;
b. Economic performance (i.e., life-cycle costs);
c. Functional performance, including user satisfaction, and:
d. Aesthetics.

The Construction Group, currently regarded as one of the leaders in this field in Australia, uses the following categories of evaluation:

a. Value for cost (cost-in-use);
b. Technical performance;
c. Fitness for purpose, and:
d. Client-user satisfaction.

The key word in POE is, of course, "occupancy," and a POE must concentrate on the assessment of a facility as a result of its occupation. Construction Group POEs study

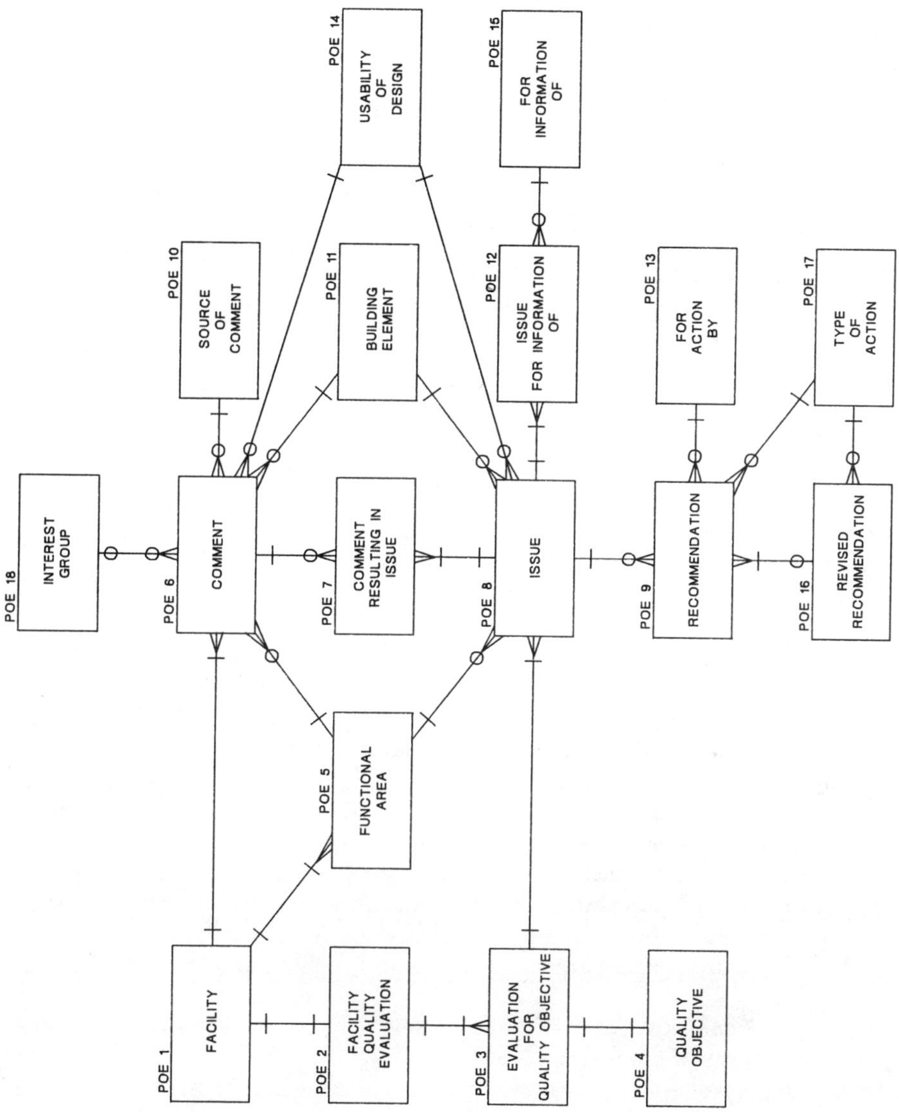
POE 1
FACILITY
POE 2
FACILITY QUALITY EVALUATION
POE 3
EVALUATION FOR QUALITY OBJECTIVE
POE 4
QUALITY OBJECTIVE
POE 5
FUNCTIONAL AREA
POE 6
COMMENT
POE 7
COMMENT RESULTING IN ISSUE
POE 8
ISSUE
POE 9
RECOMMENDATION
POE 10
SOURCE OF COMMENT
POE 11
BUILDING ELEMENT
POE 12
ISSUE FOR INFORMATION OF
POE 13
FOR ACTION BY
POE 14
USABILITY OF DESIGN
POE 15
FOR INFORMATION OF
POE 16
REVISED RECOMMENDATION
POE 17
TYPE OF ACTION
POE 18
INTEREST GROUP

Figure 6

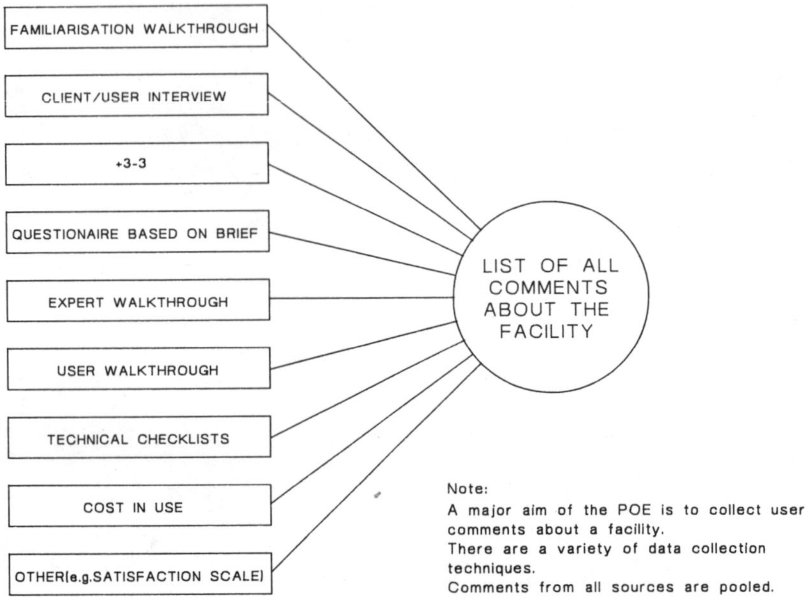

Figure 7

two different aspects related to this occupancy. The first is how the facility has responded to occupation, which involves measuring two distinct aspects of building performance: Cost-in-use – the ongoing costs of the facility, including the operational and maintenance costs, and Technical performance – the physical performance of the facility, its finishes, and its services.

Cost-in-use is measured through the examination of records such as energy usage, maintenance costs, and cleaning and servicing cost records. Actual costs are compared with targets set at the design stage, and where possible these costs are compared with costs of similar public and private sector facilities.

The systematic evaluation of the technical performance of the facility is measured by a team of building professionals, convened specifically for each POE. The team evaluates the technical performance of the facility using standard checklists and a structured inspection of the building. Information gathered from other POE methodologies such as user interviews and surveys is used to supplement the technical assessment.

The second aspect of occupancy evaluated by a POE is how the users have responded to the facility. There are two main aspects considered in this assessment:

User satisfaction – how satisfied are the users with the facility as a whole and/or with particular aspects of the facility, and:

Fitness for purpose – the extent to which the facility suits the need or function for which it was intended.

User satisfaction is measured with the standard survey form known as the Satisfaction Scale, which was originally developed by David Canter in England. The Scale has been used in the evaluation of many buildings in the U.K. (Royal Hospital for Sick Children, 1972) and in office and residential environments in Australia by Ross Thorne (Canter and Thorne, 1972). It is one of the few validated and reliable instruments

FOR ACTION

- Client
- Facility Management
- R and M
- Corrective
- Minor New Work
- Further investigations
  (i) of builiding delivery process
  (2)of building or part

FOR INFORMATION

- Facility specific info
- Client
- Facility management
- Client/Department Liaison
- Design Architects
- Design Engineers
- Eng/arch co-ordination
- Project management
- Construction management
- Life-cycle costing
- Asset Management

Figure 8

for measuring user satisfaction with buildings. The Scale is used to measure user satisfaction with both the facility overall and with specific areas and aspects.

The assessment of fitness for purpose requires several different and overlapping methodologies. Interviews with a client representative and representatives of the different user groups allows an open-ended and interactive discussion and provides an important source of information. However, interviews are time-consuming and expensive and the opinions offered are often not representative of general views about the building. It is also clearly unrealistic to attempt to interview all users of all facilities. A supplementary technique is used by the Construction Group to measure the response of a large number of users in a format that is relatively cheap and easy to obtain. A single-page survey form referred to as the +3-3 survey asks users to list the three best and three worst aspects of the facility. This instrument is not only quick and simple to use, but it also has "face validity" and the added advantage of gathering both positive and negative comments about the facility. On facilities where the total capital cost justifies the expenditure these techniques are supplemented by user inspections and user walkthrough.

Construction Group research on the use of POE techniques has indicated that some are by their nature "finger-pointing" techniques which produce an inordinate proportion of negative responses. Most of the early POEs in Australia and New Zealand relied almost exclusively on the "user walkthrough" which involved groups of users touring a building in a systematic fashion with the POE project team to gather data and opinions about the performance of the facility. In a subsequent analysis of the results of user walkthroughs, the Construction Group found that the number of negative comments outweighed the positive comments by a factor of 6:1. This imbalance between positive and negative was clearly an artifact of the methodology in which participants perceived themselves as involved in a critical inspection of the facility and hence took a negative stance to the exclusion of positive features.

The issue of potential bias was not the only problem in relying on the user walkthrough as the primary POE data-gathering technique. A more significant problem involved sampling inadequacies and the resource implications of the technique. For example, only a relatively small proportion of users could participate in a user walkthrough, which is, in essence, a mobile interview. The reliability of the findings were as a result always open to question. In addition, the user walkthrough technique involves a substantial amount of time to complete with up to six different interest groups touring a facility and participating in briefing and debriefing workshops. It is perhaps unfortunate that the walkthrough has been so influential in POE research to date, because it is a technique which needs to be integrated with other, more reliable and cost-effective, measures if it is to be useful in POE research. The Construction Group has in general balanced it with an expert walkthrough, which avoids some of the problems by employing a team of professionals to assess performance using standard checklists.

Post-occupancy evaluations are resource-intensive and without careful planning can involve substantial expenditure, particularly in the form of staff time. To be cost-effective, POE needs to show some return. A POE should at the very least result in some cost savings by reducing the extent of building failures from design faults. The Construction Group has adopted common international practice in limiting the total cost of any post-occupancy evaluation to within 1% and 1.5% of total project cost. These costs are currently being examined with a view to achieving substantial reductions and it is anticipated that the introduction of a range of carefully staged techniques for application to different facilities will reduce the total cost of the program.

The evaluations are undertaken by a project team in each State. Members of the team generally have had no previous association with the facility being evaluated and are trained in POE techniques by Central Office staff. Projects for evaluation are nominated through the Director of Architecture to a steering committee, which makes a final selection which reflects as closely as possible the ongoing works program. The first group of centrally controlled POEs was completed in 1987 and included a defense housing project, two government office buildings, two laboratories, two gymnasiums, and two post offices.

The Construction Group program recognizes the two-part process of evaluation, of which data-gathering is the more straightforward and more easily managed. Certainly there will always be questions about appropriate methodologies and there is a need to ensure that data-gathering instruments are both valid and reliable, but to a large extent the principles of social science research are directly applicable to this first step. The second part of the process, that of disseminating the findings from the evaluation in a usable and constructive way to designers and decision-makers, is a great deal more difficult. Recycling information from the completed and occupied facility offers a means for design professionals to build upon past experience to provide quality products.

For POE to be fully effective, the design process must be restructured to allow the systematic input of POE data. The effective management of evaluation findings is a key determinant of their ultimate value (Leslie, 1985:9). The 1981 report of the Experimental Building Station (Law, 1981) argued that several assumptions are necessary to develop a valid perspective of evaluation:

1. The design process is assumed to be open-ended, and the design of buildings must be looked upon as a circular process and not a linear one which ends at the construction of the building (Markus, 1973).
2. Evaluation is an integral and accepted part of the design decision-making process and not an independent activity.
3. In order to produce worthwhile feedback, the evaluation must look at the total design decision-making process, not just the results of the process, and it must use evaluation criteria that reflect issues which were in fact requirements during that decision-making.
4. The results must be presented in a form that is understood and acceptable to the designers or other users.

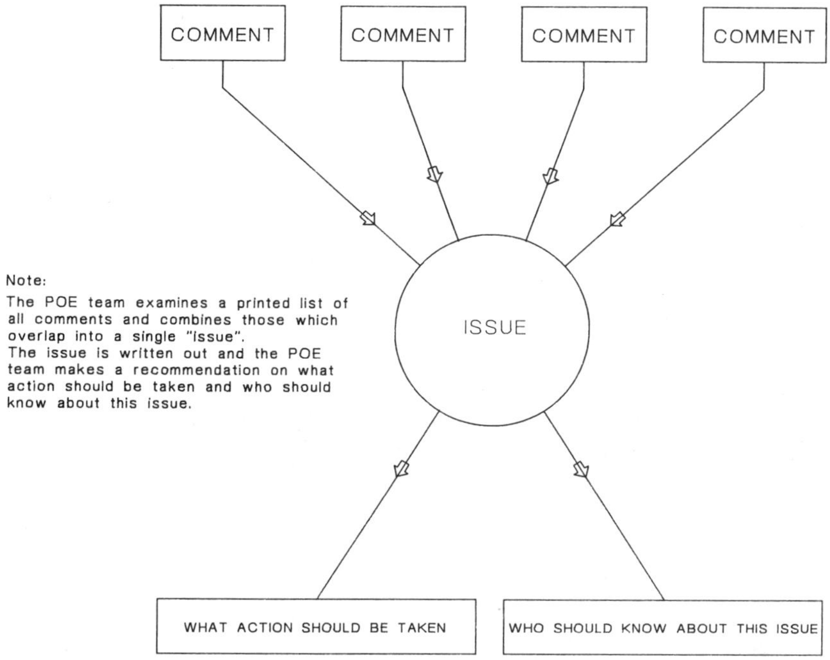

Figure 9

The Construction Group classifies POE findings for information which is then distributed to the major professional and technical groups involved in the facility delivery process. This process is known as feed-forward to emphasize its value in the design, construction, and operation of similar facilities in the future. An outline of the Group methodology is shown in figure 5. The storage of findings, measurements, comments, opinions, and issues is now carried out in a new computer system known as POETIC.

DEVELOPMENT OF THE COMPUTER SYSTEM "POETIC"

The post-occupancy evaluation computer system – POETIC (POE Technical Issues and Comments) – has been designed by the Construction Group to assist with recording, sorting, and retrieving the results of activities undertaken in POE projects. POETIC is a menu-driven single-keystroke program developed to run on an IBM XT or AT computer with at least a 10-megabyte hard disk and a 360K floppy drive. It was developed to run under version 3.2 of the MS-DOS operating system, and is written using Ashton-Tate's dBase III+ with the program modules being interpreted at run-time. The data model for POETIC is shown on figure 6. The system was specifically developed to facilitate the sorting of all data collected for each POE project and to provide a computerized and abridged version of each written POE report. In developing POETIC a logical analysis of the POE process was prepared and the system was designed to support this process.

In the POETIC analysis of POE data, a distinction is drawn between "comments" and "issues". A "comment" is a term which covers a statement, information, or

DGS HMAS Stirling Gymnasium

Print date: 03-09-87 POETIC printout

RECOMMENDATIONS FOR ACTIONING AREAS

FOR ACTION BY: CL Client

Recommendation Number		RECOMMENDATIONS recommended actions				
R11		CONSIDER REPLACING BACKBOARDS WITH CLEAR POLYCARBONATE SHEET IN ACCORDANCE WITH CURRENT PRACTICE.		Type of Action: Priority For action by:	2 3 CL	Vary Fitout/Layout Arrangement For consideration Client
Issue:	S 7	THE DESIGN OF THE BASKETBALL BACKBOARDS IS (-) UNSATISFACTORY. THE OPERATION IS DIFFICULT, THE BACKBOARD OBSTRUCTS THE VOLLEY BALL COURT, THE POSITION OF THE WINDING HANDLE IS TOO LOW.		Functional Area: Building Element: Usability:	4E1 SE 3	Main Hall Special Equipment User Safety
Comments:	C29	CLEAR POLYCARBONATE BACKBOARD PREFERRED TO SOLID FOR BASKETBALL	(-)	Functional Area: Building Element: Usability:	4E1 SE 2C	Main Hall Special Equipment Visual (lighting, glare, views)
	C30	AUTOMATIC CONTROLS SHOULD HAVE BEEN PROVIDED FOR BASKETBALL BACKBOARDS NOT MANUAL	(-)	Functional Area Building Element: Usability:	4E1 SE 1A	Main Hall Special Equipment Layout
	C32	BASKETBALL BACKBOARDS OBSTRUCT VOLLEYBALL COURT AREA	(-)	Functional Area: Building Element: Usability:	4E1 SE 1A	Main Hall Special Equipment Layout
	C51	HANDLES FOR MANUAL OPERATION OF BASKETBALL BACKBOARDS TOO LOW	(-)	Functional Area: Building Element: Usability:	4E1 SE 3A	Main Hall Special Equipment Safe/unsafe design
	C108	OPERATING HANDLES FOR BASKETBALL RINGS ARE BADLY LOCATED	(-)	Functional Area: Building Element: Usability:	4E SE 1A	Other Functional Areas Special Equipment Layout

Figure 10

observation about a facility resulting from the data collection process. The major sources of "comments" are client/user interviews, client/user surveys, and user and expert walkthroughs.

However, all data collection methods can provide "comments" about any aspect of the facility. For instance, in completing a cost-in-use survey it may become obvious that a particular design decision or choice of materials (e.g., floor coverings) is responsible for an above-average maintenance or cleaning cost. This observation is recorded as a "comment". For example, "the carpet in the foyer and passageways is excessively worn" is entered as a "comment" for further analysis. The collection of "comments" is shown diagrammatically in figure 7.

The POE team examines a list of all comments and combines those which overlap into a single "issue". The issue is written out and the POE team makes a recommendation on what action should be taken and who should know about the issue.

Within the POETIC system, an "issue" is defined as an aggregation or explanation of one or more "comments". The combining of "comments" into "issues" is the first step in categorizing or grouping similar, related, or the same "comments". It is likely, particularly with survey techniques such as the "+3-3", that there will be repetition of some comments. All comments are recorded, but duplication and repetition are avoided by combining related or similar comments into a single issue. An "issue" can also be in the form of an explanation of a comment. In the case of the comment about worn carpet cited above the issue would be something like: "The choice of low-grade carpet in the foyer and passageways is causing major wear and increased cleaning costs."

Through conventional reports and recommendations, the POE program assesses the quality of a facility in the four categories previously listed (cost-in-use, technical performance, fitness for purpose, user satisfaction) and POETIC extends these assessments to provide data which will improve future performance. There have been many attempts by other groups at sorting POE data using categories such as fine tune, corrective action, avoid in future, design item, or brief item. POETIC amalgamates these groups into the two most useful categories: ACTION and INFORMATION.

The ACTION category is self-explanatory. ACTION may include immediate or long-term maintenance or a physical change to the building or equipment. It may also suggest a more detailed study of the asset or of the practice which produced it. Those issues raised by the POE which should be incorporated into the future decision-making process are listed as INFORMATION. They are of value to one or more of a range of contributors and form the basis of continuing quality assurance for government programs. Sub-categories for both ACTION and INFORMATION are shown in figure 8. The process of organizing comments and issues into ACTION and INFORMATION is shown in figure 9.

Comments raised by a POE are individually coded by reference to the three checklists used in the data collection stages (functional area, building element, usability). Using these codes, it is possible to define the location in the building, details of the building fabric, and relevant human factors. POETIC stores this information in a number of modules or subsystems, accessed through a hierarchy of menus. The six main processing subsystems are shown below:

Subsystem	*Operations*
Facility Information	Allows the amendment of facility information.
Comments	Allows initial entry and amendment of comment data. Comment reports are also produced.

Issues and recommendations	Allows initial entry and amendment of issue and recommendation data. Reports on this data can be produced.
Evaluation of facility	Allows the entry and reporting of evaluation data.
Reporting	Allows the production of reports on all areas of data.
Backup and recovery	Allows data backup and recovery.

The four basic data groups stored and processed in POETIC are:

A. FACILITY, consisting of

- Facility descriptive information
- Functional areas of the facility

B. COMMENTS, consisting of

- Comments and other information collected in the standard categories.
- Functional areas.
- Interest groups.
- Building elements.
- Usability of design.
- Source of comments.

C. ISSUES AND RECOMMENDATIONS, consisting of

- Issues determined
- Comments resulting in issue cross-reference
- Recommendations and the standard categories for issues
- Functional areas
- Building elements
- Usability of design
- For information of
- Recommendations for action by
- Type of action

D. EVALUATION, consisting of

- Report Summary/Abstract
- Evaluation paragraph for (each) quality objective and the standard category.
- Quality objectives.

The POETIC system was designed to process each of these major data groups separately, but more importantly, to fit in with the stages of the POE projects. A typical printout from the POETIC system is shown in figure 10.

RESULTS OF THE 1986-87 POE PROGRAM

Despite the wide range of building types studied in the 1986-87 program, it has been possible to draw some valid and valuable conclusions as to the success of the

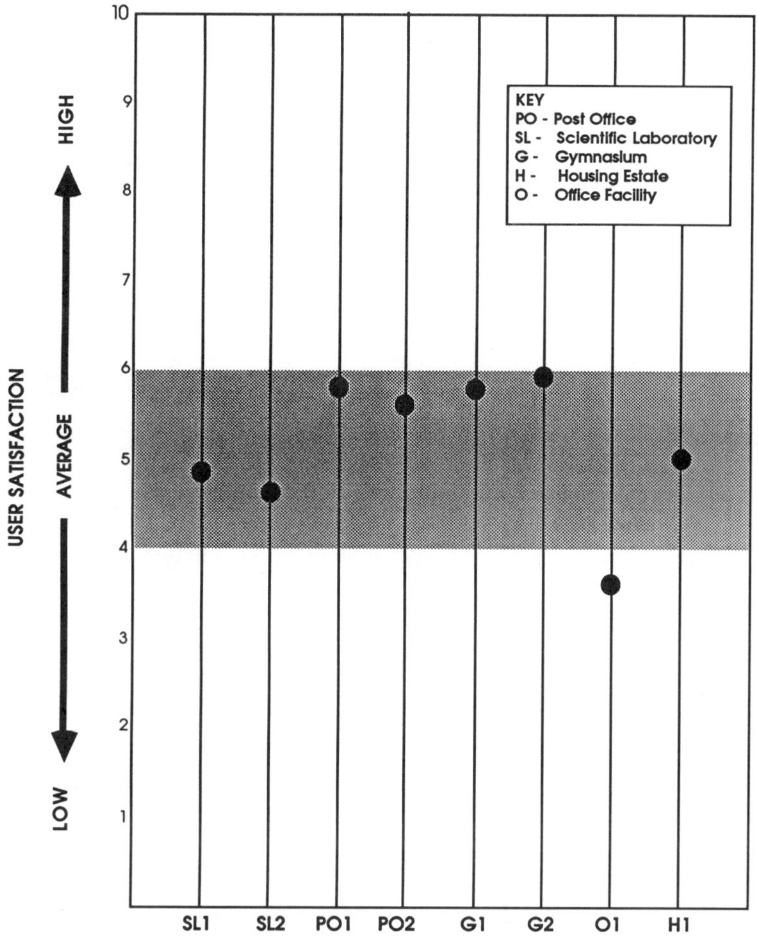

Figure 11

Construction Group in its work for several important clients. The following brief comments summarize the findings of the POE teams in the four major categories of evaluation.

Cost-In-Use

The procedure followed was that recommended by the NCRB, and each POE team collected data on in-use costs such as maintenance, energy consumption, cleaning, and security. A number of important issues have emerged from the first year's program. The first of these is that much of this data is simply not available to the POE teams or to anybody else. Despite the existence of a number of computer-based asset management systems, all POE teams had difficulty in obtaining accurate records of recurrent costs and client departments themselves were unable to provide accurate records of the operational costs of their facilities. In some cases (e.g., on defense bases) individual buildings

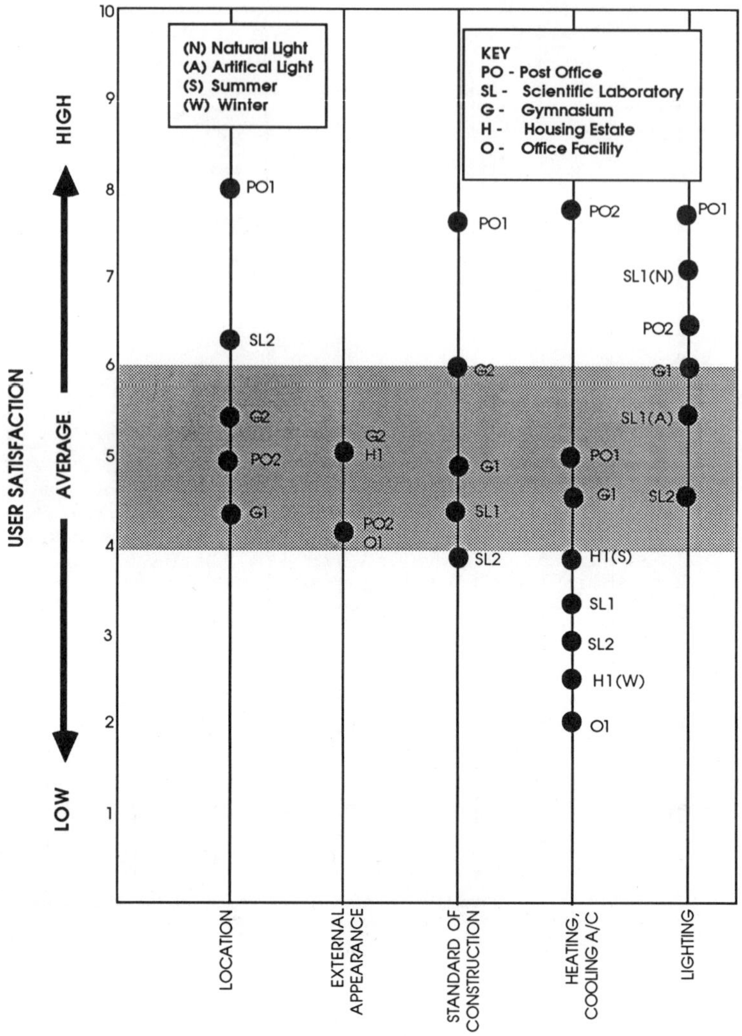

Figure 12

are not being separately monitored, nor are records being kept on an individual building basis.

As the industry generally moves towards a greater emphasis on life-cycle costs as opposed to capital costs, this state of affairs needs to be redressed. The urgency is well demonstrated by comparison of the heating and cooling costs of two gymnasiums: one for the Navy at Perth, Western Australia, the other for the Army in Canberra, ACT. The Perth gymnasium, admittedly in a milder climatic zone, uses a central mechanical plant and its energy consumption was well within normal limits for this building type. The gymnasium in Canberra uses wall-mounted fan heaters in the severe tableland winters. Installed to save the initial cost of a centralized system, the individual units have logged an astronomical annual electricity bill equal to 1.7% of the capital cost of the whole facility. If this pattern continues in the next five years of operation, it will cost 10% of capital cost to heat the building alone. Steps are now being taken in conjunction with the Defence Department to establish life-cycle cost projections for all recreational facilities, and to ensure that well-meant brief directives for lower capital costs do not result in absurd operational burdens.

Technical Performance

The assessment of facilities from the point of view of their technical performance has consistently resulted in comments on the performance of HVAC systems. The ventilation system in both gymnasiums was regarded as inadequate, quite apart from the heating costs previously noted. Zoning and performance aspects of the air conditioning were found to be problematic in the two post offices and in both scientific laboratories. The climatic performance of the defense housing project was roundly criticized, and problems with the performance of the air conditioning in the office facility studied were a major cause of user dissatisfaction prior to and during the POE project. The problems raised are not specific to the Construction Group since POEs undertaken by outside organizations regularly report the performance of HVAC systems as a major issue.

In addition, the POE projects have highlighted a number of other technical aspects of the facilities which require further attention. These aspects are specific to building type and are covered in the feedback sheets prepared for distribution to professional staff. The collection of this information, together with other technical standards previously issued for other project types, offers a firm base for improved performance in Construction Group work now in the design stages.

Fitness For Purpose

Generally there is a high correlation between fitness for purpose and user satisfaction with the eight facilities assessed in the 1986-87 POE program. The summary of user comments on the post office buildings, for example, notes that these buildings are well-regarded by users. User satisfaction with aspects of all buildings studied can be summarized by the statement that "...all buildings are well located, of good standard of construction and high standard of lighting."

These observations are also borne out in the detailed analysis of the fitness for purpose of each design. Nevertheless, a number of major issues relating to specific buildings and to specific building types have emerged from the fitness for purpose evaluation. By way of example, five key design inadequacies have emerged from the analyses of the post offices. These are:

- insufficient room for expansion;
- inadequate provision for storage;
- noise transmission problems between different functional areas;
- design and detailing of stairs and public access; and
- inconsistency in type and location of staff facilities.

The fitness for purpose aspects of the 1986-87 POEs are currently being developed into the professional feedback documents with a view to improving the quality of future designs.

Photo 1

User Satisfaction

All eight facilities in the 1986-87 POE program were assessed from the point of view of user satisfaction. The technique used a standard "Satisfaction Scale" based on the completion of a survey form containing ten bipolar descriptive adjective (e.g., adequate/inadequate, good/bad, etc.). The analysis of user satisfaction with the eight buildings involved the comparison of satisfaction rating for each facility. The values used are based on a specific request to each user to evaluate the facility overall.

Figure 11 shows the results of user satisfaction ratings of all eight facilities. Only one facility (the office building in Townsville, Queensland) is rated as below average by its users and this particular building has had a long history of problems with air conditioning performance. The POE team assessing this facility has made a particular point that this singular problem had a negative influence on the users' opinions overall. Otherwise, on the basis of these results, the Construction Group has generally provided buildings which are regarded as being satisfactory by the users, and in four of the eight cases the level of satisfaction is approaching a very high level.

The eight facilities evaluated in the 1986-87 POE program represent a number of different building types. However, in undertaking the surveys several aspects were repeated in the evaluation of each building type. Although this requirement was not applied across all eight facilities, the overlap is adequate to allow a comparison of user satisfaction with particular services across the building types. The common aspects in this group of eight evaluations is shown diagrammatically in figure 12.

Of the five aspects with common data available, only air conditioning was consistently below average. Lighting and the standard of construction were generally average to highly satisfactory. Two other aspects, external appearance and location,

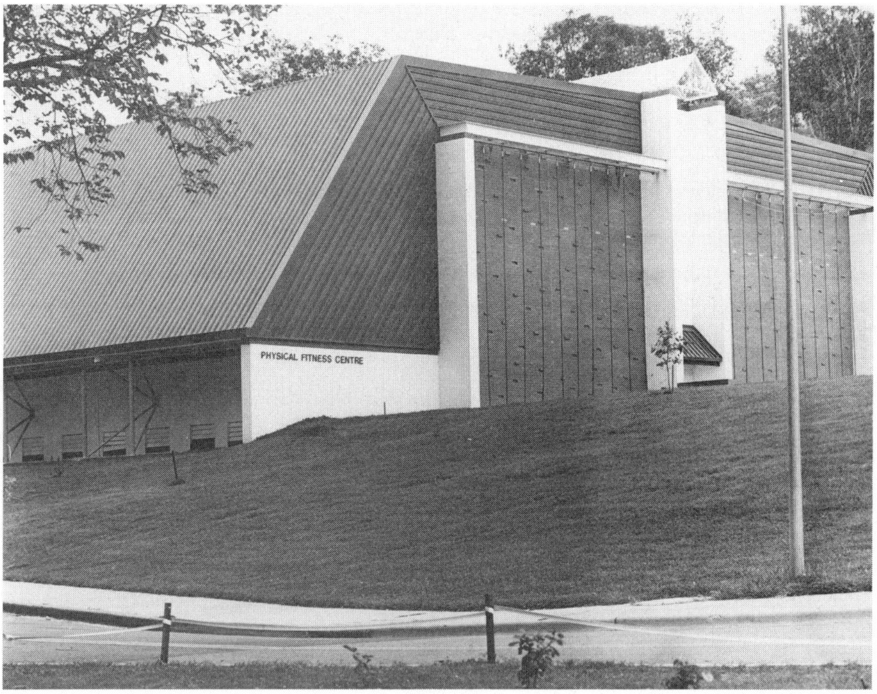

Photo 2

were common to a number of studies and are generally within the average range. Taken overall, and with the exception of HVAC systems which are marginally below average, it would seem that the Construction Group can also be well satisfied with the specific factors listed.

The user satisfaction component of this first round of POE studies has been useful in highlighting particular aspects of buildings which users have found to be unsatisfactory and which may require further attention and investigation. In particular, dissatisfaction with the thermal environment seems to be a problem for many users. As noted previously, the Construction Group is not alone in this regard. Several external studies have highlighted the recurring dissatisfaction of building users with the ability of HVAC systems to answer their requirements for comfortable working conditions wherever they might be in Australia.

FURTHER DIRECTIONS FOR POE

The original initiative for the establishment of a POE program within the Department was based on the need for an effective performance indicator for the works program. This intention was clearly reflected in the linking of POE to the existing Program Review procedures, and its identification in the Department's program budgeting proposals as a key performance indicator. However, the POE program offers more than a mechanism for assessing the finished output of the Department. It has the potential to provide direction in the following areas:

Photo 3

Market Testing

An organization producing repetitive products can improve those products through an in-use assessment of their performance. This applies to the production of buildings as much as it does to the manufacturing of products. POE provides the means for applying the same principles to the assessment of repetitive building types.

Photo 4

Brief Development

The POE program has proved to be an excellent tool for assessing client briefs and the success of particular designs. This success is reflected in client requests relating to the current (1987-88) POE program. One project currently planned for Department of Defence involves the evaluation of senior NCOs' accommodation at the New South Wales country town of Wagga Wagga. The request for the evaluation originated from the client, which is considering changing the Army Scales and Standards for senior NCOs' accommodation and were interested in basing the changes on the buildings at Wagga Wagga. The POE was seen as the tool for assessing the success or otherwise of these designs and the standards behind them.

This interest is not unique. Inquiries have also been received from the Victorian Country Fire Authority who have a number of repetitive buildings and wish to use POE to assist in the development of a standard. This role of POE is an extension of the notion of market research, and is clearly relevant to all clients who are involved in repetitive building types.

Staff Development

The POE program has potential for use in professional development where the results are used to improve the performance of staff involved in delivering the works program.

The Department has structured the POE procedures in such a way that each year different staff are trained in the POE procedures and the undertaking of the POE is part of that professional development. In debriefing sessions, post-occupancy evaluation teams have confirmed that undertaking a POE has changed the way they view their own architecture and has created a greater interest in the detail and practicality of their designs.

REFERENCES

Allan, P., and Funnell, S., July 1987, *Five Year Plans for Program Performance Reviews*, paper presented at the National Evaluation Conference of the Australasian Evaluation Society.

Baden, R., June 1987, Are You Quality Assured?, in: *The Architect*, pp. 35-36.

Bonsher, R. B. and Harrison, H. W., September 1982, Traditional Housing: a BRE Study of Quality, in: *Building Research Establishment Information Paper 18/82.*

Brauer, R. L., and Koch, M., 1981, A Method for Users to Review Facility Concept Designs, in: *C.E.R.L. Technical Report*, Illinois, p. 117.

Brauer, R. L. and Preiser, W. F. E., May 1976, Impact of Organisational Form on Identification of User Requirements in Building Delivery, in: *CIB Working Commission No. 65.*

Building Appraisal: St. Michael's Academy Kilwinning, in: *Architect's Journal* (Information Library), January 1970, pp. 10-50.

Building Performance Research Unit, December 1967, Measurement and Appraisal of Building Performance, in: *Architect's Journal.*

Canter, D. V.,and Thorne, R., March 1972, Attitudes to Housing: A Cross Cultural Comparison, in: *Environment and Behavior*, 4(1) March (3-32), 1972.

Daish, J., Gray, J., and Kernohan, D., *Post-Occupancy Evaluation of Government Buildings*, School of Architecture, Victoria University, Wellington, New Zealand.

Defects in Post-War Housing: Feedback and Prevention: Report of HTC Seminar, in: *Housing Review*, March-April 1983, pp. 63-70.

Dell'isola, A., 1982, *Value Engineering in the Construction Industry*, 3rd Edition, New York, Van Nostrand Reinhold.

Department of Housing and Construction, Canberra, 1986, *Draft POE Explanatory Manual* and *Field Manual.*

Elder, J., Turner, G., and Rubin, A. I., 1979, *Post-Occupancy Evaluation: A Case Study of the Evaluation Process*, Center for Building Technology, Washington, D.C.

Harrison, H. W. and Keeble, E. J., August 1981, Working with the Performance Concept in Building, in: *CIB Working Commission, No. 60.*

Hill, A., Spring 1986, Quality Assurance in Design, in: *Feedback Digest 54.*

Law, N. M., 1981, Evaluation of the Post-Occupancy Performance of Buildings: A State of the Art Report, in: *Experimental Building Station, 1981 Technical Record 462*, Chatswood, New South Wales.

Leslie, H. G., 1985, *Concepts of Project (Building) Evaluation: An Overview*, Canberra, National Committee on Rationalised Building.

Linard, K. T., August 1987, Management of Program Evaluation in Portfolios, in: *Practical Management Issues*, Vol. 1/1, Department of Finance, Canberra.

Markus, T. A., December 1967, The Role of Building Performance Measurement and Appraisal in Design Method, in: *Architect's Journal*, pp. 1567-73.

Markus, T. A. (Ed.), 1972, *Building Performance*, New York, Wiley.
Markus, T. A., 1973, Optimism by Evaluation in the Appraisal of Building, in: G. H. Hutton and A. D. G. Devonvald, *Value in Buildings*, Applied Science Publications, London, 1973.
National Committee on Rationalised Building, 1984, *A National System for Building Life Cycle Cost Performance Data Reading*, Melbourne.
National Public Works Conference, September 1984, *Measurement of Performance of Public Construction Agencies.*
Noble, J. and Ash, J., August 1966, Appraisal of User Requirements in Mass Housing, in: *Architect's Journal* (Information Library), pp. 479-486.
Performance Standards in Building, Draft International Standard ISO/DIS 6241, 1983.
Post-Occupancy Evaluation: The "State of the Art", in: *Research and Design*, Vol. 1, July 1978.
Royal Hospital for Sick Children, in: *Architect's Journal* (Information Library), September 1972, pp. 526-562.
Space Shuttle Challenger Incident, in: *Project Management Journal*, Vol., XVIII, No. 2, June 1987, pp. 41-68.
Standards Association of Australia, 1987, *Australian Standard 2990-1987: Quality Systems for Engineering and Construction Projects*, North Sydney.
Walters, J., December 1986, Quality Assurance Overseas, in: *DHC Technical Bulletin*.
Wells, B. W. P., Subjective Responses to the Lighting Installation in Modern Office Building and their Design Implications, in: *Building Science*, Vol. 1, 1965, pp. 57-67.

CHAPTER 18

GENERATIVE EVALUATIONS USING QUANTITATIVE METHODS: A CASE STUDY

Robert W. Marans

Ph.D Program in Urban, Technological, and Environmental Planning
The University of Michigan
Ann Arbor, Michigan

As discussed by Preiser, Rabinowitz, and White (1988) and Wener (1988), numerous advances have been made in evaluating built environments over the past 15 years. Advances in research design, methodologies and techniques, and theoretical model development have been displayed in a plethora of evaluation studies undertaken in a wide variety of settings and for different of purposes. Whereas earlier work highlighted the purpose of evaluation as providing feedback to clients and environmental designers on the impact of physical environments on occupant behavior, which could suggest directions for environmental modification (Zeisel, 1975; Zimring and Reisenstein, 1980), new purposes for undertaking evaluation in the environmental design fields have surfaced in the 1980s.

One new purpose for evaluating built environments identified by Wener is the generation of data to be used in an open-ended search for a) increased knowledge about a particular setting and b) new ideas, models, and hypotheses (1988). Generative evaluations which serve these purposes have become increasingly prevalent, particularly in efforts to increase knowledge about a particular setting as a prelude to space or building alteration (Gray, Watson, Daish, Kernohan, 1985). However, fewer examples exist of generative evaluations designed to produce data to be used to test models about particular environmental systems.

This chapter discusses a generative evaluation that produced data suitable for testing hypotheses, formulating new conceptual models linking environments to behavior, and specifying these models so as to advance theory. The generative evaluation to be discussed focuses on the particular issue of lighting in the context of office settings.

Following a brief review of past evaluative research relating lighting to behavior, conditions which prompted the evaluation of office lighting are reviewed. Next, the design for the generative evaluation is discussed, including the selection of offices and workstations, the data requirements deemed necessary, and procedures for gathering the data. Analytical steps are then outlined followed by examples of concepts and hypotheses tested with the data. Finally, model development and testing and other uses of the database are discussed.

EVALUATIVE RESEARCH ON LIGHTING – A BRIEF REVIEW

Although lighting in work places is acknowledged to be important to job performance and the well-being of workers, relatively few evaluation studies have focused on the luminous environment and human responses to it. The 1924 Hawthorne studies were perhaps the first significant attempts to determine the impact of different quantities of light on worker productivity in an actual work setting (Roethlisberger and Dickson, 1941). The results failed to demonstrate that systematically raising or reducing levels of illumination had a corresponding effect on either the performance or satisfaction of the workers. The lack of relationships between worker responses and objective conditions has been attributed to the fact that employees were aware that they were subjects in a lighting experiment.

More recently, experiments or tests of different lighting systems in an office environment have been reported by Ellis (1986). In efforts to find a lighting scheme that would substantially reduce energy costs in a large office, alternative lighting arrangements were introduced and evaluated in stages. The experiment resulted in the installation of an appropriate lighting system that was based on both objective, functional characteristics of the office lighting and on the users' attitudes toward them. Ellis notes that "these attitudes are colored by the users' perceptions of the lighting's general aesthetic qualities and also by the symbolic role the lighting plays in organizational life, in particular whether users are involved in its introduction" (Ellis, 1986:242).

Other evaluative studies of offices have examined, although not focused on, lighting conditions and peoples' responses to them (Manning, 1965; Ne'eman, Sweitzer and Vine, 1984; Boyce, 1975; Nemecek and Grandjean, 1973; Marans and Spreckelmeyer, 1981; Brill, Margulis, and Konar, 1984; Louis Harris and Associates, 1978). Each of these studies shows that characteristics of artificial lighting are associated with workers' feelings about the workspace and, in some cases, worker performance and eye problems. However, the collective research demonstrates that lighting is not the most important physical design attribute among office workers. Nonetheless, it does contribute to their well-being and, under poorly designed systems, it can adversely affect the satisfaction, health, and performance of workers. Clearly, a better understanding of the impact on people of new lighting arrangements and lower levels of illumination resulting from reduced energy loads in buildings is needed.

Background to the Current Research

Over the past 15 years, illuminating engineers, architects, and space planners have become increasingly interested in office lighting design and its impact on the office worker. The energy crisis of the early 1970s has prompted many in the design community to reconsider previously established lighting standards. In the process, planned reductions have taken place in illumination levels for offices and other types of buildings. At the same time, the emergence of new technologies, including the automated office with its problems as well as benefits, has altered the thought and activities of designers and manufacturers of lighting systems. These concerns have spawned new research in the United States and abroad which has examined the impact of lighting on people in work settings (Hedge, 1982; Ellis, 1986). More recently, illuminating engineers have shown interest in understanding relationships between lighting power density and occupant responses to the lighting in the workspace.

In late 1984, the U.S. Department of Energy (DOE) initiated a major study through the American Institute of Architects Foundation and the University of Michigan's

College of Architecture and Urban Planning. A key purpose of the study was to develop a database that could be used by researchers and standards writing organizations to better understand relationships between measures of lighting power density, photometric conditions, and occupant responses to lighting in offices. The approach was field research using methods, procedures, and orientations developed during the prior decade by architectural researchers and others as part of post-occupancy evaluation studies.

In conceiving and guiding the study, input was received from a Technical Review Committee consisting of representatives of several organizations. These included the Illuminating Engineering Society (IES), International Association of Lighting Designers (IALD), American Society of Heating, Refrigerating, and Air Conditioning Engineers (ASARE), American Institute of Architects (AIA), Building Owners and Managers Association International (BOMA), National Electrical Manufacturers Association (NEMA), National Fenestration Council (NFC), Lighting Research Institute (LRI), Oak Ridge National Laboratory, and New York State Energy Research and Development Authority.

Three broad study objectives were stated. First, the study aimed at understanding how to define the visual environment. Second, it sought to explore relationships between selected quantitative measures of lighting in occupied environments and other qualitative measures related to occupant satisfaction. Finally, it intended to examine the relationship between the quality of the lighted environment and the energy required to produce it (Gillette and Brown, 1986a). In addition to creating a database which would help in fulfilling these objectives, the study was intended to investigate the feasibility of developing a lighting quality metric or index for use in occupied office buildings.

With these objectives in mind, a study was designed involving the collection, compilation, and analysis of data from a probability sample of nearly 1,000 occupied workstations in 13 office buildings in the United States. For each sampled workstation, photometric data, lighting power density (LPD) measures, and descriptive information about the lighting system were gathered. In addition, data were obtained from the occupants of the sampled workstations using self-administered questionnaires. A third source of information from a subset of workstations was a team of professional designers that was asked to describe and assess specific workstation attributes. Prior to discussing the specific types of information gathered and the methods of gathering them, procedures used to identify and select buildings and workstations are reviewed.

Office Building Selection

With the focus of the study on lighting in office buildings, the researchers and the Technical Review Committee addressed two important questions: What particular aspect of lighting should be addressed, and which buildings would be selected for study? Although lighting is associated with several spaces in office environments including conference rooms, lounges, and hallways, lighting at the work space or workstation was chosen for detailed examination. Furthermore, interest in occupant responses as well as photometric conditions required workstations in use as units of analysis. Subsequently, procedures were developed for selecting occupied workstations and the buildings in which they were located.

As in the case of other evaluative studies covering multiple sites of the same type of environment (Zube, Crystal, and Palmer, 1976; Lansing, Marans, and Zehner, 1971; Farbstein and Wener, 1981), a judgmental sample of office buildings was selected. Several criteria were used in choosing the buildings. First, they needed to contain lighting representative of various systems currently in use in office buildings. Second, the buildings had to be geographically distributed throughout the country, or at least not concentrated in a single location. Third, variability in the quality of the overall building design as well as the quality of lighting as judged by the Technical Review Committee was required. Fourth, the buildings had to be occupied by organizations which were relatively stable. That is, spaces housing organizations which were undergoing organizational change and/or environmental change were avoided. Finally, the buildings, the organizations housed within them, and the workers within these organizations needed to be accessible. Resistance to the study on the part of personnel associated with any one would be grounds for exclusion as a candidate building.

A total of 13 buildings in 10 states were selected. These buildings were characterized by 7 major lighting systems at workstations which varied in their degree of enclosure and the extent to which they were exposed to daylighting. A summary of the main characteristics of the 13 buildings is presented in table 1.

TABLE 1
Characteristics of Sample Buildings

Building Location	Predominant Lighting System(s)	Predominant Office Type
Florida	Direct Recessed Fluorescent-Louvers Direct/Indirect Pendent Mounted	Open office with partitions
Virginia	Direct Recessed Fluorescent-Louvers Direct/Indirect Pendent Mounted	Open office with partitions
Ohio	Indirect Fluorescent-Furniture Mounted	Open office with partitions
North Carolina	Metal Halide-Direct Pendent	Private office, Open office without partitions
Washington, D. C.	Direct Recessed Fluorescent-Lens	Open office with partitions
Wisconsin	Indirect Pendent Mounted	Open office with partitions
Maryland	Direct Surface Mounted	Conventional private office
New York	Direct Recessed Fluorescent-Lens	Conventional private office
Michigan	Direct Recessed Fluorescent-Lens	Conventional private office
New York	Indirect Fluorescent-Furniture Mounted	Open office with partitions
New York	Direct Recessed Fluorescent-Lens	Conventional private office, Open office with partitions
New York	Direct Recessed Fluorescent-Louver	Conventional private office, Open office with partitions
New York	Direct Recessed Fluorescent-Louver	Conventional private office, Open office with partitions

Workstation Selection

In order to develop a better understanding of each building, reconnaissance visits were made to each site in order to lay the foundation for subsequent fieldwork. As part of a walkthrough, similarities in workstation layout and furnishings were noted. At the same time, differences were observed in the size, locations, and lighting characteristics between buildings and between floors within the same building.

To achieve variability in both workstation characteristics and occupant responses, a representative sample of occupied workstations was selected in each building. During the first year of the study, a sample of 150 occupied workstations in each of three large office buildings was selected. In large part, the number was based on the estimated number of workstations that could be observed and measured by technicians and design professionals in a 5-day period. In the second year, 10 buildings were added to the study with an average of 40 workstations per building. Workstations were selected by using probability sampling techniques. Sampling fractions varied for the 13 buildings from 1 in 6.76 occupied workstations to 1 in 1 (i.e. all workstations in the building). (See note 1 below.)

In order to facilitate data gathering, three sets of plans identifying the sampled workstations were prepared for each building. One set was used as a guide to distributing and collecting questionnaires. The second set was used by technicians to identify places where photometric and ambient environmental measurements would be made. The third set of plans guided the team of design professionals in their movements throughout the building. (See note 2 below.)

Informational Needs

Based on the objectives of the study, informational needs were determined. These needs covered occupant responses to their workstations, the responses of design professionals to the luminous environment associated with each workstation, and the actual photometric and other environmental conditions at each workstation.

Questionnaire Development and Use

A questionnaire was developed which dealt with occupants' perceptions of and satisfactions with lighting at their workstations and in other parts of the building. Questions measuring worker behaviors including the specific tasks performed on the job, the amount of time allocated to these tasks, and the amount of time spent in the building were also asked. Finally, a set of questions were designed to tap workers' feelings about their job and their health as it might be affected by the building. In preparing the questionnaire, consideration was given to the imposition placed on the office workers and the degree to which their responses could be compared with those of workers in other office buildings. Consideration was also given to the manner in which questionnaires would be administered.

A self-administered questionnaire was compiled, requiring approximately 15 minutes to complete. While it contained the same items for all study buildings, efforts were made to include questions specifically geared to the interests of management of the individual buildings. At the same time, questions used in two federal building evaluations (Elder and Tibbet, 1981; Marans and Spreckelmeyer, 1981), in two national studies of office buildings (Louis Harris and Associates, 1978) and by two Canadian consulting firms (Sterling and Sterling of Vancouver and Architectural Diagnostics of

Ottawa) were also used so that comparative data would be available to assess the representativeness of the sample.

Procedures for distributing and collecting questionnaires at each building involved several steps. First, letters were sent to occupants of all sampled workstations one week prior to questionnaire distribution. The letters described the study, its purposes, and the fact that the worker's workstations were selected using scientific sampling procedures. The letters also indicated that the occupants would be receiving the questionnaires the following week and that shortly thereafter, the workstations would be examined by technicians. Finally, the letters guaranteed the anonymity of each worker's responses and informed the occupants that their participation in the study was voluntary. These letters were reviewed and approved by the building managers who assisted in their distribution.

One week after the letters were distributed, a member of the research team, working with one set of plans, distributed questionnaires to the sampled workstations. Initially the process of distributing and collecting questionnaires was planned to take place during an eight hour day. However, the amount of time required for introductions, building orientation, and recording information necessitated two days for these activities. After the second day, questionnaires that had not been completed were identified, their workstation numbers recorded, and a list was left with the building manager who was instructed to collect all questionnaires by the end of the week and mail them back to the researchers.

Professional Designer Assessments

A second study objective was to examine lighting quality from the perspective of design professionals. Four types of design professionals were identified as relevant and used to collect information: a lighting designer, an architect, an illuminating engineer, and an interior designer. The team was assembled and worked with researchers in developing an assessment form covering their impressions of workstation lighting as it impacts on the ability of the occupant to perform work, their views on the visual environment, their ratings of specific light sources and other attributes of the workstation. As in the case of the questionnaire, the expert assessment form and procedures for carrying out the assessments were pretested and modified. The final procedures were subsequently documented in an "experts manual."

In addition to members of the professional team recording their impressions and ratings at workstations, each expert had a specific assignment. One member was designated as the team leader, responsible for determining the sequence of workstations to be visited, for informing the workstation occupant why the team was there, and for maintaining the schedule of activities while at each building. A second photographed each workstation after the team had completed their assessments. Using a 180 degree fish eye lens camera, two photographs were taken. One covered the workstation from a standing upright position; the other covered the workstation as viewed upward from the primary task location. A third team member recorded specifications for lamps associated with each workstation while the fourth recorded the presence or absence of lamps at or near each workstation.

Photometric and Environmental Measurements.

In accordance with study objectives, photometric and other environmental measurements were required for each sample workstation. Operationally, these measurements could be made directly at the buildings or indirectly using building floor plans and photographs of each workstation.

Photometric measures covered spatial luminances, work surface illuminances and luminance contrast. Specifically, these direct measurements were:

- Illuminance with body shadow at the primary work surface
- Illuminance with body shadow at the secondary work surface
- Illuminance without body shadow at the primary work surface
- Illuminance without body shadow at the secondary work surface
- Luminance contrast at 45, 90, and 135 degrees
- Contrast rendition factor at 45, 90, and 135 degrees
- Luminance at the primary work surface (white bond paper)
- Luminance of the primary work surface immediately surrounding paper
- Luminance of ceiling between luminaires
- Brightest light source in field of view
- Wall luminances at eye level straight ahead, 90 degrees to the right and 90 degrees to the left
- Brightest sky luminance in each of the four cardinal directions

Other light-related direct measurements included the degree to which daylight penetrates the workstation, the types of window covering and window glass and the types and location of supplemental task lighting and other lighting fixtures at the workstation.

Direct environmental measurements for each workstation were also made covering thermal conditions, noise levels, the type of workstation setting including predominant furniture and wall materials, predominant colors and chair style, the presence or absence of items reflecting the occupants ability to exert control over their workstations, CRT/word processing characteristics and the degree to which there were intrusive sounds.

Indirect environmental measures deemed important to the study were the size and density of workstations, the distances between workstations and the nearest glazed exterior wall, atrium, stairwell and elevator, and the window orientation of the primary work surface of the workstations. These measurements were made using floor plans supplied by building managers and the photographs taken at each workstation.

In sum, information covering workstation attributes and responses to those attributes were gathered from four sources – questionnaires administered to workstation occupants, a team of design experts, photometric and other environmental measurements made by technicians working in the field, and indirect measures of environmental conditions using floor plans and photographs.

Analytical Procedures

The compilation and coding of the data resulted in the creation of four separate data files which were subsequently merged into a database suitable for computer analysis. Several steps in the analysis process were outlined (see figure 1).

First, the process began with the presentation of basic descriptive statistics covering all the data gathered through the occupant questionnaires, the expert assessment forms, and the direct and indirect environmental measurements. These descriptive statistics, presented as means, standard deviations, ranges, and percentage distri-

ANALYSIS PROCESS	DATA SETS: OCCUPANT RESPONSES	EXPERT ASSESSMENTS	DIRECT ENVIRONMENTAL MEASURES	INDIRECT ENVIRONMENTAL MEASURES	MERGED (COMBINED) DATA
UNIVARIATE ANALYSIS (Descriptive Statistics)	X	X	X	X	
IDENTIFY KEY OUTCOMES (D.V.)	X	X			
IDENTIFY POSSIBLE PREDICTORS (I.V.) (Positing Hypotheses)	X	X	X	X	
DATA REDUCTION (Build New Variables)	X	X	X	X	
BIVARIATE ANALYSIS (Explore Relationships One at a Time)	X	X	X	X	X
MULTIVARIATE ANALYSIS (Explore Relationships Simultaneously)					X

Figure 1

butions in tabular and graphic form, served to familiarize the researchers, the sponsors, and prospective users (the technical review committee) with the data. The statistics also confirmed the initial premise that the sampled buildings and their occupants were representative of office buildings of workers elsewhere.

Following a review of the descriptive statistics, the researchers and the technical review committee identified key outcomes which could be used in understanding the luminous environment. Since quality of the luminous environment was a major theme of the overall study, two important outcomes requiring further investigation were lighting quality and visual quality – concepts used with regularity by design professionals but seldom defined and measured.

As a third step in the analysis process, hypotheses were posited suggesting the manner in which the physical environment was related to key outcomes. That is, conceptual models showing how various environmental attributes might influence each outcome were specified.

The large quantities of statistical data collected for the study was carefully scrutinized and new variables representing important concepts were constructed. The process of data reduction involved the combining of two or more variables into a single variable. As we shall see momentarily, this process was used in creating latent variables representing lighting quality and visual quality.

The next step in the analysis process involved the exploration of simple relationships between two variables for the sample as a whole or for subsets of the sample. For example, when the entire data set was analyzed, peoples' satisfaction with lighting at the workstation decreased with increased levels of illumination at the primary work surface. This bivariate analysis was also examined for workers having different lighting systems with mixed results. Whereas the relationship between illuminance and satisfaction was maintained for most lighting systems including those with furniture integrated lighting, the opposite relationship was found for workers with no task units at their workstation. That is, their satisfaction with lighting increased when higher levels of illumination (foot-candles) at their primary work surface. Eventually, relationships such as these were explored in a multivariate context. That is, several variables were examined simultaneously in trying to understand their impact on key outcomes. This last step in the analysis represents efforts to test hypotheses and conceptual models developed earlier in the analysis process.

Selected Findings

As noted earlier, one objective of the research was to develop a quality metric for subsequent use in assessing lighting in office buildings. A review of the lighting literature suggested that an operational definition of quality was indeed needed. In one article, lighting quality has been described as light meeting biological, psychological, and aesthetic needs in contrast to quantity which is intended to fulfill functional needs (Greer, 1984). McGuiness, Stein, and Reynolds (1980) define lighting quality to include all factors in a lighting installation not directly concerned with quantity. Specific items referred to are luminance ratios, diffusion, uniformity, chromasticity, uncomfortable brightness ratios where background luminance exceeds object luminance (glare), and the general notion of visual discomfort (1980: 740). The noted lighting designer, Howard Branston, has suggested that good (high quality) lighting is realized when the mood created is consistent with the function of each space, when the lighting provides spatial clarity, and when it promotes productivity (Wagner, 1985). Lam, on the other hand, simply states that a good luminous environment is one which emphasizes and highlights that which we want or need to see and downplays that which is not of interest (1977:12). Clearly, lighting quality is viewed as a concept having implications for performance (functional needs) and comfort (biological, psychological and aesthetic needs) of the individual.

Measuring Lighting Quality

In the context of the research, the multi-dimensionality of the concept of lighting quality was dealt with through a series of questions asked of the worker. That is, lighting quality was measured by combining the responses of each individual in the sample to four separate questions. Instead of assessing specific attributes of lighting such as glare and brightness, the four questions were intended to be general in nature. Subsequent analysis would determine the relative importance of the specific attributes in understanding this global evaluation of lighting quality.

The four questionnaire items were, "Overall, how satisfied are you with the lighting at your office or workspace?" "Please rate the lighting available to you for reading." "Please rate your workspace for the amount of lighting for the work you do." and "Lighting at my desk hinders me from doing my job well." For the last item, respondents indicated the extent to which the statement was true using a four point scale. Fixed response categories on four and five point scales were used to elicit responses to the first three questions. (See Note 3 below.) The four lighting questions which touch on aspects of comfort, performance, and satisfaction have an average correlation of .76 and a coefficient of reliability of .89. These figures indicate that the responses are highly, although not perfectly, related to one another in people's minds. Based upon this information, the four responses for each individual were combined to form an overall measure or index of lighting quality.

Visual Quality

Although lighting is an important factor contributing to the quality of the visual environment, it is by no means the only factor which influences our reactions to a space. Therefore, as in the case of lighting quality, visual quality was considered to have multiple dimensions which could be measured as part of the occupant questionnaire.

Respondents were asked to indicate the extent to which their workstations were attractive, pleasant, interesting, spacious, and comfortable. While some of these dimensions are purely visual (e.g.. aesthetics), others (e.g.. comfort) have physiological as well as psychological components. Nonetheless, previous research has suggested that all of these spatial characteristics are highly related statistically as well as in people's minds (Marans and Spreckelmeyer, 1981). In fact, the five items for the current study had an average correlation of .67 and a coefficient of reliability of .86. As in the case of lighting quality, responses to the five items were combined for each respondent to create an latent variable representing visual quality.

Distinguishing Between Visual Quality and Lighting Quality

Subsequent data analysis demonstrated that visual quality of the workspace is related to but distinct from lighting quality. This was done by examining relationships between each index and several other measures obtained as part of the study. As in the case of the measures used to build the two indexes, most of the selected measures were derived from the questionnaire. Other measures, however, covered actual environmental conditions at the workstation. For example, one of these "objective" measures was the type of artificial lighting system available. This measure combined the type of ambient lighting with the availability and type of task lighting. A second "objective" measure was the degree to which each workstation had been personalized. That is, the number of personal objects such as family photographs, plants, and desk memorabilia associated with each workstation was counted and recorded. Finally, a measure combining the ability to see outside and the view from the workstation was used.

It was hypothesized that some of these measures would be more strongly associated with lighting quality whereas others would be more strongly associated with visual quality. The strength of the relationship of still other measures to lighting quality and visual quality would not differ. Table 2 shows each measure and its relationship to lighting quality and visual quality expressed in terms of an eta coefficient. The higher the value, the stronger the relationship.

With two exceptions (lighting quality as it relates to personalization and to view), each of the measures is significantly related to both lighting quality and visual quality. However, measures that reflect people's assessments of specific lighting conditions are most strongly related to lighting quality (preference for improved lighting, reflective glare problem, dim/bright lights, problem with bright lights). On the other hand, attributes of the workstation such as furniture quality, the color of walls and partitions, amount of personalization and view are more strongly associated with visual quality. This analysis demonstrates that, while lighting quality and visual quality are linked in people's minds, they represent different concepts and are influenced to varying degrees

TABLE 2
Strength of Relationships between Selected Measures and Lighting Quality and Visual Quality

(Figures are eta coefficients)

Selected Measures	Lighting Quality	Visual Quality
Questionnaire Items		
Bright lights problem	.25 p<.00	.16 p<.00
Reflected glare problem	.42 p<.00	.20 p<.00
Preference for improved lighting	.58 p<.00	.09p<.01
Furniture quality	.20 p<.00	.49 p<.00
Color of walls and partitions	.16 p<.00	.48 p<.00
Dim-Bright Lighting	.60 p<.00	.45 p<.00
Soft-Harsh Lighting	.32 p<.00	.39 p<.00
Actual Conditions		
Type of Lighting System	.25 p<.00	.27 p<.00
Amount of Personalization	.00 n.s.*	.09 p<.04
View Outdoors	.24 n.s.*	.18 p<.00

* Relationships are not significant.

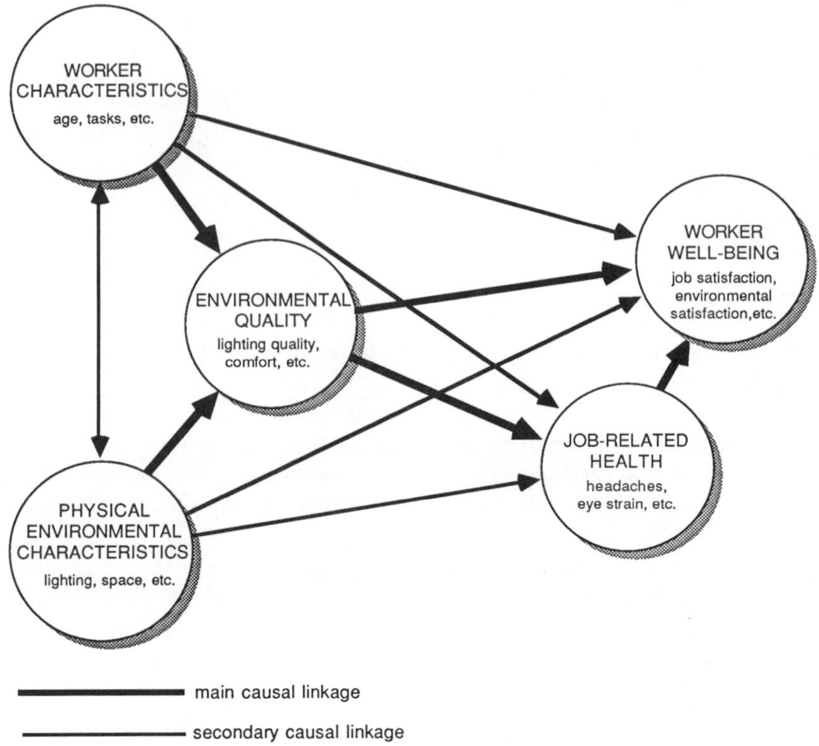

Figure 2

by different attributes of the workstation. Subsequent data analysis aims at determining the appropriate set of attributes and their qualities which contribute to high levels of both visual quality and lighting quality at the workstation.

Model Development

Figure 2 presents an overall model showing the major concepts considered in the study and their relationships. Specifically, it outlines the manner in which environmental characteristics of offices acting in concert with characteristics of workers and worker activities are linked to environmental quality assessments and ultimately to individual health and well-being at work. This general model serves as framework for the analysis and has led to the specification of detailed models showing specific predictors and key outcomes. To illustrate, the model in figure 3 suggests how lighting quality might be affected by three sets of exogenous concepts represented by measurable physical environmental characteristics, worker characteristics, lighting conditions.

The model posits a set of interrelated hypotheses that can be tested with the data collected as part of the study. Each linkage indicates a hypothesized causal relation. The entire set of causal relationships can be mathematically expressed as a set of simultaneous equations through application of the basic principles of path analysis (Markus, 1979). The data analysis estimates the value of the parameters of these equations to determine both the strength and statistical significance of the causal linkages.

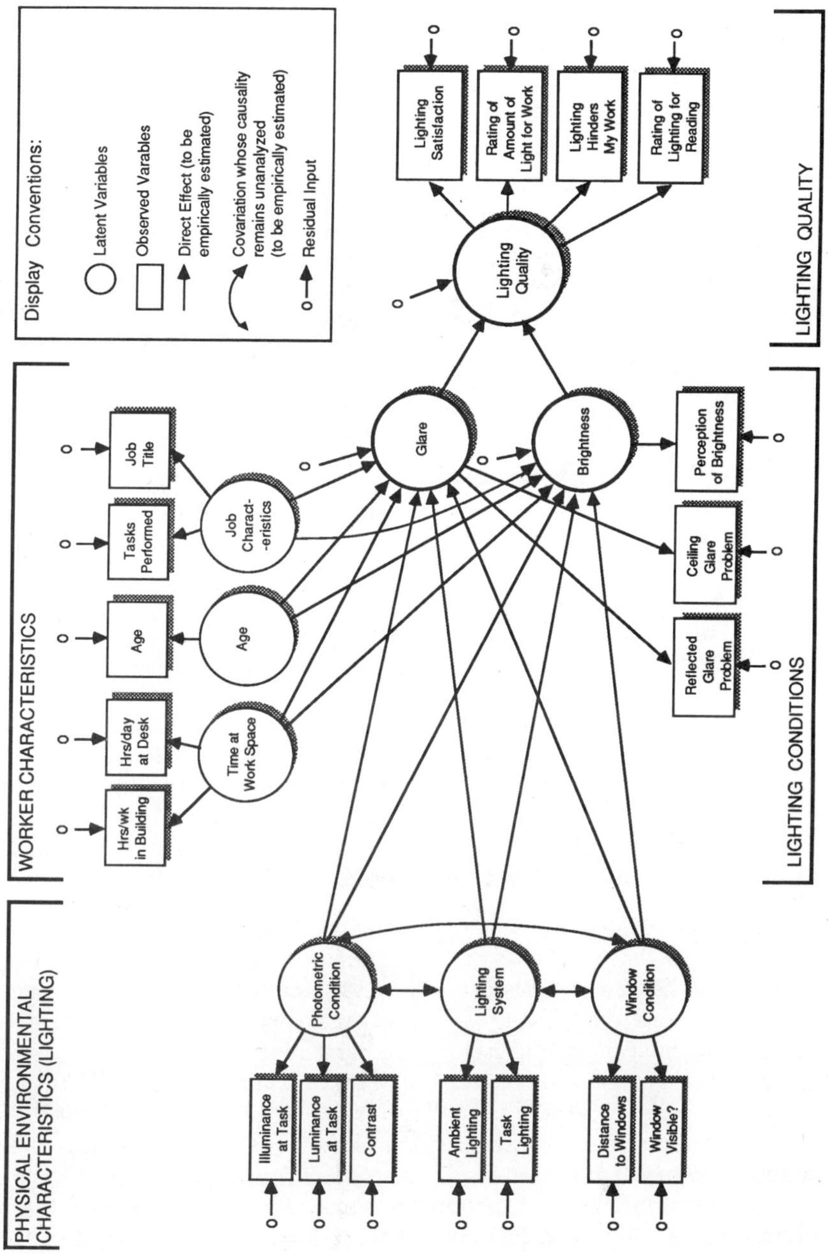
Display Conventions:
Latent Variables
Observed Variables
Direct Effect (to be empirically estimated)
Covariation whose causality remains unanalyzed (to be empirically estimated)
Residual Input
PHYSICAL ENVIRONMENTAL CHARACTERISTICS (LIGHTING)
WORKER CHARACTERISTICS
LIGHTING CONDITIONS
LIGHTING QUALITY
Illuminance at Task
Luminance at Task
Contrast
Ambient Lighting
Task Lighting
Distance to Windows
Window Visible?
Photometric Condition
Lighting System
Window Condition
Hrs/wk in Building
Hrs/day at Desk
Age
Tasks Performed
Job Title
Time at Work Space
Age
Job Charact- -eristics
Glare
Brightness
Reflected Glare Problem
Ceiling Glare Problem
Perception of Brightness
Lighting Quality
Lighting Satisfaction
Rating of Amount of Light for Work
Lighting Hinders My Work
Rating of Lighting for Reading

Figure 3

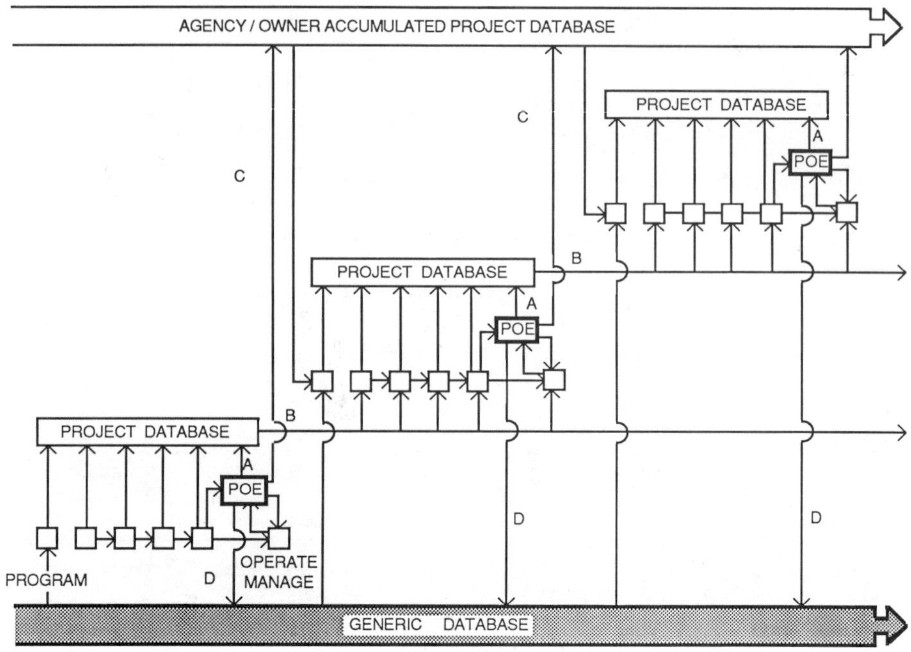

Figure 4

The primary method for estimating these parameters is the maximum likelihood fitting algorithm incorporated in the LISREL computer software (Joreskog and Sorborn, 1964). LISREL is a relatively new and powerful analytical tool that is gaining widespread acceptance and use in the social sciences and is ideally suited for analyzing data such as those collected as part of the lighting study. Several models such as that shown in figure 3 have been specified and are currently being tested with the data at hand.

Further Uses of the Data

At the onset of the office lighting study, the sponsors and researchers agreed that the data would be made available to specification writers and to other researchers interested in commercial offices and lighting in particular. Indeed, a database and a supporting database management system suitable for use on an IBM PC/XT/AT or compatible machine using MS-DOS 2.0 (or greater) are available through the National Bureau of Standards (Gillette and Brown, 1987b). At the same time, the data are available for mainframe use from the author at the University of Michigan.

It was also envisioned that this generic database could be expanded in the future by incorporating information about occupied workstations derived from post-occupancy evaluations of other commercial office buildings. That is, data from environmental measurement and occupant questionnaires covering workstations in other office buildings could be added to the database and analyzed in conjunction with the initial data. The concept of a generic database has been discussed as one of three types of databases emanating from post-occupancy evaluations in a recent National Academy of Science report (Building Research Board, 1987).

The generical database, shown in figure 4, is defined as one containing both raw data covering buildings and building occupants, and information on how buildings and their attributes interact with occupants. It would be expanded over time as data and general information (i.e. findings) from new post-occupancy evaluations become available.

Several constituent groups might use the generic database. One group might be design professionals who, in programming and designing a new facility, seek background information to aid them in their work. Other potential users of generic databases are architectural researchers interested in understanding buildings, their occupants, and the interrelationships between them. Secondary data analysis has occurred with regularity in the social sciences and, with the availability of databases derived from post-occupancy evaluations, it could become an increasingly important activity among architectural researchers.

The maintenance of a generic database could be under the domain of a professional organization such as the American Institute of Architects or the International Facility Management Association, a governmental agency such as the National Bureau of Standards (NBS), or a university affiliated data archiving organization. As noted, the database of commercial offices is available at NBS and is expected to expand as new NBS evaluations are conducted.

SUMMARY

This chapter has presented an overview of a generative evaluation covering office lighting. Unlike other generative evaluations which are designed to spark new ideas to aid in the planning or programming process, this evaluation produced large quantities of data that enable researchers to posit numerous hypotheses which can be linked together and tested using multivariate statistical techniques. Furthermore, the data produced by this evaluation are available to researchers interested in testing ideas about office lighting and other attributes of the workplace.

The chapter has emphasized the overall design of the study including the selection of buildings and workspaces, data requirements in order to achieve the study objectives, and the multiple data collection methods employed. The manner in which data analysis was carried out was then outlined and two examples of concept development (i.e. lighting quality and visual quality) were presented. Finally, analytical models representing linkages between attributes of the workstation and key outcomes were discussed.

NOTES

1. Because differential sampling fractions were used in the 13 buildings, subsequent data analysis required the weighting of the data inversely proportional to the sampling fractions. A discussion of weighting is presented in Marans, 1987.
2. The team of design professionals was not used in data collection during the second year of the study. Consequently, their views are represented on approximately half of the total number of workstations.

3. A fourth item indicating the extent to which workers wanted improvements in their lighting was also considered to be an indicator of lighting quality but was not used in the construction of the overall lighting quality index for several reasons. First, the question was confounded since it involved making choices between lighting and other workstation improvements. Responses to the lighting item intrinsically were linked to feelings about other parts of the workstation. Secondly, although the measure was correlated with the four other indicators of lighting quality, its association was not as strong.

REFERENCES

Boyce, P. R., 1975, The Luminous Environment, in: D. Canter and P. Stringer (Eds.), *Environmental Interactions: Psychological Approaches to Our Physical Surroundings,* New York, International Universities Press, pp. 81-124.

Brill, M., Margulis, S. T., and Konar, E., 1984, *Using Office Design to Increase Productivity,* Volume 1, Buffalo, Workplace Design and Productivity, Inc.

Building Research Board, 1987, *Post-Occupancy Evaluation Practices in the Building Process: Opportunities for Improvement,* Washington, DC:, National Academy Press.

Elder, J., and Tibbet, R., 1981, *User Acceptance of an Energy Efficient Office Building: A Study of the Norris Colton Federal Office Building,* Washington, D.C., National Bureau of Standards.

Ellis, P., 1986, Functional, Aesthetic, and Symbolic Aspects of Office Lighting, in: J. Wineman (Ed.), *Behavioral Issues in Office Design,* New York, Van Nostrand Reinhold, pp. 225-250.

Farbstein, J. and Wener, R. E., 1982, Evaluation of Correctional Environments, in: *Environment and Behavior,* 4, 671-695.

Gillette, G. and Brown, M., 1986a, *Occupant Evaluation of Commercial Office Lighting: Volume 1, Methodology and Bibliography,* Oak Ridge National Laboratory Report TM-10264/VI.

Gillette, G. and Brown, M., 1986b, *Occupant Evaluation of Commercial Office Lighting: Volume III, Data Archive and Database Management System,* Oak Ridge National Laboratory Report TM-10264/V3.

Gray, J., Watson, C., Daish, J. and Kernohan, D., 1985, in: S. Klein, R. Wener, and S. Lehman (Eds.), *Environmental Change/Social Change: Proceedings of the 16th Annual Conference of the Environmental Design Research Association,* Washington, DC, EDRA, pp. 275-287.

Greer, N. R., 1984, Lighting Design: State of the Art, in: *Architecture,* 73(10), 64-67.

Hedge, A., 1986, Enclosed Workspaces: The Impact of Design on Employee Reactions to Their Offices, in: J. Wineman (Ed.), *Behavioral Issues in Office Design,* New York, Van Nostrand Reinhold, pp. 139-176.

Joreskog, K. G. and Sorborn, O., 1986, LISREL - Analysis of Linear Structural Relations by the Method of Maximum-Likelihood, Users' Guide, Version VI. Chicago, National Educational Resources.

Lam, W. C., 1977, *Perception of Light as Formgivers for Architecture, New York,* McGraw-Hill.

Lansing, J. B., Marans, R. W. and Zehner, J. B., 1970, *Planned Residential Environments,* Ann Arbor, MI,Institute for Social Research, University of Michigan.

Louis Harris and Associates, 1978, *The Steelcase National Study of Office Environments: Do They Work,* Grand Rapids, MI, Steelcase.

Louis Harris and Associates, 1980, *The Steelcase National Study of Office Environments, No. 11: Comfort and Productivity in the Office of the 80s,* Grand Rapids, MI, Steelcase.

Manning, P. (Ed.), 1965, *Office Design: A Study of Environment by the Pilkington Research Unit,* Liverpool, University of Liverpool Press.

Marans, R. W., and Spreckelmeyer, K., 1981, *Evaluating Built Environments: A Behavioral Approach,* Ann Arbor, MI, Institute for Social Research, University of Michigan.

Markus, G. G., 1979, *Analyzing Panel Data,* Beverly Hills, CA, Sage.

McGuiness, W. J., Stein, B., and Reynolds, J., 1980, *Mechanical and Electrical Engineering for Buildings,* New York, John Wiley.

Ne'eman, E., Sweitzer, G. and Vine, E., 1984, Office Worker Response to Lighting and Daylighting Issues in Workspace Environments: A Pilot Study., in: *Energy and Buildings,* 6, 159-171.

Nemeck, J. and Grandjean, E., 1973, Results of an Ergonomic Investigation of Large-Space Offices, in: *Human Factors,* 15:111-124.

Preiser, W. F. E., Rabinowitz, H. A., and White, E. T., 1988, *Post-Occupancy Evaluation,* New York, Van Nostrand Reinhold.

Roethlisberger, F. J., and Dickson, W. J., 1941, *Management and the Worker: An Account of a Research Program Conducted by the Western Electric Company, Hawthorne Works, Chicago,* Cambridge, MA, Harvard University Press.

Wagner, W. F., 1985, Lighting: An Art Supported by Technology, in: *Architectural Record,* 173(4), 156-163.

Wener, R. E., In press, Advances in the Evaluation of the Built Environment, in: E. H. Zube and G. T. Moore (Eds.), *Advances in Environment, Behavior, and Design: Volume 2,* New York, Plenum Publishing Corp.

Zeisel, J., 1975, *Sociology and Architectural Design,* Russell Sage Social Science Frontiers Series, No. 6, New York, Free Press.

Zimring, C. M. and Reisenstein, J. E., 1980, Post-Occupancy Evaluation: An Overview, in: *Environment and Behavior,* 12:429-451.

Zube, E. H., Crystal, J. H., and Palmer, J. F., 1976, *Visitor Center Design Evaluation,* Amherst, MA, Institute for Man and Environment, University of Massachusetts.

CHAPTER 19

BIG BUILDINGS: HOW THEY CHALLENGE EVALUATION THOUGHT AND PRACTICE

Francis T. Ventre

College of Architecture and Urban Studies
Virginia Polytechnic Institute and State University
Blacksburg, Virginia

INTRODUCTION

The size distribution of the non-residential building stock in the United States is skewed, with a very large number of small buildings and a small number of very large ones. As figure 1 reveals, just two percent of the nation's nonresidential buildings account for 32% of the nation's nonresidential floor area; five percent of the buildings account for 47% of the floor area (U. S. Department of Energy, 1981). This conformation of the nation's physical inventory of buildings – of which many building professionals and their research cohorts are but dimly aware – will be the primary influence on the evolution of the theory, method, and practice of both "post-construction evaluation" and "post-occupancy evaluation" (PCE/POE), transcending the otherwise useful distinction drawn by Anderson and Butterfield (1980). This influence will be pervasive, affecting what attributes and effects are to be measured, to whom the measurements are to be reported, and how the measurements are to be made. Moreover, many of the building evaluation practices and perspectives employed to date and oriented to custodial environments provide little precedent or preparation for what is now required for evaluating discretionary or entrepreneurial environments. Herein lies the challenge to evaluation thought and practice.

These challenges are induced only in part by the size of the buildings themselves: the buildings, after all, are but an effect of more basic changes in what Knox (1987) describes as the "social production of the built environment." These larger structures are often units of multibuilding inventories or of investment portfolios, collectivities that span regions and continents and that now cross national boundaries. This last is no exaggeration, either: two of the largest multicity developers in the United States are based in Canada and "flagship" properties in America's most glamorous cities are in foreign ownership (*National Real Estate Investor*, 1987). This "globalization" is a remarkable change for an industry – real property development – that until just yesterday was the preferred vehicle for independent (even maverick) operators who plied a strictly local market. These inventories, and the larger structures within them, tend now to be professionally managed and operated against specific control regimens that are, themselves, increasingly universal, abstract, and calculable. The "control revolution" (Beni-

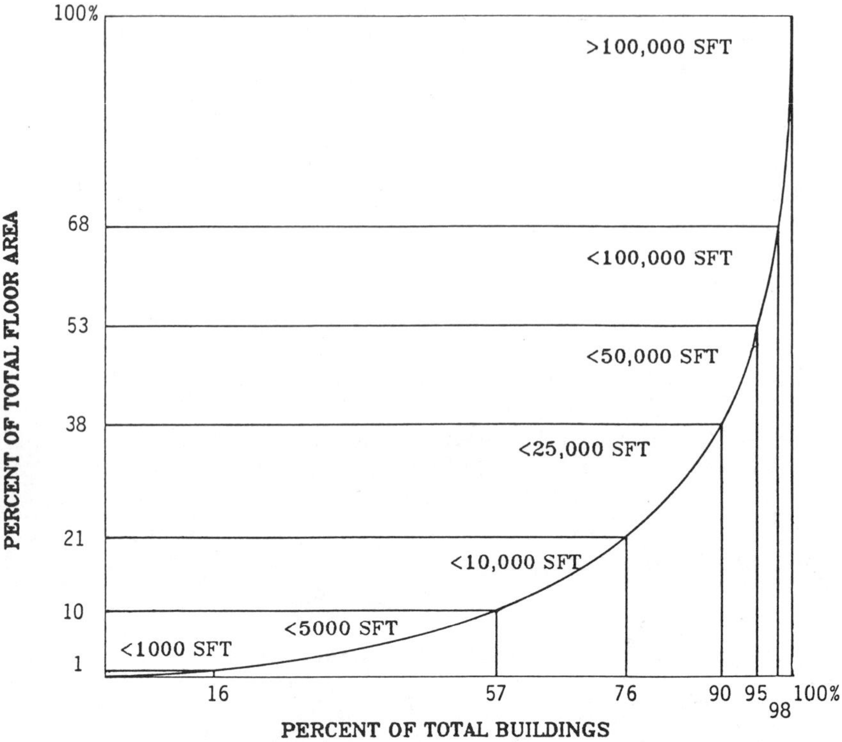

Figure 1. Cumulative distribution of nonresidential buildings by size class.

ger, 1986) that ramified through the world's industrial economies earlier in this century is now coursing through the building industry with consequences for the evaluation movement.

Controls are made universal, abstract, and calculable to facilitate the more effective deployment of the financial assets of the larger economic entity that owns the assets. The managers at higher levels of responsibility are able, through superior knowledge of investment alternatives, to substitute higher yield investments for lower yield ones allowing for greater capital accumulation. To consider buildings, objects so long celebrated as being highly differentiated and even unique, as substitutable or fungible may offend cultural sensibilites and, specifically, the received view of architecture. (I recall wincing when I first heard real estate professionals referring to their respective buildings as "products".) But it is happening all around us, an instance of the "commodity fetishism" that Karl Marx (*Das Kapital*) warned of over a century ago.

This substitutability of specific units in a physical inventory results from the increasingly precise language for characterizing the economic performance of investment vehicles. Contemporary commerce and industry, the entities that own and manage the assets, extends its linguistic hegemony and intellectural influence even into the realm of environmental design research. It is the owners and operators who manage the assets that have succeeded the designers as the constituency for evaluation studies. This is not what the pioneers in environment-behavior research had expected, as Robert Gutman attested in his remarks on receipt of the 1985 Environmental Design Research Association (EDRA) Award (*Design Resarch News*, July/August 1985). As a consequence, the research priorities and field methods developed for the initial PCE/POE work seem now to be quaintly out of scale with the environments and out of sympathy with the interests of the entrepreneurial owners and managers likely to commission formal evaluations in the future.

The governments and eleemosynary institutions that sponsored the pioneering evaluation studies and whose custodial settings continue to draw the attention of the methodologically more venturesome researchers find themselves in reduced political circumstances and facing tighter budgets, leaving fewer resources for evaluation efforts. But the traditional ideology of the PCE/POE movement persists, reflected the deep-seated ethical committments of the researchers to:

> "foster 'innovative forms of life-praxis' that would, in turn, foster self-determination and 'help to keep large-scale political organs responsive to local situations.'" (Knox, 1987, citing Knesl, 1984).

Take, for an illustration, the question of who "owns" an evaluation. Shibley and Schneekloth (1988-89) assume a figurative meaning of the term and propose as owners those involved in or occupying the setting being evaluated; Sommer (1984) assigns ownership to the "questioning community" of those who can challenge and change their own setting. These sentiments resonate still with the emanations of another time, that of People's Park, and other building types, such as public housing projects, correctional facilities, and health care settings: these are the domains of the dependent who are under one or another form of social custody. But these kinds of environments are a smaller fraction of the inventory of physical assets than previously (Bowlby and Schriver, 1984; Ventre, in press). Look now to the entrepreneurial or private sector for a contrast. There the idea of ownership retains its common-sense meaning and extends even to intellectual property like evaluation findings themselves. As an illustration of this, consider the selective promulgation of the results of the mammoth BOSTI (1981) study on the effects of office environments on productivity of the workers sheltered therein.

New Uses for Evaluation Research

Owners and operators of industrial and commercial facilities face intense competitive pressure from domestic and overseas rivals. These owners and operators seek greater effectivness for the one quarter of their total corporate assets devoted to land and buildings (Zeckhauser and Silverman, 1982). This has created two kinds of opportunities for PCE/POE researchers. Private firms seeking new facilities are willing to undertake more deliberate and thorough pre-design planning. In the words of an executive of the fifth-ranked engineer/architect firm in the United States (only one tenth of whose revenues derive from institutional or custodial clients), "We find that clients are taking . . . productivity seriously. . . . These buildings are truly designed from the inside out." (*Building Design and Construction*, July 1988) The second opportunity lies with firms that need no new buildings but, responding to the same competitive pressures are reducing costs by downsizing their in-house corporate facilities staffs and turning instead to specialty consultants (*Engineering News-Record*, 1988).

The opportunities cited occur particularly at the larger end of the building-size distribution illustrated in figure 1, the market where professional design and consulting firms earn their livlihoods. And it all hangs together: larger buildings, professional management, formal control regimens, portfolio-wide economic investment criteria, inventory-wide performance standards, professional evaluation services. Are the building evaluators equipped to assist? The evaluation pioneers, and their students who are, themselves, now imparting skills and attitudes to another generation of protegés, may find commercial and industrial environments exotic territory indeed. In partial consequence of this, evaluators manifesting a less-pronounced social agenda have come forward. Professional designers, both consulting engineers and architects, are increasingly retained to evaluate the performance of both older existing and newly-constructed buildings. A veritable movement, which the National Research Council's Building Research Board (1986) labelled "building diagnostics," has emerged to deal with this pervasive concern. Large diagnostics efforts have been undertaken by corporate owners in government and industry and by fiduciary institutions contemplating investments in real property assets. Indicative of the breadth of the American building industry's concern and need for reliable measures of both the performance of buildings and the serviceability of facilities is the decision by the American Society for Testing and Materials (ASTM) Committee E-6 on Performance of Buildings Contructions to establish Subcommittee EO6.25 on Overall Performance of Buildings (Davis and Ventre, in press).

To be sure, environmental design researchers – psychologists, sociologists, and architects prominently among them – have pursued building performance assessments for upwards of 20 years (Moore, 1987). But the published behavioral science-based evaluations manifest two limitations that have diminished their utility to the professional managers responsible for the lion's share of the Nation's nonresidential building inventory. First, these studies, for reasons adduced by Marans and Ahrentzen, have slighted "collecting data about environmental reality" (1987, p. 261). This neglect is the more surprising, given the large growth of interest the representation of reality both in science studies (Latour, 1986) and in design theoretic circles (Hillier and Leaman, 1974; Benedikt and Burnham, 1980; Akin and Weinel, 1982; Crowe and Hurtt, 1986;) and in the greater than three-fold growth in the literature comparing the "physical qualities of a building with some acceptable physical reference" since Mattar and Fazio's (1973) bibliography (Johnson and Bosworth, 1987). The second shortcoming is that published studies have addressed environments so small that exhaustive (i.e. "100%") surveys were feasible. But the buildings under professional management that the emerging building diagnostics movement now addresses are too large for exhaustive survey because of the great outlay of skilled effort required. How, then, to deal with large buildings?

Measuring physical performance. The problem of whole building post-construction appraisal persists despite the existence of fairly precise and amply documented evaluation methods and standard test procedures for single aspects of building performance, for example: office lighting; thermal conditions; visual task lighting; structure, heating, ventilating, and air conditioning (HVAC), plumbing, and electrical systems. (A representative listing is included at the end of the reference section of this chapter.) But whole building assessments remain a problem because (a) the standard methods address performance within single zones or compartments: and (b) the selection of single performance attributes for measurement typically occurs without a preceding and systematic whole building survey. But until a valid and reliable whole building survey has been completed, the significance of the single performance aspects cannot be confidently established. The current situation, that of measuring single attributes in single spaces, puts second things first and is, in the literal sense of the word, preposterous (pre-post-erous) and analogous to a physician ordering neurosurgery before the first-time patient's medical history has been elicited.

Buildings, large and small, are usually examined by means of "judgment sampling," that is observers select for measurement a finite (and usually small) number of spaces from within the building as a whole. Rarely is the selection completely random: since many assessments are undertaken in response to specific complaints, examiners tend to converge on the complainants' locations within the building. Yet randomness is the *sine qua non* of representative sampling (Kalton, 1983). Clearly, aggregate methods are needed for preliminary performance assessments to determine when the specialized evaluation methods (such as those cited above) should be invoked.While production engineering has long employed statistical procedures of product sampling for quality control, counterpart methods for assessing the spatially distributed performance attributes of large buildings need further development. The remainder of this chapter describes how such counterpart methods might be developed for use by the building community and provides criteria by which candidate methods may be evaluated for appropriateness.

A SAMPLING APPROACH

The problem confronted in the performance assessment of buildings is, in many ways, similar to the problem of product acceptance. In both cases measuring all items is infeasible. This is an opportunity for (a) rational sampling, and (b) inference from sample observations to the total population. In building performance sampling, however, there are three significant differences: (a) the building universe to be sampled (i. e., the spaces in a large building) is, itself, spatially distributed: (b) overall building performance is based on the simultaneous evaluation of several factors; and (c) the evaluation of those factors are strongly affected by temporal (e.g. time of day and season of year) considerations.In light of these differences, statistical quality control techniques used for product acceptance need be adapted (and not merely adopted) to the specialized problem of measuring building performance, facility serviceability or occupant satisfaction.

Paraphrasing the language of product acceptance: The building may be considered a "lot," a number of things. Subdivisions within the building – whether floors, structural bays, air-handling zones, or arbitrary (but fixed) space-units – are items within this lot. Some of these units are sampled and inspected. Based on inspection reports and previously specified acceptance criteria, a decision is reached: either the whole lot is accepted or it is not. On the other hand, items (or space-units), may deviate from the test criteria for different attributes: e.g., one space unit is rejected on thermal comfort grounds, another on acoustical privacy criteria. Multivariate analysis may be used to combine the information on these two types of deviations: the combined score versus the single scores. Similar judgments may be made about larger spatial subdivisions of the total building, but not about individual attributes across the whole building.

Procedures for measuring building performance and/or facilities serviceability and/or occupant satisfaction would need to be applied in rapid succession at selected points (sample values) within the building. Repeated use of such measurement can generate a time series of performance, serviceability or satisfaction at a sample point. Aggregating the sample point values into an estimated distribution over the entire building would be the principal objective of the sampling effort. The principal components of the sampling procedure are now named and their specific intellectual challenges briefly described.

Elements of a Sampling Procedure

Strata definition. Both building researchers (Marans and Ahrentzen, p. 262) and architectural historians and theorists (Journal of Architectural Education, *passim*), though each for different reasons, have realized great cognitive economies by typologizing. Statistical sampling in large buildings is also drawn to the definition of significant subdivisions as a first step because it would ensure that different types of environments will have a higher probability of being represented in a sample. Richer, more subtle environmental descriptors make it possible to stratify environments in many different ways. The fineness of the stratification reflects, of course, the purposes for which the building is to be surveyed and the theoretical or technological orientation of the investigator: an environmental psychologist may choose to partition a building of the basis of the tasks being performed or the characteristics of the occupants themselves; a mechanical engineer may need to segregate perimeter from interior workstations.

The validity of any stratification scheme is established when it can be shown that variation in the attributes being measured are smaller within strata than between them. Where this criterion is not met, then strata may be – should be – collapsed. This can only be done after measurements and preliminary data analysis are completed.

Multicharacteristic estimation. Building perfomance evaluation involves the simultaneous assessment of many environemtal attributes. These are not independent of each other. Ventre (1983, table 2.1) summarized Canadian, British and American efforts to isolate a smaller number of performance attributes to represent completed buildings. The largest such number was seven. Hence it would be necessary to estimate the joint distribution throughout the building of up to seven characteristics rather than the distribution of each characteristic individually. Naturally, the larger number of attributes considered simultaneously, the more confidence we may have in the evaluation's validity. The "open-endedness" of this concept argues for a robust statistical methods of a general character. That's on the empirical side. On the theoretical side, the interactive simultaneity of environemental effects must proceed apace and Wener (in press) reports movement by Winkel and by Stokols in this direction.

Error management. Statistical sampling, while offering advantages of economy and efficiency, entails error. That is, we expect a systematic discrepancy between values generated from sampled observations and those for the whole building. Responsible users of sampling methods always state forthrightly the size of the error and their confidence in the resulting estimate. But what is an acceptable sampling error for a building evaluation: 5%, 10%, 20%? All are achievable providing one is willing to pay for larger samples. This issue, like all questions of measurement quality, cannot be resolved *a priori* for, as Aristotle advised, quality in measurement inheres not in the measurement itself, but in the fitness of the measurement to the purpose for which it is intended. If a building's performance is being addressed, then the precision of the measurements (including the allowable sampling error) must accord with the resolution, so to speak, characteristic of the technologies employed in delivering that level of performance or service or derived from the perceptual acuity of the building's occupants. The designers of lighting, for instance, stipulate target values – the quantity and quality of lighting on a visual task, for instance – and are willing to accept a limited deviation above or below the target values. Target values and tolerances around them are either expressly stipulated in technical specifications or they may be inferred from the technical descriptions of the service systems incorporated in the building. The quality of measurements required for gauging occupant satisfaction, on the other hand, must accord with the perceptual acuity of the building's occupants (Halldane, 1978). With the foregoing in mind, we can state that the optimal sampling error must be no greater and should be no less than those tolerances characteristic of the pertinent technologies and human perceptual systems they stimulate. Optimality is at issue because the commercial purchasers of evaluation services will pay for no more precision than they can use. For instance, a sampling error of +/- 10% may be too coarse for lighting measurements but too fine for acoustical measurements. Photometry and visual psychobiology are relatively exact sciences employing elegant instrumentation. The acoustician, on the other hand, will have paid too much for the survey because it generated data of a resolution too fine to be achieved with present day design and construction practices.

Recording physical performance. As mentioned, methods for measuring single attributes in single spaces are well developed for both instantaneous readings and for measurements integrated over an interval of time. Several vendors provide meters that report out single figures-of-merit that integrate measures of thermal stress, environmental acoustics, or visual discomfort. These are marketed to owners and regulatory officials seeking rapid means for accepting or rejecting ambient environments and are suitable for rough diagnostics. But researchers seeking to devise their own indices, or, equally important, testing the validity and reliability of indices in wide use, need to devise their own means of simultaneously measuring and electronically recording several environmental attributes. Whatever means are chosen must be mobile so that set-up time in the field can be minimized. All building owners and managers – entrepreneurial or custodial – are likely to be more accommodating to evaluators who operate inobtrusively.

Pertinent indicators of building performance, facility serviceability, and occupant satisfaction. Improvements in measurement are usually associated with advances in science (and vice versa); little appreciated – except by historians of metrology – is how much the demands of commerce have influenced the evolution of our measurement system (Ventre, in press). Who decides what attributes of environments need to be better characterized? I nominate the user of measurements and not their supplier: who, after all, is in the better position to identify what measures are needed? Again, remember Aristotle's advice: the quality of a measurement inheres not in the measurement itself, but in the fitness to the purpose for which it was intended. It is the users of measurements, the managers of facilities, owners, designers and occupants who are in the advantaged position.

PUTTING SAMPLING TO WORK: A CASE STUDY

Researchers at Virginia Tech seeking analytically tractable methods for statistically representing large buildings, enabling more rigorous, reliable and repeatable assessments of their spatially-distributed performance attributes, developed and are field-validating many of the ideas presented thus far. Faculty and students from the College of Architecture and Urban Studies and the College of Engineering identified and selected procedures of product sampling for quality control and combined them with standard procedures used heretofore for measuring single aspects of building performance in single spaces.

The project measured 11 performance characteristics at 596 workstations in the

A. CONFIG. TYPE		B. PLAN TYPE	C. SPATIAL TYPE	D. DEPTH TYPE
1. Internal Service Core 2. Perimeter Service Core	OPEN	1. Open Layout	1. Open without Screen 2. Open with Screen	1. Very Deep - Over 65 ft. 2. Deep - 35 ft to 64 ft.
3. External Service Core	ZONE	2. Single Zone 3. Double Zone 4. Triple Zone	3. Cellular 4. Group	3. Medium - 20 to 30 ft. 4. Shallow - Less than 19ft.

E. BOUNDARY TYPE	F. SCREEN TYPE	G. ENCLOSURE TYPE	H. FLOOR COVERING
Perimeters 1. North 2. East 3. South 4. West 5. Northeast 6. Southeast 7. Southwest 8. Northwest 9. Interior	1. Less than 4 ft./No gap 2. Less than 4 ft./4" gap 3. 4 to 5 ft./No gap 4. 4 to 5 ft/4" gap 5. Over 5 ft/ No gap 6. Over 5 ft/4" gap 7. No screens	1. One side 2. Two sides - continuous 3. Two sides - non continuous 4. Three sides 5. Four sides 6. Non glazed wall with 1 screen 7. Non glazed wall with 2 screens 8. Non glazed wall with 3 screens 9. No screens	1. Carpet 2. Resilient 3. Others

Figure 2

210,000-sq. ft. high-rise portion of the 13-story Richard H. Poff Federal Building and Courhouse in Roanoke, Virginia between July 6 and 14, 1987. (The dates were chosen to capture the effect of the routine week-end shutdown). Each workstation was described by its spatial location and its immediate surround classified according to the schedule shown in figure 2. The characteristics measured were: lighting (both luninance and illuminance); acoustics (A- and C-weighted sound pressure levels); thermofluid conditions (air and radiant temperatures, relative humidity and air movement at 4" and 42" above the floor); and indoor air quality (CO and CO_2). The measurements were completed within 30 to 60 seconds at each workstation, resulting in minimal disruption of occupants' tasks. The specially-built mobile measurement and recording device illustrated in figure 3 made this speed possible. (The Poff Building occupants, with evident good humor, referred to the the data collection team as the "Ghostbusters.")

Tentative results and future plans. Preliminary analysis indicates that large buildings may, paradoxically, be "in" and "out" of control at the same time. This may account for the discrepant reports from building occupants and building managers about the adequacy of a building's performance. Most of the 11 attributes were found to vary from one location to another across the building but none of the variation could be associated with the specific strata derived from the environmental descriptors listed in figure 2. This suggests that a simple "random walk" may be an effective device for quickly characterizing large buildings. One stratification rule was found serviceable: time of day, suggesting that temporal sampling (time of day, season of year) approaches may be worth pursuing.

Although variation was found across the building, the performance characteristics at any single workstation were highly correlated. These results, and the absence of explanation other than the principles guiding the designer, indicate the desirabilty of simultaneous observations of all characteristics and their joint evaluation. Work on the method of "multicharacteristic evaluation," i.e. simultaneous evaluation of a large number of quality attributes, must continue. The project team is working on the development of a single evaluation statistic that combines information about all characteristics monitored at a single point. Theoretical insight will likely set the pace of that development.

A word to colleagues. To collect reliable performance data in a fully-functioning office building during a typical workday requires not only the ingenuity of field workers but also the goodwill of building occupants and managers who are, after all , fully occuppied by and committed to their own pursuits. Consequently, care must be taken to keep all project participants fully apprised of the building occupants' sensitivity and curiosity. Not only must measurement methods and instruments be minimally disruptive, the research team must be as inobtrusive as possible without appearing stealthy. Selection of instruments and the design of data collection methods must show a committment to inconspicuousness.

Generalizability and standardization. Although spontaneous complaints from occupants do occur, the Poff Building whose performance is described here may be an exceptionally well-engineered and well-operated building, making generalizations suspect. This preliminary report is offered not to broadcast results as much as to recruit other evaluators to replicate our methods in office buildings designed and operated differently. The project's internal reports provide copious detail on the methods used and are available. The project has already reported early results to ASTM Subcommittee 06.25 on Overall Performance of Buildings to accelerate the development of standard methods for conducting evaluations of this type.

Your participation in the Subcommittee's work is earnestly solicited.

ARE BIG BUILDINGS "DIFFERENT"?

This chapter began with an assertion that the great size and professional management of the buildings in the Nation's nonresidential inventory portends a shift in the orientation of the PCE/POE specialists. The professional managers of buildings built for entrepreneurial purposes will have greater influence in the evolution of PCE/POE practices and the methods and instruments employed. Is this surprising? Not really, for although improvements in measurement are usually associated with advances in science (and vice versa), it is generally little appreciated – except by historians of metrology – just how much the demands of commerce have influenced the evolution of our measurement system (Ventre, in press).

The first would be some agreement on what attributes of environments need to be better characterized.

It is the users of measurement services and not their suppliers who are in the better position to identify what measures are needed. But better measurements are surely needed at every point in this reticulated model of facilities acquisition and

Figure 3. Key to items: 1. Photometer. 2. dl-714 Data Logger. 3. db-308 Sound Level Analyzer. 4. Air velocity meter. 5. Humidity transmitter. 6. Infrared gas analyzer. 7. Heat stress computer. 8. Carbon monoxide monitor. 9. Power supply.

maintenance process that this chapter proposes. What constitutes better measurement? It is not precision, necessarily, nor reproduceability necessarily, or even absence from bias. Aristotle warned that the quality of a measurement inheres not in the measurement itself, but in the fitness to the purpose for which it was intended. It is the users of measurements: the managers of facilities, owners, designers and users who are in the advantaged position. They know what their measurement needs are. Whether they are balancing investments between buildings and equipment, buildings and buildings or country and country. They also know what precisions are required to meet those needs. Rather, put it another way, the users of measurements services are better equipped to identify the upper limit of what they will pay to get a good measurement. In other words, price/performance ratios need to be calculated in this realm of business enterprise.

It is virtually certain that once users of measurements assume their responsibility of dictating to instead of being dictated to by the metrologists one of the first changes to be made will be replacing the highly aggretated and gross measures of buildings and their performance. We are all innured to such crudities as overall building volume or costs described only in terms of unit area. Could you image purchasing automobiles on the basis of relative cost per pound? Or audiences measured only in costs per thousand of readers or millions or viewers? Yet building are routinely described by buyers and sellers alike in terms as crude as dollars per square foot. Designers and builders employ language hardly more exact than that, although it seems ridiculous consideration of the intricacies of mechanical performance or esthetics. or user satisfaction.

One might argue that we lack indicators of building performance. It is true, we may lack indicators of building performance at the present moment, but the inability to measure something precisely does not mean that we leave it out of our deliberations. A cautionary tale on the measurement challenge we face comes from Japan. I participated in the first annual meeting of the Intelligent Buildings Institute (IBI) in 1986. One of the high points of the meeting was the joining of an agreement between the IBI and its counterpart in Japan. In the course of the two-day meeting in Washington, the IBI stipulated three characteristics a building must possess in order to be qualified as "intelligent." These were fairly straight-forward, foreseeably measureable attributes: capability for flexible, expandable in-building data communication; for interconnection with other buildings and sites; and advanced building management systems integrated with the first two features. The Japanese representatives then described their four-part definition of intelligent buildings. Three of the four attributes were identical to those enunciated by the IBI. And what was the fourth? Amenity! The Japanese are undaunted by the relative incommensurability between the first three items and the quality of amenity. Undaunted, our Japanese peers were reminding us of what we already knew: if an attribute is important, we will find a way to gauge its magnitude and its effects.

While the American building industry ratifies its three-part definition and perfects the associated measurement methods, will we see a replay of the stranded motorist looking for his lost key not near his car, but under the lamppost across the street merely because the light was better there? Will we see Japanese designers, builders and facilities managers bringing into the stream of commerce the amenities that we ignore because of their now nonmetrical character or their analytic intractibility?

ACKNOWLEDGEMENTS

The financial support of the National Science Foundation (Grant Number: MSM-8420540) and the cooperation of the General Services Administration's Public Buildings Service both in Washington and in Roanoke is gratefully acknowledged.

NOTE

The data for the preparation of Figure 1 came from the Department of Energy, *Nonresidential Buildings Energy Consumption Survey: Building Characteristics*, DOE/EIA-0246, March 1981.

REFERENCES

Advisory Board on the Built Environment (ABBE), 1983, *A Report from the Workshop on Building Diagnostics*, National Academic Press, Washington.

Akin, O., and Weinel, E. F. (Eds.), 1982, *Representation and Architecture*, Information Dynamics, Silver Spring, MD.

Anderson, J., and Butterfield, D., 1980, *Generalization and Post-Occupancy Evaluation*, Housing Research and Development Program, Department of Architecture, University of Illinois-Urbana, Illinois.

Aristotle (M. Ostwald, trans.), *Nichomachean Ethics*, Bobbs-Merrill, Indianapolis, 1962, p. 18.

Benedikt, M. and Burnham, C., 1980, Describing and perceiving the physical world: research with isovists, Presented at the 88th Annual Convention of the American Psychological Association, Montreal, Canada.

Beniger, J. R., 1986, *The Control Revolution: Technological and Economic Origins of the Information Society*, Cambridge, Harvard.

Bowlby, R. L. and Schriver, W. R., (1986), Observations on productivity and composition of building construction output in the United States, 1972-1982, *Construction Management and Economics*, 4:1-18.

Buffalo Organization for Sociological and Technical Inovation, Inc, 1981, *The Impact of Office Environment on Productivity and Quality of Working Life*, BOSTI, Buffalo, New York.

Engineering News-Record, Owners and engineeers try long-term 'partnering', May 12, 1988, p. 13.

Gutman, R., 1985, Environmental design as a social institution, *Design Research News*, 16 (4): 1.

Halldane, J. F., 1978, Performance specification – a quandary of measurement, *The Construction Specifier*, 30(1): 20-35.

Johnson, V., and Bosworth, F. M., 1987, Measuring building performance: a bibliography, unpublished.

Kalton, 1983, *Introduction to Survey Sampling*, Sage, Beverly Hills, CA.

Knox, P. L., 1987, The social production of the built environment: architects, architecture, and the post-modern city, *Progress in Human Geography*, 11(3): 354-377, citing Knesl, J.A., The powers of architecture, *Environment and Planning*, 2: 3-22.

Latour, B., 1986, Visualization and cognition: thinking with eyes and hands, in: *Knowledge and Society: Studies in the Sociology of Culture Past and Present*, vol. 6.

Marans, R. W., and Ahrentzen, Sherry, 1987, Developments in Research Design, Data Collection, and Analysis, in: *Advances in Environment, Behavior, and Design*, Vol. 1, E. H. Zube and G. T. Moore, (Eds.), New York, Plenum Publishing Corp.

Marx, Karl, *Das Kapital*, Vol. 1.

Mattar, S. G., and Fazio, P. P., 1973, A Bibliography and Reviewof Building Evaluation Schemata and Practices, *Council of Planning Librarians Exchange Bibliography 470*.

Moore, G. T., 1987, Environment and behavior research in North America: history, development, and unresolved issues, in: *The Handbook of Environmental Psychology*, D. Stokols and I. Altman (Eds.), New York, John Wiley and Sons.

Olsen, C., 1988, Top firms are buoyant despite industry's rough weather, in: *Building Design and Construction*, July: 42-43.

Shibley, R., and Schneekloth, L., 1988-89, Risking collaboration: professional dilemmas in evalution and design, *Journal of Architectural and Planning Research*, 5:4.

Sommer, R., 1984, Action research is not business as usual, in: D. Duerk and D. Campbell (Eds.), *The Challenge of Diversity*, Environmental Design Research Association, Washington, D.C.

U. S. Department of Energy, 1981, *Nonresidential Buildings Energy Consumption Survey: Building Characteristics*, DOE/EIA-0246, U.S.G.P.O., Washington, D.C.

Ventre, F. T., Establishing Appropriate Measurements for the Performance of Buildings and the Serviceability of Facilities, in: Gerald Davis and F.T. Ventre., (Eds.), *Performance of Buildings and Serviceability of Facilities*, in press, American Society for Testing and Materials, Technical Publication 1029, Philadelphia, PA.

Ventre, F. T., 1983, *Documentation and Assessment of the GSA/PBS Building Systems Program: Background and Research Plan,* (NBSIR 83-2662), National Bureau of Standards, Washington, D.C.

Werner, R., in press, Advances in the Evaluation of the Built Environment, in: E. H. Zube and G. T. Moore (Eds.), *Advances in Environment, Behavior, and Design*, Vol. 2.

Zeckhauser, S., and Silverman, R., 1983, Rediscover Your Company's Real Estate, *Harvard Business Review*, 61 (1): 111-117.

CHAPTER 20

SIMULATION TECHNIQUES IN DESIGN RESEARCH

Robert B. Bechtel

Department of Psychology
University of Arizona
Tucson, Arizona

ADVANCED SIMULATIONS BY COMPUTER

The last few years have seen an explosion of computer simulation. Once mathematical models can be developed for any natural process, they can be programmed for computer simulation. In this way weather prediction has improved (Tribbia and Anthes, 1987); tidal power can be simulated for the design of power plants (Greenberg, 1987); the design of trophy-winning yachts (Letcher, et al, 1987) can be pretested without touching the water; ecological systems (Luker, 1987) can be modeled; nuclear war can be programmed as a computer game (Dewdney, 1987a); and even a cocktail party simulation can be a computer game (Dewdney, 1987b). But the most sophisticated and expensive simulations are those used in training military and airline pilots (Haber, 1986; Chiles, 1986). Flight simulators have now become so reliable that they can be used to train pilots to react to hazardous situations that have not heretofore been experienced.

An example of this degree of sophistication is the recent knowledge obtained from wind shear accidents. In August of 1985 a Lockheed L-1011 crashed at Dallas-Fort Worth airport as it was attempting to land. Data from this and other accidents were compiled to develop a mathematical model of wind shear forces. This model was then programmed into the flight simulator to train pilots how to fly through such encounters. It was discovered that the way to fly through a wind shear was to increase the throttle and lift the nose of the plane. Most pilots instinctively lower the nose to help increase air speed, but this is a sometimes fatal mistake under wind shear conditions. Although the primary instructions are to avoid wind shear whenever possible, all pilots trained on the simulator will have experience flying through wind shear, even though the average pilot seldom encounters such conditions during an entire career.

Flight simulators illustrate the new and prominent position that simulation of environments has taken in ever increasing aspects of modern life. Simulation saves a great deal of money. Airlines estimate that using simulators is ten times less expensive than training pilots in real aircraft (Haber, 1986).

Simulation also enables encounters, in short sessions, with emergencies and conditions that would normally require a lifetime to experience. As one pilot explained, "It puts twenty years of experience in one hour."

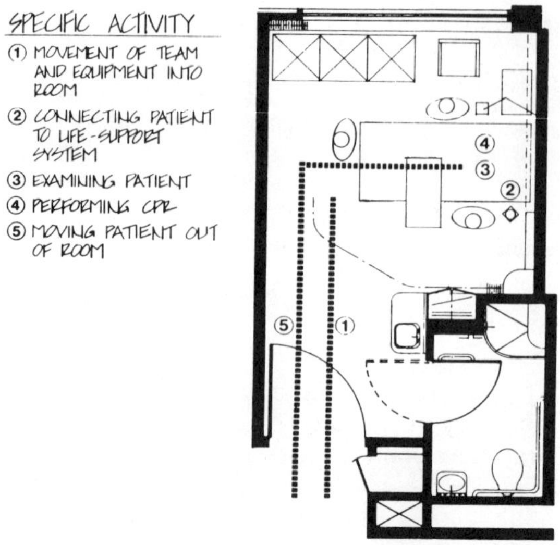

Figure 1. Emergency treatment of patient.

An important fact for environmental simulations in general is that these sophisticated reproductions of the environment are free from the static models used in earlier simulations. The new environments can be projected on screens using film taken from any place on the globe, and computer graphics can be used to simulate photographs. This is extremely helpful in freeing the simulation from its earlier reliance on realistic hand-made models.

Unfortunately, computers have not so greatly influenced the design process. Computer-assisted design (CAD) remains a promise that has yet to reach completion. MacCallum and Duffy (1987) point out that even though this form of simulation, which greatly assists the design process, has reached a high level of sophistication, comparatively few designers take advantage of it. MacCallum and Duffy feel the reason for this underutilization is that few designers realize the richness of material available through CAD. Perhaps we are experiencing a technology lag, where it will take another generation of computer-trained designers to take advantage of CAD.

SIMULATION BY MODELS

One of the most intriguing uses of architectural models was described by Delong (1976, 1981). Delong and his students at the University of Tennessee Dept. of Architecture constructed three different architectural scale models of a building and had subjects simulate behavior within the models, with the use of appropriately scaled dolls, for thirty minutes. The subjects were then measured for how they behaved and how long they estimated those thirty minutes to take. Indications were that the subjects tended to collapse time to a degree consonant with the scale of the models: i.e., smaller model, shorter time. Earlier studies (1976) also indicated that subjects moved the dolls in such a way as to maintain the kinds of social distances described by Hall (1959) as Proxemics.

Unfortunately, attempts to replicate these studies in other locations have not succeeded. It would open up a new era of investigation if, indeed, doll play in scale models could replicate full-scale life.

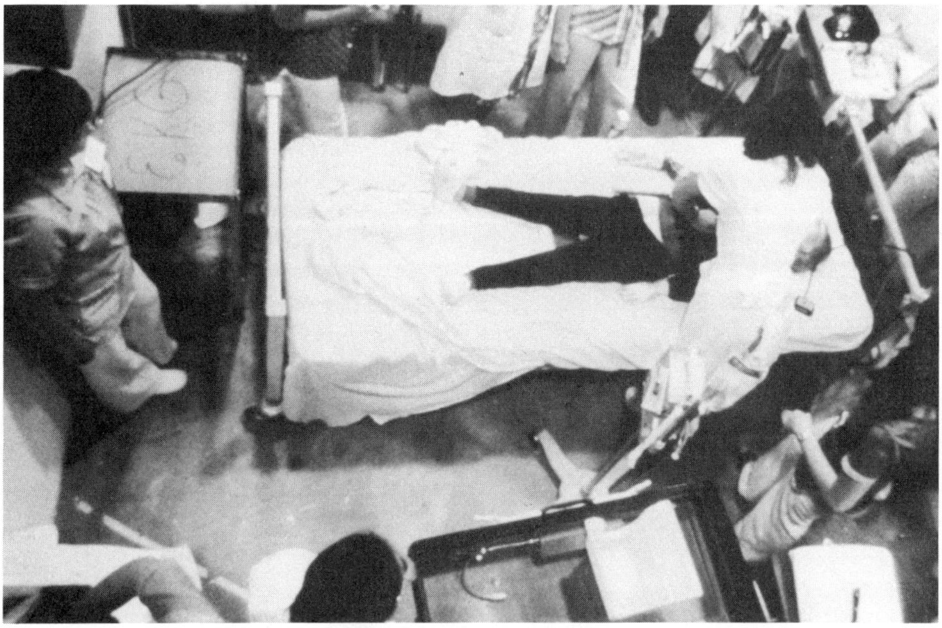

Figure 2. When the bed was pulled away from the wall, there was not enough room for the staff and equipment at the foot of the bed. (Photograph courtesy of Dr. Robert W. Marans.)

Hardie (1988) describes a more conventional use of scale-model simulations, used to involve community residents in participation in the design process. In this process Hardie describes how illiterate Third World subjects manipulate models of houses and street plans in order to solicit preferences and choices. What is especially important about the use of models for these purposes is that it helps bring out information that would not be forthcoming by any other method. Not only illiterate people, but even many sophisticated middle-class persons find it hard to visualize street plans and housing designs that they have not experienced. The models permit them to visualize many variations and then to explore the consequences of different designs through manipulating the models. It is hard to see how this process could be replaced by computers, however, because it represents not only three-dimensional objects but also the physical sensations of touching and moving those objects.

FULL-SCALE MOCK-UP SIMULATIONS

Simulation in architectural design research has not yet reached the sophistication of the flight training simulators. Most attempts at full-scale simulation still involve photographing models and presenting them on film (Bosselman and Craik, 1987). However, the Architectural Research Laboratory of the University of Michigan (King, Marans and Solomon, 1982) did a full-scale mock-up of hospital rooms for pre-design research. At first it may seem that this is too expensive a simulation to consider practical. But it becomes clear, given the complexities of hospital routines, that only a full-scale

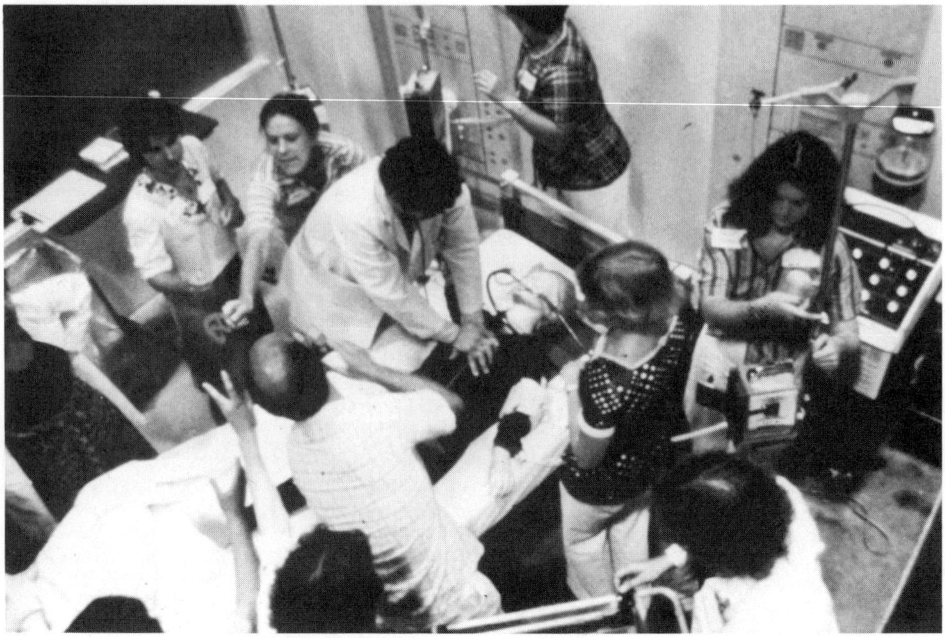

Figure 3. There was enough room for the staff and equipment to move on the sides of the bed, but the area was cramped. (Photograph courtesy of Dr. Robert W. Marans.)

mock-up can reveal the kind of spaces needed for the many procedures that must be performed in patient rooms, nursing stations, and intensive care units.

Figures 2 and 3 show the mock-up under full simulation conditions. The principle involved in the use of a full-scale mock-up was to simulate all hospital procedures that would take place in the room. What was particularly valuable in this case was the finding that merely adding a few inches to the room space – the very amount eliminated by the original design to save construction costs – increased the efficiency of the rooms considerably!

Perhaps the largest and most elaborate full-scale mock-up ever conceived will take place near the town of Oracle, Arizona in the very near future. Construction has already begun on Biosphere II (Allen and Nelson, 1986) which is a full-scale mock-up of a simulated Moon or Mars colony. The purpose, of course, is to test the efficacy of present technology, to determine whether a completely self-contained colony is possible outside the earth. Figure 4 shows a cutaway plan of the colony, which will have three different biomes – rain forest, savannah, and desert – in a completely self-contained structure that will have to function for two years sealed off from any inputs of atmosphere, water or food. Crew quarters can be seen behind the biomes.

What will be tested in this simulation is whether the biomes can sustain an oxygen-rich atmosphere and provide enough food to support human life indefinitely. The scheduled date for sealing in the occupants is December of 1988. The structure, which must be airtight, covers about 2 1/2 acres of space and has about 5,000,000 cubic feet in volume. This is a truly environmental simulation, but it is much more compre-

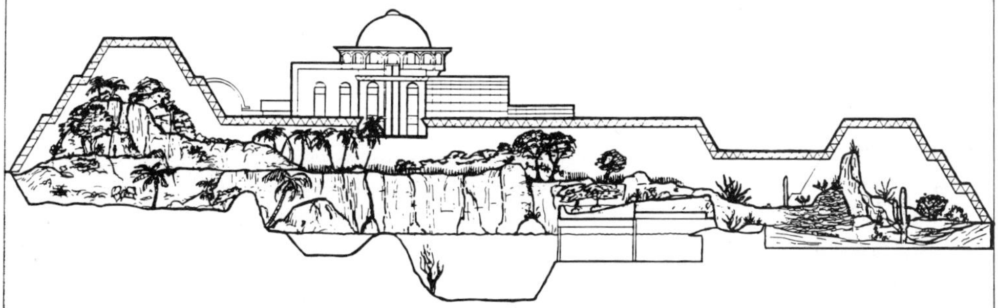

Figure 4. From Allen and Nelson, 1986, p. 60.

hensive than the kind usually thought of to test building designs. This experiment may greatly influence or even determine the final design of a Mars colony.

VIDEO SIMULATIONS

Video cameras, the VCR, and television have greatly affected simulations and will continue to do so. The Environmental Simulation Laboratory (Bosselman and Craik, 1987) has largely switched from film to video tape, even though this of course means some sacrifice in detail.

The hospital study of Reizenstein Carpman et al. (1985) is already a classic in the use of video techniques to study pre-design problems. (See chapter 15.) The problem in this case was whether to construct a visitor parking entrance close to the patient drop-off entrance, or to put the entrance further away, so as to redirect traffic away from the drop-off and reduce congestion.

Both conditions (near and far entrance) were simulated by videotaping an entrance sequence to models and asking subjects who watched the video where they would turn. The results were unequivocally in favor of placing the entrance farther away, because it was found that subjects would use the drop-off zone to get to the parking entrance, even when they had no patients to drop off, and even with signs telling them to go elsewhere. Thus, placing the parking entrance nearer would result in congestion. An important theoretical finding of this study was the predominance of physical environment clues over signage.

The practical advantages of this simulation were that it was very cheap, costing only about $200, and very quick, taking 14 days from problem identification to results. This means that the potential for quick and easy solutions to pre-design problems is now more available than was even thought possible only a few years ago.

THE FUTURE OF SIMULATIONS

The future direction of simulations is already clear. It will involve the exploding technology of computer graphic simulations. Already computer simulation experts are talking about simulation of "artificial environments" (Luker, 1987). Single chip graphic displays (Peled, 1987) will permit more flexible and higher-speed manipulations.

Computers will be able to display 256 colors or shades of gray simultaneously, and may accept voice or handwritten commands. Designs will be previewed on the screen and subjected to numerous manipulations based on PDR and POE data.

The particular utility of these future computer simulations is that they can replace the scale model, or many versions of a scale model, until its final form has been tested by engineering, aesthetic, and even behavioral science performance criteria. The problem will be developing the software for such testing. An example of how such criteria can be utilized in developing the programming for a design that cannot be pre-tested in real life is the development of criteria for NASA's proposed space station. In 1986, Stuster developed 14 behavioral issues around which to develop design guidelines:

- sleep
- clothing
- exercise
- medical support
- personal hygiene
- food preparation
- group interaction
- habitat aesthetics
- outside communications
- recreational opportunities
- privacy and personal space
- waste disposal and management
- on-board training, simulation, and task preparation
- behavioral and physiological requirements associated with microgravity

Since the space station is a unique project, only situations that approximate the isolation and other conditions of the station can be used to arrive at guidelines. Stuster systematically evaluates these "analogous conditions" by first evaluating the appropriateness of each, using a panel of experts to rate them on 14 dimensions. This produces a matrix which converged on overall fidelity to the space station vs. the 14 specific dimensions.

The importance of the Stuster study is that it provides the raw data whereby future "simulations" can be made for testing space stations which cannot be otherwise simulated except in highly expensive full scale mock-ups like Biosphere II. Unfortunately, Stuster stops short of the mathematical modeling which would have permitted computer simulation.

To date, no Meta analysis has yet been done on design research. Meta analysis is a term familiar in social psychology where a series of studies done on a particular issue is collected and all the old data is re-analyzed from a new perspective. It is a way of recycling old studies, but more importantly, it is a source of data for mathematical modeling to program computer simulation.

For example, Mullen (1983) collected studies involving group size, to test out a model which he had developed to explain the inverse relationship between group size and performance (the so-called "too many cooks spoil the broth" situation). He found most studies agreed with his model, thereby finding powerful support for it.

Such data are sorely needed for design research; this is a plea for researchers to begin collecting various kinds of POEs and other design research to do Meta analyses.

The final expectation of simulations is that computer graphics (Bosselman and Craik, 1987; Haber, 1986) will reach the point where photographing models will become obsolete and designs can go directly into computer simulation based on a large accumulation of data from analogues or similar designs. It is also assumed that the hardware will advance to the point where all these simulation capabilities will be available in the form of desktop workstations.

REFERENCES

Allen, J., and Nelson, M., 1986, *Space Biospheres*, Synergetic Press, Inc., Oracle, AZ.

Bosselman, P., and Craik, K., 1987, Perceptual Simulations of Environments, in R. Bechtel, R. Marans and W. Michelson, (eds.) *Methods in Environmental and Behavioral Research*, Van Nostrand Reinehold, N.Y. pages 162-190.

Chiles, J., 1986, How pilots learn to cope with the worst, *Smithsonian*, 17:78-87.

Delong, A., 1976, The use of scale models in spatial behavior research, *Man Environment Systems*, Vol. 6, 179-182.

Delong, A., 1981, Phenomenological Space-Time: Toward an experiential relativity, *Science*, Vol. 213, 681-683.

Dewdney, A., 1987a, Computer recreations: Diverse personalities search for social equilibrium at a computer party, *Scientific American*, 257:112-115.

Dewdney, A., 1987b, Computer recreations: "After MAD": A computer game of nuclear strategy that ends in a prisoner's dilemma, *Scientific American*, 257:174-177.

Greenberg, D., 1987, Modeling Tidal Power, *Scientific American*, 257:128-133.

Haber, R., 1986, Flight simulation, *Scientific American*, 255:96-103.

Hall, E., 1959, *The Hidden Dimension*, Doubleday.

Hardie, G., 1988, Community participation based on three-dimensional models, *Design Studies*, Vol. 9, 56-61.

King, J., Marans, R., and Solomon, L., 1982, *Preconstruction Evaluation*, Architectural Research Laboratory, University of Michigan, Ann Arbor.

Letcher, J., Marshall, J., Oliver, J., and Salvesen, N., 1987, Stars and Stripes, *Scientific American*, 257:34-40.

Luker, P., 1987, Computer simulation: A developer's perspective, *Academic Computing*, 1:18-21.

MacCallum, K. J., and Duffy, A., 1987, An expert system for preliminary numerical design modelling, *Design Studies*, Vol. 8, pages 231-237.

Mullen, B., 1983, Operationalizing the effect of the group on the individual: A self attention perspective, *Journal of Experimental Social Psychology*, 19:295-322.

Peled, A., 1987, The next computer revolution, *Scientific American*, 257:56-65.

Reizenstein-Carpman, J., Grant, J., and Simmons, D., 1985, Hospital design and way finding: A video simulation study, *Environment and Behavior*, 17:296-314.

Stuster, J., 1986, Space Station Habitability Recommendations Based on a Systematic Comparative Analysis of Analogous Conditions, NASA.

Tribbia, J. and Anthes, R., 1987, Scientific basis of modern weather prediction, *Science*, 237:493-499.

CHAPTER 21

ADVANCES IN POST-OCCUPANCY EVALUATION APPLICATIONS: AN OVERVIEW

Jay Farbstein

Jay Farbstein & Associates, Inc.
San Luis Obispo, California

POE APPLICATIONS: AN OVERVIEW

The chapters in this section of the book deal with recent advances in applying building evaluation to a variety of settings and situations, especially work environments.

Joiner and Ellis describe the effort of the Ministry of Works and Development in New Zealand to institute a building evaluation program with a focus on the organizational problems. Using evaluations effectively over a period of eight years, another goal was to improve the design quality and standards for public buildings by establishing an appropriate data base.

Serra, in his chapter, "POE at the Urban Scale in Brazil," reports on an innovative application and expansion of evaluation techniques to the urban design scale, including such issues as the quality of public housing sponsored by the Brazilian government over the past 20 years, urban infrastructure, quality of roadways, bus shelters, and so on.

In the chapter on "Buildings in Use: Analysis of Office Buildings," Vischer proposes a "building-in-use assessment system" which is inexpensive and quick to use. It utilizes rating scales to assess such things as worker job satisfaction, building complaints, and other items. In an ever-changing building environment, especially in offices, these items should be important to facility managers in their resource planning for maintenance and repair of facilities.

Evaluation as an organizational intervention is the topic of the chapter by Jay Farbstein and Associates as well as the representatives of his client. The U.S. Postal Service has adopted a POE program as a multi-year funded activity which has had significant impact on the process of facilities design, construction, and management. The lessons learned in this project are presented for the benefit of other organizations, including effective ways to state goals, choose personnel, communication, and involvement of members of the client organization, as well as the presentation of project findings in practical, positive terms.

ADVANCES IN POE APPLICATIONS

Advances in post-occupancy evaluation (POE) applications are looked at from two points of view. First, recent developments are observed, and second, those developments that may represent advances for the field are identified. Developments in POE applications include an apparent increase in the volume and acceptance of POEs, shifts in the sponsorship of POEs, and changes in the types of POE programs that are run by some sponsors. Possible advances include the integration of behavioral and technical assessments, moving toward the application of "total building performance." Other changes due to sponsor demands include a growing awareness of "bottom line" measures of performance in POE. Possible changes due to experience with large scale POE programs which may also advance the field include the development of greater sophistication in dealing with organizational issues and the clearer discrimination of multiple levels of POE.

INTRODUCTION

First, some definitions. The term "POE applications," as used here, is meant to describe POE as it is practiced for (or by) clients who make use of the results. POE may be used to change and improve the way clients program, design, build, manage, or use their environments. The title of this chapter, "advances in POE applications," suggests not only change, but progress; not only doing things differently, but doing them better.

How can we measure the advance in POE as it is applied in real world settings? This is not a trivial question and reminds us that our knowledge of POE applications is limited in terms of how much we know about what is actually happening out there. Even more critically, our thinking about the application of POE and environment and behavior (E&B) research in general proceeds without an accepted conceptual framework which would explain the activity of applied research. That is, theory in E&B tends to be viewed as explaining phenomena, rather than describing or explaining the process of developing information and acting on the person-environment system.

Other observers have commented on the application of E&B research – which is the essence of POE. Kantrowitz (1985) asked the rhetorical question "Has E&B research made a difference?" and answered that it has, as research has become more applied in its orientation, as clients have recognized its potential contributions, and as research programs have become more institutionalized.

Shibley, in discussing building evaluation programs in large organizations, proposes at least two criteria for their success. First, that evaluations be in the "mainstream" – that is, be embraced as the usual way of doing things – in relation to long-range planning and policy decisions within an organization; in other words, that their results "count" to decision makers. Second, that they be shown to be cost-effective in generating this information. Zimring and Wener suggest that if the field is to advance, standards are needed to ensure the quality of information generated by even the "quickest and dirtiest" of POEs. This chapter will suggest a number of other measures in addition to these, and will attempt to report on the movement in each of the following areas:

- Are more POEs being done?
- Are POEs being applied in "new" contexts or facility types?
- Are "new" people doing or supporting POEs?
- Are new techniques or technologies generating new POE applications, or vice versa?
- Are POEs being done better?

- Are POEs measuring new kinds of outcomes?
- Are POE results being communicated more effectively?
- Is POE having more impact?

ARE MORE POES BEING DONE?

Perhaps the most difficult question to answer is this quantitative one. No new data on the volume or monetary value of work being done in this field has appeared since Bechtel conducted his survey of housing POEs for the United States Department of Housing and Urban Development in 1978. However, there is anecdotal evidence which can be cited on both the positive and negative sides of this issue, and conditions seem to vary greatly among countries. On the positive side, one can cite two factors. The first is the institutionalization of POE in several countries (Australia, Canada, New Zealand, and perhaps others). The second is the emergence and continuity of private firms able to subsist by doing (at least in part) POE work. Some of these firms have been active for a number of years, but there seem to be more now, and, with a larger number of firms, many have specialized in particular user or facility types (e.g., the Project for Public Spaces, which studies plazas and airports, and other firms which specialize in offices, health care, and so forth).

On the negative side, it is possible to identify governmental agencies (particularly in the United States) which sponsored POE studies in the past, but do so no longer, or have reduced their support. The National Science Foundation and National Endowment for the Arts provided money for early conceptual work on POE, but appear to be less involved at this time. This may be because they see their role diminishing as POE becomes better understood and accepted, leaving sponsorship to the agencies and organizations which can benefit directly from use of research results.

Also on the negative side is the level of sponsorship by U.S. federal agencies such as Housing and Urban Development and the Department of Energy, both of which had been major sponsors of directed POE studies and both of which appear to have greatly curtailed, if not abandoned, their programs.

While the magnitude of this reduction and the reasons why it has taken place are unclear, it may reflect shifts in priorities or perceptions which are particular to these organizations. Because, at the same time that some U. S. agencies have reduced funding for POE, many others have stepped in to fill the gap. Examples include the General Services Administration, Postal Service, Veterans' Administration, some state governments (e.g., Massachusetts), and many private clients and voluntary organizations.

It appears that POE is becoming better understood and more accepted. Staff and managers in many client organizations are familiar with the term and are more likely to see its value and to include it as an integral part of the facility management cycle. POE seems to be shifting into a "demand-driven" mode, where clients are defining the need for research, as compared to the prior "supply-side" mode, where researchers had to convince reluctant clients that they might get something of value from a POE.

ARE POES BEING APPLIED IN "NEW" CONTEXTS OR FACILITY TYPES?

Environment and behavior research began mainly with studies of public sector housing and institutional facilities. More recently, POEs have been performed on private sector facilities, including office, retail, and "hospitality" (restaurants and hotels). Kantrowitz (1985) reviewed recent developments in several of these areas. Clearly, for private clients to support POE, it must prove itself useful – and apparently it is doing just that. At the same time, increasingly effective POEs (and resulting design guides) are being developed for more traditional public sector clients, e.g., Welch, Parker and Zeisel, as well as Regnier and Byerts.

ARE "NEW" PEOPLE DOING OR SUPPORTING POES?

POEs have been done in the past by environment-behavior researchers and some design professionals. In discussing the experience of their firms in conducting design research, Symes, Duffy and Ellis (1985) conclude that a new generation of design researchers whose education and experience span social science and design will be required to bring about an effective synthesis of skills needed in this field. Could it be that improvements in POE cited elsewhere in this chapter are, in part, a result of the emergence of this hybrid practitioner?

An additional emerging category, that of "facility manager," may also have an impact upon POE practice. Facility managers are taking responsibility for carrying out or managing tasks that span from strategic planning to janitorial and maintenance duties (Margulis, 1987). Somewhere in this spectrum falls POE, and facility managers recognize that evaluation and feedback are essential to their ability to manage effectively. One sees POE results reported in facility management trade journals (e.g., Gere Picasso's article in *Facilities Design and Management* which briefs facility managers on how to develop an in-house POE) and discussed at their conferences (e.g., International Facility Management Association [IFMA]).

It can be argued that the existence of the discipline of facility management will have an impact on the diffusion of POE; as large organizations develop in-house skills and integrate all aspects of planning and control, evaluation will become a more obvious thing to do. And, with professionals inside of organizations whose job it is to look after the facilities, demand for POE may increase.

ARE NEW TECHNIQUES OR TECHNOLOGIES GENERATING NEW POE APPLICATIONS?

The availability of new technologies may influence how POEs are applied, their cost, and the time they require for completion. Conversely, the contexts in which POEs are performed may require new techniques (not necessarily hardware). We have seen POEs which utilize such technologies as videotape (to record walk throughs, interviews or focus groups, and to present results), lap-top and hand-held computers (e.g., the "datamite," for field data collection), on-line questionnaires (where the respondent can connect to the researcher's computer, answer questions, and have results processed instantaneously), and the use of electronic bulletin boards (where remote sponsors of a POE project can keep informed and trade messages). Widespread and affordable "desk top publishing" (with personal computers driving laser printers) is allowing researchers to generate more attractive, readable and effective POE reports. The development of increasingly sophisticated, interactive, and realistic image processors and computer aided design (CAD) simulations (allowing one to "walk through" an unbuilt project, change colors at will, and inexpensively simulate alternatives) will probably have an impact on the ability to do "preconstruction evaluations" of alternatives (see Molloy, chapter 5).

"New" low-tech POE methods have been stimulated by the demands of applications (many from or inspired by the organization development field). An example is the "walk through" or touring interview (Shibley, 1985; Gray et al., 1986), which entails the simultaneous, directed, in-person assessment of an environment by several parties who see the place from different perspectives.

ARE POES BEING DONE BETTER?

Are we learning to do POE applications better or smarter? I would not argue that evaluation methods, per se, have been improved, or even that there may not be poor quality POE work being done in some instances. But there is evidence that POE practitioners are thinking more about the information needs of their clients and of designers, how to gather that information most efficiently, and how best to present it, as exemplified by Tim White and his illustrated publications. In this sense, POEs may be being done "better": with more direct, obvious and useful payoff to the client. This is evident in several relatively recent developments which are reviewed below.

First, there is a growing awareness of the role that POE plays in organizational change (see Farbstein and Kantrowitz, chapter 25; Joiner and Ellis, chapter 22). Some observers feel that POE consultants are more sophisticated about the POE's organizational context and what characteristics may be required to get results used effectively (Richard Wener, personal communication). Craig Zimring (personal communication) points out that we can observe an increasingly better match of POE methods and products to our clients' goals and needs.

An example of this improved match is the recognition of needs for differing levels of POE in what may be an ongoing research program (Shibley, 1985). The first level may be diagnostic, intended to identify the range of problems and issues which need to be looked at in more depth. Often, this stage will involve a wide range of people from within the organization, as well as a broad spectrum of expertise on the evaluation team. This approach should not be denigrated as "quick and dirty," but rather seen as laying the foundation for further research. Results of this initial stage range from immediate feedback and fine-tuning (see Kantrowitz, 1986, p. 123 for an example of fine tuning to get optimal energy performance) to problem definition, to "buy-in" of the organization.

The next level of POE may include detailed assessments of a single building or a set of similar buildings. A single building study may serve to test programmatic assumptions. If POEs are completed on a sufficient sample of similar buildings, conclusions may be drawn which are applicable to other buildings of the type, and to the documentation of that knowledge as design guidelines. Thus, POE results are used in the next generation of building designs.

The final level of POE would focus in greater detail on a particular functional area or technical issue which had been identified as needing further study. This operational approach to defining levels of detail is being utilized in such POE programs as those of the New Zealand Ministry of Public Works, Health and Welfare Canada, Public Works Canada, the California Department of Corrections (Zimring, 1987), and the U. S. Postal Service.

ARE POES MEASURING NEW KINDS OF OUTCOMES?

Environmental design researchers struggled for many years to get performance-based notions accepted in facility programming applications (Preiser, 1988). The notion was that POEs would measure environmental performance. The initial performance measures were largely related to such outcomes as habitability, satisfaction, and task effectiveness. More recently, new concepts and measures of performance have been integrated into POE practice.

The first is the notion that POE should measure "bottom line performance" which can be expressed in terms of dollar costs. Brill (1984, 1985) began use of the term in discussing office design, showing many of his findings to affect worker productivity (while also reporting other findings which were felt to be important but could not be measured in dollars). POEs of correctional facilities have pointed to such bottom line costs as replacement of vandalized equipment, provision of health care services, staff time lost, and public liability (Wener et al., 1987). The U. S. Army's Construction Engineering Research Laboratory (CERL) demonstrated a return of 77-to-1 on the cost of research which lead to productivity and morale improvements in an office building. Recently the Postal Service has embarked on a program to measure productivity in relation to design, not only in terms of worker productivity, but also in terms of marketing effectiveness (USPS). These concerns demand new conceptual and methodological approaches to POE and require that environmental design researchers coordinate efforts and findings with economists and financial analysts.

Another emerging direction for POE is the concept of total building performance, which demands the integration of behavioral and technical factors in a more comprehensive evaluation scheme (see Loftness et al., chapter 12). The total building performance concept is consistent with the notions of environmental systems and holistic approaches which have been with the field since its beginning.

Assessing total building performance requires expertise from a range of disciplines in addition to environment and behavior research. The need for such an approach arose from problems which were confronted by POE researchers examining health and behavioral effects in sealed buildings (required by recent energy conservation standards). The effects could only be understood by looking simultaneously at the totality of factors, including behavioral, managerial, chemical, epidemiological, technical (lighting, HVAC, etc.), and so forth. All disciplines had to be present together to begin to discover the complex interactions that were occurring. The concept has been shown to be relevant to general situations as well, where impacts and outcomes need to be looked at broadly. Inherent limitations can be built into behaviorally oriented POEs that ignore technical or physical design issues, just as much as with technically based building research which ignores user/behavioral interactions and effects.

Good examples of POE work within the total building performance paradigm include the Department of Energy-sponsored studies of non-residential passive solar buildings (Burt Hill et al., 1987; Kantrowitz, 1986), and Public Works Canada's office studies. Other examples are cited in a publication of ASTM (Davis, 1986), which has a task group working on defining total building performance.

ARE POE RESULTS BEING COMMUNICATED MORE EFFECTIVELY?

There are many examples of POE results which are being more effectively applied to feed forward into future generations of building designs in the form of design programs, guidelines and standards. This is not necessarily better research, but better "packaging" and communication of results. While some practitioners have been working for years to make effective and readable reports, their recent work shows much more sophisticated use of typography and illustrations. Compare, for example, Zeisel (1975) with Welch, Parker and Zeisel (1984).

It is important to recognize that many designers and managers may not be very interested in research results per se, but rather in being shown what the research demonstrates that design should do. When Farbstein and Kantrowitz asked a convocation of very busy and results-oriented design managers from U.S. Postal Service offices around the country whether they wanted to perform POEs themselves, to read about the results, or to be given design direction based on the results, they almost unanimously asked for the latter only. They were glad that the research was being done, insisted that the design direction be well founded in that research, but did not feel that they or the architects whose work they managed had time to read research reports.

An intermediate approach is taken by Carpman et al. (1986), who provide brief summaries of research findings juxtaposed to clear design direction. Similarly, Brill et al. (1984 and 1985), in reporting on results from a large scale empirical study of offices, took great pains to make the data immediately imageable and understandable to lay people. More recently, Brill produced a users' manual and instructional videotape to reach office users directly (Dixon, 1988).

IS POE HAVING MORE IMPACT?

There is some evidence that POE results are beginning to have a greater impact on building design, management, and policy. Despite the fact that some previous sponsors of POE studies are no longer as active as they once were, the past five years or so have seen a growing number of examples of high visibility, high impact POE studies, many of which have already been mentioned. Other facts that contribute to this conclusion are the wider dissemination of findings through such client- or user-dominated outlets as the American Hospital Association, Building Owners and Managers Association, American National Standards Institute, National Institute of Corrections, Gerontological Society, and many others. A book on how to conduct POEs has been published by a major press (Preiser, Rabinowitz and White, 1988). Many articles about POE or reporting results have appeared in the design, trade and popular presses. A design research awards program sponsored by the National Endowment for the Arts in 1985 supplemented *Progressive Architecture* magazine's annual program in giving recognition and publicity to some of the best POE work. Finally, the on-going POE programs in large governmental organizations, though only examples and not yet the dominant mode, provide a number of models for others to emulate.

THE RELATIONSHIP BETWEEN PRACTICE AND THEORY

Kurt Lewin's statement to the effect that there is nothing so practical as a good theory is often quoted at E&B conferences. Based upon the review of research applications, researchers' experiences in the "real world" can also have a stimulating effect on theory – sometimes leading to the development of theoretical constructs, and at other times pointing out the need to explain observed phenomena or processes and to integrate them within a new theoretical synthesis (Kantrowitz, 1985). An example of this stimulating contribution from the applied to the theoretical can be seen in the recognition of "bottom line measures" which require the identification and quantification of environment-behavior effects in terms of their economic value (dollars gained or saved). Total building performance, which in itself is a quasi-theoretical construct (or, at least, one with implications for theory), required a new synthesis in relation to problems observed in field – especially indoor air pollution and energy utilization.

FUTURE DIRECTIONS FOR POE APPLICATIONS

The purpose of this chapter has been to discuss emerging trends in POE. While we would not simply extrapolate these trends into the future, among the possible scenarios which may unfold are:

- further acceptance of POE within private and government user agencies;
- a greater degree of understanding about how to intervene in complex organizations;
- technologies which allow more effective user participation and more efficient data gathering; and
- a greater penetration of research findings in use (as design and policy guidance).

A more controversial potential development is toward a degree of standardization of the process of rating buildings, which is a variant of POEs. This is likely to be embraced by client organizations as making POE more accessible and efficient, but may be resisted by researchers.

NOTE

An earlier version of this chapter was published in Lawrence, D., and Wasserman, B. (Eds.), *Proceedings of the Environmental Design Research Association Conference*, Pomona, 1988.

REFERENCES

Bechtel, R., 1978, *Post Occupancy Evaluation of Housing*, Washington, D.C., U.S. Department of Housing and Urban Development.

Brill, M. with Margulis, S., Konar, E., and BOSTI, *Using Office Design to Increase Productivity*, Volume One (1984) and Volume Two (1985), New York, Westinghouse.

Burt Hill Kosar Rittelmann Associates and Min Kantrowitz Associates, 1987, *Commercial Building Design: Integrating Climate, Comfort, and Cost*, New York, Van Nostrand Reinhold.

Carpman, J., Grant, M., and Simmons, D., 1986, *Design That Cares: Planning Health Facilities for Patients and Visitors*, American Hospital Publishing.

Construction Engineering Research Laboratory (CERL), 1982, *Information Exchange Bulletin*, R5:1, Urbana.

Davis, G. (Ed.), 1986, *Building Performance: Function, Preservation and Rehabilitation*, ASTM STP 901, American Society for Testing Materials, Philadelphia, PA.

Davis, G., et al., 1985, *ORBIT-2: Executive Overview*, Norwalk, CT, Harbinger Group.

Dixon, J. M., January 1988, Research Awards, *Progressive Architecture.*

Gray, J. M., Daish, J. R., and Kernohan, D. Q., 1986, A Touring Interview Method of Building Evaluation; the Place of Evaluation in Building Rehabilitation, in: *Building Performance: Function, Preservation and Rehabilitation*, Davis, G. (Ed.), ASTM STP 901, American Society for Testing Materials, Philadelphia, PA, pp. 46-68.

Jay Farbstein & Associates and Min Kantrowitz and Associates, May 1986, *Image Study Final Report*, Washington, D.C., United States Postal Service, Headquarters Facilities Department.

Kantrowitz, M., January 1985, Has Environment and Behavior Research 'Made a Difference'?, *Environment and Behavior*, 17:1, pp. 25-46.

Kantrowitz, M., 1986, Energy Past and Future, *Progressive Architecture*, 4:86, pp. 118-123.

Lewin, Kurt, *Resolving Social Conflicts*, NY, Harper Brothers, 1948.

Margulis, S., 1987, Facilities Management: An Insider's View, workshop reported in J. Harvey and D. Henning (Eds.), *Public Environments*, proceedings of the Environmental Design Research Association Conference, Ottawa, Canada, p. 271.

Picasso, G., November/December 1987, How to Develop an In-House POE, *Facilities Design and Management*, pp. 64-67.

Preiser, W. F. E., Rabinowitz, H. Z., and White, E. T., 1988, *Post-Occupancy Evaluation*, Van Nostrand Reinhold, NY.

Project for Public Spaces, January 1978, *Plazas for People: Seattle Federal Building Plaza, a Case Study*, NY.

Regnier, V. and Byerts, T., 1985, Applying Research Findings to the Planning and Design of Housing for the Elderly, in T. Vonier (Ed.),*Proceedings of the Research and Design '85 Conference*, Washington, DC, American Institute of Architects.

Schneekloth, L., 1987, Advances in Practice in Environment, Behavior and Design, in E. Zube and G. Moore (Eds.), *Advances in Environment, Behavior and Design*, Vol. 1, Plenum Publishing Corp., New York, pp. 307-334.

Shibley, R., January 1985, Building Evaluation in the Main Stream, *Environment and Behavior*, 17:1, pp. 7-24.

Symes, M., Duffy, F., and Ellis, P., January 1985, Research and Practice: A Case Study, *Environment and Behavior*, 17:1, pp. 119-132.

United States Postal Service, 1987, *Request for Proposals for the Store of the Future*, Headquarters Facilities Department, Washington, DC.

Welch, P., Parker, V., and Zeisel, J., 1984, *Independence Through Interdependence: Congregate Living for Older People*, Department of Elder Affairs, Boston, MA.

Wener, R., Frazier, W., and Farbstein, J., June 1987, Building Better Jails, *Psychology Today*, pp. 40-49.

Zeisel, J. and Griffin, M., 1975, *Charlesview Housing: a Diagnostic Evaluation*, Architectural Research Office, Harvard University Graduate School of Design.

Zimring, C. and Wener, R., January 1985, Evaluating Evaluation, *Environment and Behavior*, 17:1, pp. 97-117.

Zimring, C., Goldman, M., and Fuller, C., 1987, Getting it Done: Post-Occupancy Evaluation in the Public Sector, workshop reported in J. Harvey and D. Henning (Eds.), *Public Environments*, proceedings of the Environmental Design Research Association Conference, Ottawa, Canada.

CHAPTER 22

MAKING POE WORK IN AN ORGANIZATION

Duncan Joiner	and	Peter Ellis
Works and Development Services Corporation (NZ) Ltd. New Zealand		Organizational Psychology Consultant London, England

INTRODUCTION

Works and Development Services Corporation (NZ) Ltd, (WORKS) started developing techniques for post-occupancy evaluations of government-designed buildings in 1979. It was at that time the New Zealand Ministry of Works and Development (MWD). Under the direction of the first author, who was then the Assistant Government Architect (Design), the initiative was in response to pressures from the New Zealand Government for improved accountability of the former Ministry's architectural services. A system for conducting POEs was designed with the help of researchers from the School of Architecture of the Victoria University of Wellington. Advice was taken from other experts, notably Professor Robert Shibley, who was at that time directing a POE program on behalf of the US Army Corps of Engineers, one of the few government agencies in the world then successfully operating a POE program (Shibley, 1985).

The research input to the WORKS POE program is well documented (Daish, Gray, and Kernohan, 1983). In the eight years since the first initiative, some twenty building evaluations have been completed, and the program has become firmly established. Apart from developing and modifying the techniques for POE, the MWD (and subsequently WORKS) also learned a good deal about the organizational aspects of operating a POE program, which is the subject of this chapter. Generally there has been a considerable amount of research over the past twenty years on POE techniques, but very little on the organizational problems of using POE effectively, for the simple reason that so far there is little experience to draw on in this area. Yet in a previous publication we concluded that in applying POE effectively to the improvement of building design quality, these organizational factors required as much attention as the POE techniques themselves (Ellis and Joiner, 1986).

In Section 2 of this chapter we describe the early period of using POE within MWD, and the problems associated with establishing and maintaining a POE program within an organization. Section 3 discusses POE as an organizational process and the contribution it can make to improving design quality and standards. In Section 4 we go on to consider recent and current changes in MWD's mode of operation as an agency producing public buildings, and the role of POE in the growing field of facilities management. Then in Section 5 we discuss some of the organizational problems of building a data base out of POE for use in future design. Finally, we attempt to draw some conclusions from MWD's eight years of experience as an organization employing POE.

ESTABLISHING AND MAINTAINING POE

MWD in New Zealand was, until April 1, 1988, the government department responsible for undertaking major public works projects. Recent restructuring has converted MWD into a state-owned enterprise (WORKS) which is now operating like a private sector corporation. Commercialization means that the state-owned enterprise now provides design and property management services for private sector clients. Equally, most government clients are now free to commission a private sector firm as an alternative to WORKS. The operating units of MWD consisted of seven district offices, with a Head Office in Wellington. In its restructured form as WORKS, there has been a decentralization of control to district offices, with the central office now providing co-ordination and technical support to the districts.

The POE program was initiated before restructuring, with the twin aims of improving accountability of MWD's design services, and generating information from evaluating occupied buildings, that would contribute both to the improvement of existing buildings and to policy for future design. An innovative feature of the POE process was its participative nature. The research team from Victoria University showed in a comprehensive literature review (Daish, Gray, and Kernohan, op. cit.) that most building evaluation had been done against physical rather than use criteria. Although they found an extensive body of theory on behavioral methods of evaluation, most of these were untested and at any rate did not meet their need for methods which were quick and pragmatic, but also scientifically rigorous. Since user participation had been established as a key principle in MWD's POE strategy, the researchers had to develop their own process and techniques for user involvement.

Two techniques in particular are characteristic of POE in New Zealand. One is the "Walkthrough" or "Touring Interview" method, whereby small groups of users, designers, builders, managers and other concerned parties walk through the building observing, commenting and evaluating. Walkthroughs are planned, managed and facilitated by a Task Group, which is responsible for putting together a report on the Walkthrough POE. The second technique is the Focus Study, a more concentrated evaluation which examines in detail, again with user participation, a particular problem or success identified in a building.

Three other principles have guided the development of MWD's POE techniques. Firstly, techniques were intended to be robust, that is, capable of being used by a variety of people, generalists as well as specialists. Secondly, POE was to be a "bottom-up" rather than a "top-down" process, which could be initiated locally according to need. Thirdly, POE needed to be able to clearly demonstrate its own cost benefits.

Reviews in 1983 by the Victoria University research team, and in 1984 by the second author, whose firm was then appointed as consultant to MWD to assist in various aspects of organizational development, assessed progress on the POE program. It was clear at that stage that in some ways the POEs done to date had been extremely successful. They had yielded good information for upgrading or fine-tuning the buildings under evaluation, and they had generated positive attitudes and increased awareness among participants with respect to environmental issues and aspects of building use. Where POEs had been less successful was in generating information of long-term value in the design of future buildings, and in stimulating demand for an increasing number of POEs to be done.

We shall discuss the issue of long-term information and establishing a data base in Section 5. Here we wish to discuss the problem of creating and sustaining demand for POE in a organization where its utilization is a matter of choice. While the POE program was actually set up by the former Head Office of MWD, the system allowed individual evaluations to be initiated either by the client for a particular building, or by the district office which designed it. With the slow growth in demand for POE after the first three years or so, it was realized in Head Office that if the program was to be sustained effectively, POE would have to be marketed in a more positive way than before to its potential "consumers."

The authors devised two possible marketing strategies. A "supply" strategy would be based on the idea of maintaining the POE "product" in its present form, while making greater efforts to "sell" it to the consumers. A "demand" strategy, on the other hand, would consider how the existing product could be modified, or indeed how new products might be developed, to appeal more to the consumer. A demand strategy was suggested by "market research" indicating that the cost-benefits of POE in its present form were not being perceived by everyone. Reasons given for not initiating POEs were that it was too expensive and time-consuming.

MWD decided that both strategies were needed. The existing POE product was thought to be well-founded and essential to the longer-term objectives of building a data base, and making an input to policy. At the same time there was an argument for adding to the range of POE techniques some which were even quicker and cheaper than the participative walkthrough, and which would have greater appeal, particularly to the busy architect. For it was the architects in the district offices who were showing the most reluctance to use POE.

In implementing these strategies, four types of organizational change measure were devised:

Structural

A structural measure involves a permanent or semi-permanent change. Two structural measures were decided on by MWD. The first was to establish a special POE unit at Head Office. The second was to re-define in a formal way the services offered by MWD to its clients, to make POE a structural part of those services. A structural measure provides a continuing stimulus to the result it is designed to bring about.

Motivational

Motivational measures attempt to change people's attitudes through persuasion, reward, incentive or disincentive. MWD had been doing this in relation to POE since the program was introduced, in the form of "brushfire" tours of the district offices. But the problem with motivational measures is that the results do not necessarily last. Attitudes may revert if the desired change is not compatible with individual's work aims as they perceive these, or if individuals are replaced by others.

Informational

Informational measures present rational arguments which are expected to make sense to those receiving them. Their effectiveness is limited if the logic of these arguments does not appeal to or fit with current objectives of the audience. Nevertheless, full information is essential, and MWD decided to spell out more clearly the aims and benefits of POE by improving the literature they produce for potential users.

Technical

Technical measures involve developing new or improved techniques which will make it easier for those concerned to achieve the desired aim. The research team at Victoria University was commissioned to develop new techniques better suited to the needs of working architects. These are described in Kernohan et al., 1987.

In relation to these four types of measure, it is clear that a supply marketing strategy for POE, which aims to sell an existing product more effectively, will gain more from the first two, while a demand strategy will benefit most from new techniques and from information about those techniques. However, MWD's experience suggests that all four types of measure are most effective when acting in concert.

A final consideration in relation to maintaining a POE program is the question of who pays for POEs. In MWD it was the client or district office that paid; hence the marketing analogy. However, as discussed in Section 5, there was a move towards additional POEs commissioned by Head Office. If the entire program was centrally funded, the problem of sustaining POE would no doubt be reduced. In several other organizations known to us, ownership and cost of POE is shared by the supply organization and its clients (e.g. Health and Welfare Department, Canada). Given the multiple levels of application recommended by Shibley (op. cit.) for POE to be sustained successfully in an organization, this arrangement is likely to be a common one. In that case, the above marketing considerations will be relevant to most organizations.

POE AS ORGANIZATIONAL PROCESS

Joiner (1984) and Ellis & Joiner (1985) developed the idea of design quality being negotiable. We argued that design quality depends on the effective management of resources: not just physical and technical resources, but human resources too. Furthermore, the whole notion of design quality lacks relevance if not applied to a building in use; in such cases it depends on the definitions of quality held by all those involved, users as well as planners and designers. Establishing design quality through the processes of designing, occupying, and using a building is in this sense "political" (Ellis, 1984), and is characterized in varying degrees by social negotiation about standards and about perceptions and definitions of quality. "Good physical design is a necessary but not a sufficient condition for design quality" (Ellis and Joiner, 1986).

Joiner's assertion that design quality is negotiable was both descriptive and prescriptive. It was descriptive in accepting as valid the overwhelming evidence that quality is perceived by users not purely in physical terms: their ideas about quality are strongly related to their social experiences in relation to a building. It was prescriptive in arguing that even physical design standards should ideally not be fixed or laid down in advance, but should be subject to negotiation to suit changing circumstance.

With both these purposes in mind, MWD has harnessed POE to serve the organizational process of negotiating design quality. In managing the resources of building design and production, participative POE has the particular quality of linking technical and human resources. It brings users and other parties together in communication with each other in direct relation to the physical environment. It is quite clear from MWD's experience that this process is essentially one of negotiation. Each individual or group enters into a POE with a particular perception of design quality. During the POE these perceptions are influenced and changed, for better or worse. These changed perceptions are then communicated to others in the organization, either by word of mouth, or by written report. POE affects design quality not only through the formal process of changing the physical environment – either of the building under evaluation, or of future buildings through revision of standards and procedures – but through the informal processes of communicating and negotiating those attitudes which are an essential component of design quality as we have defined it.

POE AND FACILITIES MANAGEMENT

MWD was traditionally an organization which produced and delivered buildings on behalf of clients. However, in the last few years MWD's market orientation shifted, with a greater emphasis on building and facilities management and a continuing consultancy service to clients. This came about partly from pressures to be more commercial and competitive with the private sector, and partly from changing attitudes about buildings in the market as a whole. The recent advance of the Facilities Management movement, seen throughout the Western world, has resulted at least in part from an increasing awareness, particularly among building owners and managers, of the need to manage the process of supplying and maintaining physical accommodation more professionally.

The growing use of POE is connected with this market shift. Evaluation is a logical component of the process of managing facilities; in order to make good decisions about the future allocation of facilities, you are obliged to take stock of their present allocation. This applies as much to buildings in use as to any other resource. As organizations which supply whole buildings, furniture for buildings, or design services for buildings respond to market changes by shifting towards a consultancy mode of operation, so are they more likely to introduce POE into the range of services they offer.

There is plenty of evidence of this happening, both in North America and Europe. While the POE movement has been led by large public sector organizations like MWD (Shibley, op. cit.) the private sector is gradually following. In the office field, several furniture companies are now offering some version of POE as a service to their customers, as are a few interior design and space planning practices. On the whole, the architectural profession is slow to follow this trend. As MWD found in its own organization, architects appear to be fixated by the notion that their business is the production of new buildings, and are either unable or unwilling to shift to the idea of providing a continuing service to their clients.

That this should be so is ironic, since evaluation is an inherent part of most architects' mode of operation. In the early stages of the design process, it is very common for an architect to seek out buildings of similar type, and to evaluate them informally through reading journal articles, making a brief visit or having discussions with colleagues. That this activity should be seen as different in principle, rather than in degree, from "POE" is as much a reflection on the POE and those who designed it as on the architects themselves. In MWD, the response to this perceived gulf between architects' normal evaluative activity and POE was to re-design POE to bring it closer to the everyday needs of architects. As a result, POE has been turned into a more flexible set of tools which can be used in a variety of situations. In this form it is more closely integrated with the design process, and its market appeal both to architects and to their clients has increased.

BUILDING A DATA BASE

One of the basic market questions underlying MWD's recent restructuring into a Corporation has been: what can a large public sector design practice offer to its clients that the private sector cannot offer? In the case of MWD, and now WORKS, one of the answers to this was: the advantages of scale. And one of those advantages is the scope a large organization has for maintaining and deploying specialist sources, such as a unit for accumulating POE data and translating it for future use by the organization.

It is arguable that the immediate application of POE data – to the upgrading or fine-tuning of the building under evaluation, or to an architect informing himself in relation to an immediate design problem – does not require a specialist resource. The principles and techniques can be incorporated in guide books, with perhaps a central co-ordinator monitoring and managing the POE program. But MWD's experience is that the longer-term task of building a data base is a specialist operation, requiring dedicated people with suitable skills.

Reviews of some of the POE reports produced by MWD have shown that the way information is recorded and presented is crucial to its usefulness in future situations. For example, the knowledge that the users of a particular building did not like the quality of the floor covering is of limited use to a designer in another situation. What he would need to know is why they did not like the floor covering; the uses to which it was normally put; the reason it was specified in the first place; and some technical details about its specification, performance and maintenance. Ensuring that this quality and depth of information is collected consistently during the POE process itself requires, in our view, a specialist overview, preferably from the same specialist who will be responsible for accumulating and translating that information into a form for future use.

The newly formed WORKS organization includes such a POE specialist. The job involves internal and external marketing of POE, advising district offices and their clients on its use, and compiling a data base. The POE specialist is part of the "Practice Support Group," the label given to the revised former Head Office operation, after restructuring had redefined its role as one primarily of support rather than of control. As discussed earlier, from its beginning MWD's POE program was designed on this model, with the aim that POEs should be initiated at local level according to need, with only guidance and support from the center. The need to foster and sustain the POE program led to the internal marketing activity described in Section 2, which is still compatible with the supporting role now prescribed for the former Head Office. However, the specialist requirements for establishing a data base from POE do require a more pro-active role for the central resource. In order to acquire the necessary data for this, the POE specialist may want to initiate POEs and specify particular methods of data collection which are not in the immediate interests of the owner or users of the building concerned.

WORKS are at present in the early stages of establishing a data base from POE, and still exploring the organizational issues arising. These include the issue of who should pay for a POE initiated by the Practice Support Group. Some regard it as unreasonable that the client or district office should pay if they are deriving no immediate benefit; on the other hand, a product in the form of useful design guidance should benefit them in the long term. If the Practice Support Group pays, it is in a better position to obtain the data it requires.

The form that such design information should take is another issue under debate. One possibility is to follow the example of the US Army Corps of Engineers, with their very thorough Design Guide Publication Series (Shibley op. cit.). Another is to integrate POE data into a computerized information system currently being developed.

CONCLUSIONS

An increasing number of organizations in both public and private sectors are likely to be establishing POE programs as part of their increasing concern with facilities management. In discussing MWD's experience of the past ten years, we are conscious that there can be few universal rules for the successful organization of POE. But we believe that those aspects of MWD's experience which we have singled out will be relevant to many. POE, like other institutional implants, must be sympathetic to the host culture. Participative POE, while contributing to the organizational process for democratic organizations, may be alien to others.

We believe it worthwhile to reiterate the principles laid down by Shibley (op. cit.). First, start simply. In retrospect MWD might have started even simpler than it did, by building on architects' day-to-day evaluative activity, rather than introducing a concept perceived as something distinct from this. Second, know your organization. Here we believe MWD has been quite successful, mainly because the POE program was designed "from the grass roots," rather than being imposed from above. Shibley's third principle states the need to develop multiple levels of application. From MWD's experience, it is apparent that this a long-term process, but a very necessary one.

POE will only prove its long-term value to the organization and become fully institutionalized when it has developed an operational data base, which is being successfully used to influence new design and design policy. While WORKS has made a lot of progress in demonstrating the immediate and short-term benefits of POE, it has only just started on the road to realizing the longer-term benefits. The biggest problem in the near future is going to be the integration of the very different demands which these short and long-term interests make on the POE effort.

REFERENCES

Daish, J., Gray, J., and Kernohan, D., 1983, *Post-Occupancy Evaluation of Government Buildings: Final Report*, Wellington, New Zealand, School of Architecture, Victoria University.

Kernohan, D., Daish, J., and Gray, J., Building Evaluation – Data Gathering and Review, paper given at the Conference on People and Physical Environment Research, Perth, Western Australia, 15-19 June 1987.

Ellis, P., 1984, Making Buildings Work: Should Designers get Involved in Politics?, in: Powell, J.A., Cooper, I., and Lera, S., *Designing for Building Utilisation*, London, Spon, pp 217-229

Ellis, P. and Joiner, D., Organisation Development as a Means of Improved Building Delivery, paper given at 9th International IAPS Conference, Israel, July 1986.

Ellis, P. and Joiner, D., Design Quality is Negotiable, paper given at People and Physical Environment Research (PAPER) Conference on "Place and Placemaking," Melbourne, June 1985.

Joiner, D., 1984, Quality in Design, *Proceedings of the 1984 Senior Architect's Conference*, Commonwealth of Australia, Department of Housing and Construction, Canberra, April 4-6, 1984.

Shibley, R. G., January 1985, Building Evaluation in the Mainstream, *Environment & Behavior*, 17:7-24.

CHAPTER 23

POST-OCCUPANCY EVALUATION AT THE URBAN SCALE IN BRAZIL

Geraldo G. Serra

Faculty of Architecture and Urbanism
University of Saõ Paulo, Brazil

INTRODUCTION

During the years between 1968 and 1973, Brazil experienced the so-called "Brazilian miracle," which eventually resulted in a hundred-billion-dollar debt. The population was growing fast and urbanization even faster. The GNP will soon reach $300 million and the annual per capita income will reach something like $2,000. There is not much meaning in this "per capita" figure, because income is concentrated within a relatively small population of the very rich. During the past 20 years Brazil was ruled by an authoritarian government: indeed, many people are describing the regime as "authoritarian centralism," which is a good but somewhat euphemistic characterization.

This chapter describes the creation and subsequent evaluation of the National Program for Medium-Size Towns, as well as the CURA program on urban improvement, which utilizes an adapted post-occupancy evaluation methodology. As a result, several implications of the research can be noted:

1. POE can help in the creation of minimum standards for such facilities as public amenities, day nurseries, schools, etc., in a situation where no standards exist at this time.
2. POE can be used as a means to assess cultural differences in relationship to such standards and the human requirements they are based on.
3. POE methodology, if applied to a larger urban context, can produce a valuable set of data, not only in regards to the specific buildings and neighborhoods concerned, but also on a broad urban and regional scale.
4. This kind of research could serve as a tool in assessing the programs of international aid agencies in developing regions.

POPULATION TRENDS

From 1950 to 1980 the total population increased from 52 to 119 million, an increase of 128% in 30 years. However, during the same period the urban population increased from less than 19 million to more than 80 million, growing by 321%! The urban population, which was 45% of the total in 1960, grew to 67% of the total by 1980. The trend indicates that 75% of the population will live in urban areas by the end of this century, at which time the total population will be more than 165 million. The number of rural area inhabitants at the end of this century will be the same Brazil had in 1950. This implies a need to build at least 10 million new houses, or perhaps even more than that, since there is already a large housing deficit.

During these years there was a feeling of urgency, not only to build houses, but also to supply urban infrastructure and amenities, social services, and most importantly, employment. It is true that the authoritarian centralist regime did a lot of building. But now, the "new republic" has a mandate to evaluate, as there is a need to know what the authoritarian centralist regime built, where it built, how much money it spent, and so on. Why is public housing of such poor quality? Why are our cities rapidly deteriorating? What happened to our environment? Who won and who lost? These are some of the questions that need to be answered.

In the past few years a significant amount of research has been carried out on such topics as public housing, urbanization, planning, water supply, sewage systems, etc., constituting a critical review of those 20 years of authoritarian centralism. This research is using several different methodological approaches. If one tries to find a common denominator among them, it is the critique of the positivistic vision of the process: i.e., to look for the basic cause of the problems and conflicting different interests, and not merely for technical errors.

Some of the government agencies tried to control this criticism by sponsoring research based on positivisitic premises. Such was the case with the evaluation of the "National Program for Medium-Size Towns," in which the concepts of "efficiency" and "efficacy" were imposed a priori, but even so, the critical attitude prevailed.

It was during this period of establishing appropriate evaluation approaches that the cooperation between the University of Saõ Paulo (USP) and the University of New Mexico (UNM) was established. Based on the mutual existing agreement, several

TABLE 1
Brazilian Population
Total, Urban and Rural

Year	**Inhabitants** (in millions) Total	Urban	Rural
1940	41.2	12.9	28.2
1950	51.9	18.8	33.1
1960	70.1	31.3	38.8
1970	93.1	52.1	41.0
1980	119.0	80.4	38.6
1990	150.4*	112.8**	37.6**
2000	179.5*	152.6**	26.9**

Source: FIBGE, 1987
Projections by: *FIBGE, 1987 **Author

faculty members began to collaborate on post-occupancy evaluation (POE) methodology. In the past, the general methods and characteristics of POE were mostly used in the evaluation of individual buildings. The methodology had not been applied to broader urban subjects, such as urban renewal projects, urban infrastructure, and urban social services. This chapter deals with a research effort aimed at expanding the original POE concept in order to address projects at the urban scale.

EVALUATION OF THE NATIONAL PROGRAM ON MEDIUM-SIZE TOWNS

At the end of the 1950s, the Brazilian urban population was 70% larger than it was at the beginning of that decade. As the momentum of this increase was maintained during the 60s, the growth process produced a severe shortage of urban amenities, i.e., housing, infrastructure, and social services. The entire Brazilian society was shaken by a crisis which eventually ended with the installation of the authoritarian centralist regime.

From the very beginning, the new rulers were concerned with the so-called urban question. There was a drive toward urban reform, new urban development laws, public housing projects, new financing systems, etc. The administration, in response, created the National Housing Bank (BNH) and the National Commission for Urban Development (CNDU). The first was created to finance public housing and urban infrastructure, using money from the Guarantee Fund for Employment Time (FGTS), which is a kind of unemployment insurance. The second had the task of preparing plans and proposing laws dealing with the explosive urban growth. During the 20 years, from 1967 until its closing in 1986, the BNH financed the building of some 4.5 million housing units. As for the CNDU, it proposed some urban reform laws which were never approved by Congress; but it also devised some programs and policies. One of these was the National Policy for Urban Development.

The main directive of this document (Resolution n.003, of September 11, 1979), was the decentralization of the economic activity from the Saõ Paulo and Rio de Janeiro metropolitan regions to other, undeveloped areas, mainly in the northeastern region of Brazil. This was in sharp contrast with the actual economic policy, which was concen-

TABLE 2
National Program for Medium-Size Towns
Investments

Region	Number of Towns	Investment (dollars)	Percentage of Total
North	13	6,555,897	0.9
Northeast	44	243,422,634	34.5
West-Central	16	59,548,128	8.4
Southeast	86	326,164,807	46.2
South	35	70,523,554	10.0
Total	194	706,215,020	100.0

trating economic activity and the income in few hands and in few places. With this began an even more profound separation between plans and actions, a separation which was typical of the authoritarian centralist regime.

One of the programs of this policy was the National Program for Medium-Size Towns (PNCPM). The theoretical base of this program, as well as of the entire policy, was the theory of the "metropoles d'equilibre." The general goal was to improve the quality of life and the opportunity for employment in the medium-size towns. This was done in order to increase their attractiveness for migration and to discourage in-migration from the metropolitan areas. Although the National Policy was implemented in 1979, the Program for Medium-Size Towns had already begun in 1976. Table 2 shows the investments of the PNCPM, in the various regions. Unfortunately, the influence of local politicians ended up having more to do with the criteria for selection than did, for example, the size and population of the towns.

In regards to table 2, it is worth noting that:

1. The greatest expansion of the urban population happened in the Northern, West-Central and Southern regions, exactly the ones which received the least funds from the program over a period of six years.
2. The towns included in the program represent less than 5% of Brazilian towns.
3. Although the average amount invested per town ($364,028) was significant, most towns received too little to have any meaningful impact. Also, unfortunately, many of the towns could not be considered medium-size towns by any criteria. Table 3 shows the criteria used to select towns for the evaluation program.

In the beginning, it was intended that the program be supported by federal

---

TABLE 3
National Program for Medium-Size Towns
Selection Criteria

Spatial Criteria

- Regional Relevance
- Location in Relation to Principal Axis
- Existence of Special Programs in that Field
- Distance from other Urban "Agglomerations or Centers"
- Strategic Situation

Intra-Urban Criteria

- Demographic Size
- Recent Performance
- Great Proportion of New Immigrants
- Structure of Economically Active Population
- Urban Poverty
- Recent Urban Evolution

funding only. However, in July 1979, a loan agreement was signed with the International Bank for Reconstruction and Development (BIRD), and in 1984 and 1985, CNDU sponsored a research program with the Faculty of Architecture and Urbanism of the University of Saõ Paulo (FAU/USP) to evaluate the National Program for Medium-Size Towns. As the scope of this research was very large – it covered the entire national territory – it was decided to organize the research effort with separate research teams for each of the regions: North, Northeast, West-Central, Southeast and South. Hereafter, the focus of this chapter will be on the evaluation of the Northeast region.

It was decided at the very first meeting with the government that the main categories of analysis should be "efficiency" and "efficacy." The efficiency concept referred to correct timing and budgeting in both program planning and execution. The efficacy concept was the capacity of the program to reach its goals, i.e., to promote a change in the general trend of economy and migration. Table 4 shows the priorities of the PNCPM.

Had the researchers kept the focus of research within the limits of these categories, the result would have been a clear example of a "positivistic" approach to evaluation. In that case they would have presented the conclusions as a set of recommendations merely dealing with complex technical matters. This kind of result would have introduced confusion into the evaluation, the relevant social and economic questions would not have been addressed, and the program would have looked like nothing more than a technical inquiry.

METHODOLOGY

The five regional research groups conducted frequent meetings to discuss and

---

TABLE 4
Program for Medium-Size Towns
Priorities

Decentralization Centers

- Transportation and Regional Communication
- Space and Infrastructure for Industries
- Incentives for Industries
- New Industrial Districts
- Protection for Local Industries
- Environment

Centers With Dynamic Function

- Regional Transportation
- Incentives for Trade and Primary Products
- Credit Facilities
- Managerial Help and Training
- Transportation and Telecommunication
- Urban Social Services

---

unify the methodology and the main categories of data that would be collected. During these meetings a more critical approach toward evaluation evolved. The two data categories of efficiency and efficacy were maintained, and other concepts were added. Risking the future sponsorship of their research, the teams confronted the Government with the contradictions inherent in the program.

First of all, documents from CNDU and demographic and economic data from the Brazilian Institute for Geography and Statistics were obtained and analyzed. Then the investment plans for each town were examined and key problems and deficiencies were identified. These investment plans were documents which contained a general appraisal of a town's infrastructure and social services, including pending proposals. The plans also provided for certain public works projects and investment estimates. They had been prepared by local staff following the guidelines and instructions of the program. Unfortunately, the guidelines included serious misunderstandings, especially during the phase financed by BIRD. For example, they recommended reinforcing the traditional producers of handicrafts, traditional fisheries, laundries, etc. People, however, wanted to change their condition of low productivity and not to reinforce it! The result was that fish from an industrialized fishery in Santa Catarina – 4,000 km from the town in which the investment was made – was cheaper than the locally financed and produced product.

Each of the investment plans contained items that were carried out by federal and state, as well as local, authorities and agencies. The main conclusion referring to efficiency was that local agencies were more efficient than the state ones, and in turn state agencies were more efficient than federal ones.

The program's efficacy was evaluated comparing the population of the towns included in the program with the population of the metropolitan areas of Saõ Paulo and Rio de Janeiro, between the census dates of 1970 and 1980. The evaluation took into account the projections for these populations in 1984, prepared by the Brazilian Institute for Geography and Statistics (IBGE). The conclusion was that the major part of the Northeastern medium size towns increased their population proportionally more than the Saõ Paulo and Rio de Janeiro metropolitan areas. At this point it was hoped to verify that the performance was a consequence of the program; however, this proved impossible to show. Indeed, there was a relationship between the population increase and the investments in new plants and other urban economic activities not financed by the program. The major part of these investments was accomplished with income tax incentives by private investors, supervised by a regional development agency.

The main conclusions of this research are:

1. The intentions of the program were in contradiction to the actual economic policy of the government. In fact, this policy did not have any components dealing with spatial or personal distribution of income;
2. This trend was reinforced after 1979 in the special projects developed by the federal government to increase export surpluses;
3. If the program had as a goal to reorient the migration, it had to deal with spatial location of employment and industry, but the program accomplished nothing in this regard;
4. The improvement of living conditions in these towns did nothing to attract migration and industrial investment and therefore new employment;
5. The federal government could have promoted the reorientation of

migration by deciding to locate the new plants of federally owned enterprises (which controls a significant part of the GNP) in the towns the government considered more convenient for attracting migrants, and to do that the government would not have needed the program;

6. If this could not be done because it was in flagrant contradiction to economic policy, then the program had other objectives, which could have included the will to interfere in local and regional politics.

The evolution of the evaluation methodology, from a mere technical approach to other more sensible, social and economic aspects, showed its advantage in this case. During the research the need was felt for an evaluation methodology that would incorporate the opinions of the people who lived in the towns included in the sample.

EVALUATION OF THE "CURA" PROGRAM

In 1973 the National Housing Bank instituted a program for urban improvement called "CURA." The name stands for Urban Communities for Accelerated Upgrading. CURA, in Portuguese, means "cure," which suggested that the city was ill and that the program was the medicine. Conceptually, CURA was based on two main principles: deficiencies and underdevelopment. The theory was that deficiencies in and underdevelopment of urban infrastructure and social services exist in both cities and towns, due to the accelerated urbanization process and the actions of real-estate speculators. Indeed, this theory presupposes a certain minimum standard of urban space, including paved streets, curbs and sidewalks, trees, electricity, light, water supply, garbage collection, telephone, drainage, squares and parks, schools, health care centers, child-care centers, community centers, leisure and sports facilities with equipment, and so on. If some of these items were lacking in a certain district or neighborhood, but at the same time there were many unoccupied parcels there, then the infrastructure was undeveloped.

CURA's answer to this problem was to finance the building and installation of all the lacking items, in the hope of thereby eliminating the deficiency. An annual increase of municipal taxes on unoccupied lots was imposed in order to force the owners of those lots to build on them or sell them. It was expected that through this plan the availability of urbanized lots would increase. The plan was to cost less money than it would cost to build an entire new housing complex. The program would take advantage of the underutilization of the already installed infrastructure and service and would thereby improve the urban standards of these districts and neighborhoods.

By 1984, several contracts were made between the BNH, the universities, and other institutions, in order to evaluate the CURA program. The institutions began to collect and code investment and physical data in preparation for statistical analysis. Summaries of the main aspects of each case, including maps and designs, were also collected and analyzed. By 1986, the opportunity arose to include post-occupancy evaluation (POE) methodology in the research plan. A team of 8 researchers from the post-graduate courses of the FAU/USP began to apply POE methodology to evaluate the CURA I in Osasco, which is one of the municipalities in the São Paulo metropolitan area. The final reports of the Osasco evaluation were completed by the end of July, 1987. After that, a similar research program, with a larger team of 10 architects and engineers, was developed in Cotia, which is another municipality in the Saõ Paulo metropolitan area. The final conclusions were ready in December of 1987.

Presently, the conclusions of the evaluation of the CURA Program include the following:

1. Since the main objective of the program was to increase the density of the selected area, there is enough evidence to confirm that the program was a success. Thus, in most of the cases analyzed, the increase in the number of occupied lots inside the CURA Area was greater than the increase outside the CURA Area.
2. Up until August of 1985 the program was implemented in 136 towns; now this number has increased to near 150. This comprises less than 4% all the Brazilian cities and towns, thereby very much reducing the importance of the program in the country's accelerated urbanization program.
3. The total amount of investment in the program, from 1973 to the present, is near 400 million dollars. This is very little money, considering the time span of 15 years and the number of 150 towns. It is even less than the amount invested in the Medium-Size Towns Program.

Nevertheless, in order to get a more detailed insight at the community level, a new methodological tool was needed. A suitable approach to the evaluation of urban infrastructure and services was developed from the general POE process model and principles. The following description of the research approach in the Osasco CURA evaluation will give a better idea of the CURA performance.

POE AT OSASCO

Osasco has an area of 67 square kilometers and a population (in 1980) of 473,856; its population density is 7.0 inhabitants per square kilometer. The entire Osasco area is urbanized, but there are also many open, unused spaces within the area. The city is heavily industrialized, mainly in the electrical equipment and building materials fields.

The planners chose an area of 400 hectares divided by the Tiete River. A program was developed for the area, including 37,000 square meters of street pavement, with sidewalks, curbs and drainage. The program also included the building of 5 squares, 4 child-care centers, and 3 sports facilities, and the installation of 2,302 street lights. The loan from the BNH to the Osasco municipality was approximately 3 million dollars, or 90% of the estimated project cost.

The evaluation of the streets and drainage was done by two civil engineers. They obtained the design plans and documents from the technical offices of the municipality. They inspected the streets, the drainage system, and bus shelters. They interviewed people living on those streets, taxi drivers, bus drivers and passengers, owners of shops and stores, garbage collectors, and other municipal employees. They also talked with the designers of buildings and with the municipal technical staff.

The conclusion from the street evaluation was that the asphalt pavement, the walks, the bus stop shelters, and drainage system had performed well and that people were satisfied with them. Opinions did differ between drivers and non-drivers about speed bumps built on the major arterial road, but that was quite understandable. Some local streets have pavement of a different specification called "simple superficial treatment." It is in very bad shape, full of holes, and without any maintenance. The main recommendations are to pave the local streets with asphalt, paint the bus stop shelters, and clean and maintain the drainage system.

It is important to note the effort that was undertaken to adapt the POE questionnaire technique to such a large and differentiated sample of respondents. It proved both possible and useful.

As for the child-care and sports facilities, the POE methodology was applied by using the the standard format. Several problems of construction and design were noted and recommendations were made. The most interesting case was of the Saõ Jose Sport Center. The entrance is about 4 meters (13 feet) below street level, from where it is possible to see its metal roof. Children threw stones on the roof, which became dented and in some places even perforated. Thieves then vandalized several parts of the building, including the guard house itself. Despite that, people continue to use the center for meetings and sports events.

In this case, the recommendation was for complete restoration of the building and its surroundings, including the grounds and sports installations. It was also recommended that a new wall be built, that security guards be posted 24 hours per day, and that a metal protective shield be constructed on the side from which the stones are thrown.

CONCLUSION

It is interesting to note how the collaboration between the University of Saõ Paulo and the University of New Mexico has evolved from the evaluation of buildings to the evaluation of entire urban renewal programs. This required certain methodological adaptations which were made during the process of POE research in response to practical requirements. In the Osasco research project some questionnaires proved to be not so successful. They made respondents suspicious because it is not easy to interview people in the streets or in public places.

A new project is now underway in Cotia, a town near Osasco. The technical staff is trying to perfect the research tools, especially the questionnaires. Thus has the advantage of utilizing POE methodology in the assessment of the efficiency and efficacy of public investment become evident.

BIBLIOGRAPHY

Banco Nacional da Habitacaõ, 1974, *BNH 1974*, Rio de Janeiro.

Banco Nacional da Habitacaõ, 1974, *BNH Documenta*, Rio de Janeiro.

Banco Nacional da Habitacaõ, 1973, *Manual CURA*, Rio de Janeiro.

Dickenson, J.P., 1984, in: *Geografia del Tercer Mundo*, Barcelona, Omega.

Durhan, Eunice R., 1978, *A caminho da cidade*, Saõ Paulo, Perspectiva.

Faculty of Architecture and Urbanism, University of Saõ Paulo, 1984, *Avaliacaõ do Programa Nacional de Cidades de Porte Medio e Revisaõ de Criterios de Selecaõ das Referidas Cidades*, Saõ Paulo.

FIBGE, 1973, Ministerio do Planejamento e Coordenacaõ da Presidencia da Republica, *Anuario Estatistico do Brasil - 1972*, Rio de Janeiro.

FIBGE, 1986, Secretaria de Planejamento e Coordenacaõ Geral, *Anuario Estatistico do Brasil - 1986*, Rio de Janeiro.

CHAPTER 24

BUILDING-IN-USE ASSESSMENT: ANALYSIS OF OFFICE BUILDINGS

Jacqueline C. Vischer

Division of Capital Planning and Operation
State of Massachusetts
Boston, Massachusetts

INTRODUCTION

Going back to evaluate a building once it is occupied to see how people are using it, and whether the assumptions made by designers and other team members during its planning and construction bear out, seems like an idea whose time has come. No one can disagree with it; it ranks right up there with cleaning out the refrigerator and sending money to those organizations for handicapped children that mail you colored stamps shortly before Christmas.

Convincing cases for the usefulness, ultimate value, importance, and essential goodness of Post-Occupancy Evaluation (POE) have been made repeatedly in the research and technical literature. No one can refute the logic of a practice that promises a solution to building management problems, an increase in overall knowledge about buildings, an improvement in the quality of future buildings, and a more rational planning and design process. But evaluating buildings is far from being a commonplace activity, and POE is still the exception rather than the rule. A POE might be employed when a building has a serious problem and attempts to solve it have not been successful. It is rarely applied routinely to normal, unexceptional buildings, or to buildings that work well for users.

Office buildings have become popular candidates for POE studies, in part because there are so many, and in part because many of them provide workers with dysfunctional interior environments. Office environment studies have provided a new direction for POEs and solved some of the problems of doing them. In this chapter, we review some of the reasons why POE has not fully emerged from its academic cocoon to become a practical and systematic activity, and we offer some alternative ways of thinking about evaluating office environments.

Reasons Why POE is Not Popular

1. *Mistrust of surveys.* Almost without exception, POEs involve surveys of building users. These are time-consuming and often raise users' expectations for building

improvement. If these are not met, workers complain to management. Survey results can reveal to management unwelcome information about problems and failures in the building for which they are responsible. A badly-run survey can result in no usable information at all.

2. *Changing uses.* An evaluation of something must have a frame of reference. It is customary in evaluating buildings to refer back to the design objectives and/or program goals to determine whether or not the building as it is being used has met these goals and objectives. The uses of the building change over the lifetime of the building – a process that can begin during design and construction. It is not unusual to find that the problems and failures found in a building are a direct result of changes in use, changes that are not in the control of the designers, the evaluators or the occupants. As a result, using the customary framework for POE, it is easy to conclude that evaluating buildings is pointless.

3. *Professional insecurity.* It is sad but true that what is, on the face of it, a logical way to conduct building research – that is, to examine design objectives and program goals and determine whether or not they have been met in building design – cannot avoid being a commentary on the performance of the professionals involved. If for example, users of an open plan office complain about poor acoustic privacy, and the program specified a sound-masking system that was not installed, then the project management team look bad. If the density and uses of an open plan office create a voice privacy problem for which no provision has been made, then the programmers look bad. It is not a surprise that these same professionals are reluctant to evaluate the quality of the service they delivered, and to pay for it yet!

4. *Sense of futility.* Some of the results of a POE recommend changes to a building or an environment to make it more acceptable to users. Other results are so negative that nothing can be done about them at all. For example, a recent courthouse POE found that the lack of daylight in the windowless courtrooms increased the sense of stress felt by participants in trial proceedings, and resulted in numerous complaints about the various lighting systems that were installed. However, there was no feasible way of changing the design of this courtroom to accommodate windows. Although this POE could be applied to a recommendation that future courtrooms should have some daylighting, this is a small consolation to the users of the present courtroom.

5. *Immediate pay-off.* People who advocate POEs (who are not always those who pay for them) assert that the building user is the beneficiary of a POE because changes can be made to improve buildings for users. This assumes that if the users are better off, then so are the building owners (who may well be those who pay for POEs) because the organization will work better in a building that suits its functions. This is for many owners an indirect benefit. The contribution to knowledge that accumulates over the long term informs building professions and ultimately improves the quality of buildings is not a pay-off in the minds of most building owners. Unless they are large corporations, developers, or government, owners are unlikely to build new buildings more than once in the lifetimes of the corporate executives who pay the bills. Unless the POE can solve an obvious building problem, there is no immediately obvious pay-off from doing a POE to those who pay for it.

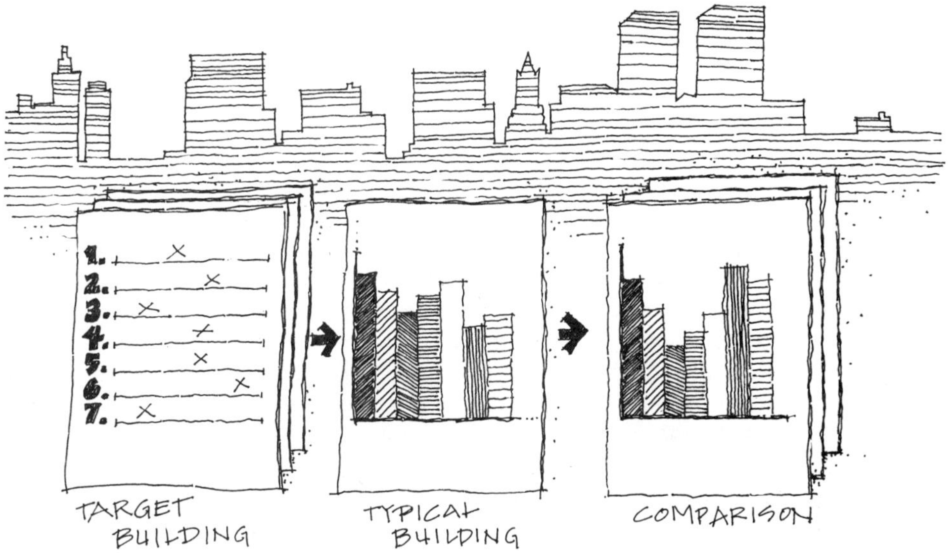

Figure 1

6. *Cost-effectiveness.* The essential value of POE is that it may lead to a more efficient, more effective environment that is better for people using it. However, it is important for those bearing the cost of the POE to understand how "better for people" translates into value for money. Housing that is better for people means less tenant turnover, faster sales, less vandalism, easier management. Offices that are better for people mean higher levels of worker productivity. The ultimate rationale for doing POEs is to improve buildings, but unless this is tied to increased worker productivity, less vandalism, or some other bottom line measure, this is not a strong argument for undertaking them. Most people know that the majority of buildings provide less than optimum environments for users, and that people manage just fine in them. Unlike POE, most services are sold on the grounds that they can produce a bottom line improvement: better quality for less cost. The improved quality that may result from POEs is a long-term proposition, and the reduced cost is not easily calculable.

Designing POE as a System

For POE to become commonplace and affordable, it must become a natural part of an organization's operation. This organization does not need to be an architectural firm, as used to be thought. Architects and other professionals benefit from the fact that POEs are done, because new knowledge is generated to apply to their projects. But architectural firms feel they do not get enough out of them to pay for them. Building owners and managers stand to gain most from POEs, followed by the design professions, and developers who have more in mind than short-term ownership and rapid turnover.

There are two kinds of POE. One type of POE is a one time only effort, usually directed at a single building. This "case study" approach is useful if information is required about a particular building, for a variety of reasons. A new building may be planned and there are lessons to be learned about the design, the detailing, the costs, and the overall effectiveness of an existing example of the same building type.

Or the POE may focus on building performance problems. In this type of POE, the building may have multiple occupancy problems, to which no obvious solution has emerged from previous analysis. The causes of building problems are often more complex than they appear: complaints about indoor air quality or access to windows from the workspace often turn out to be tangled up with flickering lights, uncomfortable furniture, poor acoustics, or other building conditions not apparent at first view. Evaluating buildings with big problems is helpful to the owner and to the industry; understanding the story of the building helps inform everyone concerned on how the mistake occurred and how to avoid it in the future.

A second type of POE is concerned with systematic upgrading and improvement of complex indoor environments. This is an ongoing operation, and not limited to a single building study. For years, POEs were done on a "one off" basis, in which findings from a case study were of interest in themselves and abstracted into recommendations for likely future buildings. More recently, building owning and/or operating organizations have collected data from several buildings and created databases on building performance and use which are only beginning to be mined. An important conceptual leap is implied by the database approach, which is the value of being able to compare environments with each other. These comparisons might be before and after a change or a move, between two comparable but different buildings, or, as in the building-in-use assessment system (see below), between a particular building and a "normal" set of comparable baseline building environments.

By comparing buildings with each other, one eliminates the uncomfortable element of evaluation, namely, the imposition of value judgments on the performance of a building. In the comparison approach the term assessment is used in preference to evaluation. A building is assessed when its performance is examined in the context of other comparable buildings; it is not being "evaluated" as something good or bad according to some larger and more absolute system of judgement.

As building performance should not be studied separately from people's activities in the building, the concept of the building-in-use incorporates the interaction of users with the environment and the effects they have on each other. The building-in-use approach to evaluating buildings is not the sum of measurements of building performance and surveys of people's attitudes and behavior, but an analysis of users and building as an interactive system. Building-in-use assessment is a conceptual approach to evaluating buildings which assumes that an occupied building is more than the sum of its parts. Building-in-use assessment is an example of the comparative approach to POE.

Building-In-Use Assessment

The building-in-use assessment methodology depends on the acquisition of small amounts of critical information from a building which can be compared to an existing database in a preestablished and proven way. A profile of office building performance that is quick and inexpensive to create can be put to use by managers in resources planning, maintenance budgeting, and trouble-shooting building problems.

The building-in-use assessment system uses the judgments of building occupants to provide an overall measure of the quality of their office environment.

Their judgments are elicited in a questionnaire survey that is standardized, short, and simple to complete. A sample of 30 can be used to test even the largest building, making this a cost-effective approach to evaluating buildings. If an organization wants to spend more money, it can develop its own database and compute its own norms. If not, then existing office environment norms will serve.

The questionnaire asks building users to assign ratings from 1 (poor) to 5 (comfortable) to a set of 22 environmental conditions. These conditions include heat and cold, ventilation and air freshness and circulation, lighting and daylight, acoustics, and furniture lay-out and comfort. Each person is asked to circle one answer or rating for each scale. The ratings are summarized into seven building-in-use dimensions: Air Quality, Thermal Comfort, Noise Control, Lighting Comfort, Privacy, Spatial Comfort and Building Noise Control. The building-in-use profile is based on the scores on each of the seven dimensions, which are computed from the 22 ratings received from occupants.

Each of the seven dimensions is a particularly salient aspect of occupants' experience of the interior of the office building, and together the scores on the seven dimensions provide a comprehensive indicator of interior environmental quality. Each building tested yields its own score on each of the seven dimensions, and these are compared to the "norms" or average scores for all office buildings, generated by the office building database.

By systematically comparing the profile of one building with the normative profile on these dimensions from a number of buildings, it is possible to evaluate how the building compares to known buildings on each dimension of the user experience of environmental quality.

Application of Results

On each of the seven building-in-use dimensions, the target building's score is either below, above or the same as the database norm for that dimension. When a building's score is below the norm, it is an indication that remedial action is required: the building needs improvement on that dimension. If the assessment shows, for example, that the lighting dimension in an office building is rated worse by the occupants than lighting is normally rated in office buildings, then building managers should improve the lighting environment in that building.

On the other hand, managers may have been receiving complaints about ventilation conditions in a building, but they determine from the building-in-use assessment that the Air Quality dimension is rated about the same as the norm. As a result, managers know their building is no worse than "normal" buildings, and they can choose whether or not to act. Each comparison between a building's scores on a dimension (its profile) and the norm for that dimension provides information on whether the building is normal, above normal or below normal on that dimension. The degree of deviation from normalcy provides a basis for action: a large deviation suggests action is urgent, whereas a smaller deviation means action can be delayed. No deviation means the building is normal, even if its scores are not very high. On dimensions where a building scores better than the norm, it may be useful to gather more information to understand better the source of its excellence.

Once the building-in-use assessment results are in, there are several possible courses of action open. One is to measure the dimensions that deviate most from the norm. If, for example, Lighting Comfort scores well below the norm, it is necessary to know whether this is due to a problem of window glare, to non-functioning overhead fixtures, to heavy VDT use and lighting glare, or to something else. Measurement using light meters and contrast meters to measure aspects of the visual environment, and analysis of the electrical system and power supply, is necessary in order to understand how to solve the problem. Additional information may be gathered through interviewing and observing workers in the problem area.

Another possible follow-up action is to develop a plan for maintenance and repair which corresponds to the priorities that have emerged from the building-in-use assessment. Those building-in-use dimensions which score lowest indicate areas of building performance that are most disturbing to users, and should therefore be taken care of first. If several of the scores are lower than normal, those dimensions that have most influence on worker productivity, satisfaction and ill-health – as explained below – are the ones to be addressed first.

Interpretation of the Building-in-use Assessment

Use of the building-in-use assessment system does not indicate whether performance of a building is acceptable in an absolute sense. The fact that a building is normal on one or several dimensions does not imply that the building is "good" or "acceptable" on that dimension. The normative rating on Air Quality, for example, is towards the lower end of the 5-point scale. The Air Quality score for a specific building, therefore, is "normal" when it is low. The normative approach to building assessment is by definition relative, and does not indicate acceptability. The acceptability of a building's in-use profile is a function of organizational goals and policy.

Not all the seven building-in-use dimensions are equally important. Although all the building-in-use dimensions have an effect on workers' comfort and their productivity, some have a greater effect than others. A regression model was constructed to test the 7 dimensions in predicting worker-rated productivity, satisfaction and occupational health. The results are summarized below.

Productivity (R^2=0.39)	*Satisfaction* (R^2=0.25)	*Illness* (R^2=0.21)
1. SPATIAL COMFORT	1. SPATIAL COMFORT	1. AIR QUALITY
2. NOISE CONTROL	2. PRIVACY	2. THERMAL COMFORT
3. PRIVACY	3. LIGHTING COMFORT	3. LIGHTING COMFORT

The item that is most important for both worker comfort and productivity is Spatial Comfort. This means aspects of furniture comfort and workstation design and lay-out, including storage, size of workspace, availability of meeting rooms and lounges, circulation, and type of furniture. The next most important dimension for productivity is Noise Control, meaning the unwanted sounds caused by equipment and people's voices that disturb workers. This is followed by Privacy, meaning the amount of visual and speech privacy workers feel they have. The second most important item for people's satisfaction is their Privacy, followed by Lighting Comfort. Good lighting, freedom from glare, and adequate daylight will increase people's satisfaction with their workspace.

The most important dimension in predicting people's health in the building is Air

Quality. This dimension includes ventilation, air freshness, feeling warm, and the experience of odors. Feeling ill causes workers to be both dissatisfied and unproductive, and may increase absenteeism, so although Air Quality is not directly important to comfort and satisfaction, it is still an important in-use dimension.

CONCLUSIONS

Of the six problems of post-occupancy evaluation (POE) identified at the beginning of this chapter, most are addressed by the building-in-use assessment approach to office building evaluation.

Organizational mistrust of surveys is offset at least in part by the short standardized questionnaire form. Unlike conventional questionnaires, this is inexpensive and quick to administer, and yields standardized information the usefulness of which is guaranteed as long as it is used in the context of a building-in-use assessment.

Building-in-use assessment deals with changes in building use by focusing on the users' experience of the environment, regardless of how they came to be there. It may be argued that length of time in one place, or impending moves, or other circumstantial factors may affect users' judgments, and this is true. Users' ratings incorporate and reflect all the circumstances that impinge upon their judgments of the workspace, just as their circumstances affect their experience of the space. Thus to try to separate users' judgments of the building from their experience in it by controlling variables affecting their ratings does not provide useful information about the building-in-use. The actual occupancy of a building is the basis for assessment; issues of past occupancy, or of occupancy that might have been, although these might impact on the current state of the building, are mainly of historical interest. The results of a building-in-use assessment are oriented to what the building IS rather than why it is.

The building-in-use assessment system sidesteps the issue of whether or not a building is generically good or bad by using workers' own experience of the environment as a basis for determining environmental quality. Regardless of the "objective" qualities of a building environment, if the workers feel comfortable and estimate that they work well in it then it is by definition a good office environment. This puts some expensive, elegant new offices at risk of being poor quality because people do not like working in them. Telling users that it is an expensive building and that they should not complain does not improve the quality of the office environment for these workers. Using their ratings to improve building performance does improve the quality of the office environment for workers.

Design professionals who are concerned about their performance may never be comfortable with building evaluation. However, building-in-use assessment is less threatening than conventional POE because the present state of the building is all that is being studied, and no criticisms or judgments are involved except those that the workers make of their own environment.

The building-in-use system is such that the person or agent who authorizes and pays for the assessment is the one who uses the results. The feedback from an assessment is directed to the concerns of the management, who uses it as a basis for planning and allocating maintenance and repair resources. Doing an assessment allays the concerns of owners or managers who need to understand the variety and range of building complaints they are receiving from building users, and who want a systematic way of responding to them.

From a planner's point of view, if workers are uncomfortable in a building, managers need a systematic way of knowing what complaints they have about the building. The random way in which individuals customarily telephone in complaints about the building is not a reliable source of planning information for managers.

Building-in-use assessment is cheap compared to conventional building evaluation, and can be carried out rapidly and conveniently. It also results in a more efficient and effective way of solving building problems than reacting to users' complaints. Building-in-use assessment represents the first step towards solving complex problems of building use, and provides a basis for taking action for building improvement. In conventional POEs, a large commitment of funds is often required at the outset, and the study maintains an open-ended strategy that makes it difficult to perceive the pay-off for each stage of investment, and to control the overall study cost.

The effort to improve an environment is vain if it is not managed, directed and controlled. Building improvement for the sake of improvement suggests an open-ended and often uncontrolled situation where dollars can be wasted on undirected efforts for environmental change. For example, complaints about indoor air quality can result in costly changes to the air handling systems, an expensive balancing contract, and lengthy and complex air quality testing procedures that provide equivocal results. By using occupants' ratings as the indicator of what is most wrong and most right about their environment, and by comparing these to a baseline of "rightness" and "wrongness" in users' judgments of environmental conditions, the improvements that result can be matched to people's concerns, to the dollars available, and to the priorities of decision-makers.

Future Directions

In the future, the building-in-use assessment of modern offices is likely to continue. More organizations will develop databases, and develop their own norms. The system is already in use in Public Works Canada for its 400 office buildings and for client agencies and others who use its norms to get a building-in-use profile of their building. As it continues to be used, the questionnaire will become refined and improved, and databases will be up-dated as new data become available.

The future also indicates the expansion of the building-in-use assessment system to other building types. The State of Massachusetts is implementing a building evaluation program for buildings that include prisons, hospitals, mental institutions, colleges and universities, police stations and park buildings as well as office space. It is not yet known whether separate databases need to be developed for each building type, or whether the environmental quality categories will hold across building types. In the latter case, the same questionnaire could be used; and it may be sufficient to compute different norms to reflect the different building types. In fact, it is possible that the norms themselves may not vary that much across building types, because users' sensitivities and preferences do not vary that much even when tasks and activities differ.

A third direction for the future of building-in-use assessment is to use it to plan for short-term and immediate changes to buildings that will improve their performance for users. Building-in-use data are valuable to help organize long-term industry-wide improvements in building quality. The technology of the building industry is constantly changing, and design and construction processes need tools to maintain the quality of environments for people. Feedback on building-in-use performance over the long term will affect the way buildings are built, and improve human health and comfort while they are at work.

BIBLIOGRAPHY

Dillon, R. and Vischer, J. C., 1988, *Technical Report on the Building-In-Use Assessment Mathodology*, Ottawa, Canada, Public Works Canada.
Vischer, J. C., 1989, *Environmental Quality in Offices*, New York, Van Nostrand Reinhold.

CHAPTER 25

POST-OCCUPANCY EVALUATION AND ORGANIZATIONAL DEVELOPMENT: THE EXPERIENCE OF THE UNITED STATES POSTAL SERVICE

Jay Farbstein
Jay Farbstein & Associates, Inc.
San Luis Obispo, California

Min Kantrowitz
Min Kantrowitz & Associates, Inc.
Albuquerque, New Mexico

Brian Schermer
Jay Farbstein & Associates, Inc.
San Luis Obispo, California

John Hughes-Caley
U.S. Postal Service
Washington, D.C.

INTRODUCTION

This chapter discusses post-occupancy evaluation (POE) as an opportunity for organizational development. That is, it looks at POE not only as a technical exercise in assessing facilities, but also as an adjunct to – and even a potential catalyst for – organizational decision making. POE, used this way, is a management tool integral to a systematic, organization-wide feedback system for assessing progress toward larger organizational objectives. In order to examine this function of POE, the chapter discusses the role of facilities in organizations, the ways organizations mobilize to manage facilities, the place of POE in facilities management, and the types of organizational analysis that can be helpful to facilities researchers who find themselves operating within large, complex organizations.

In exploring these issues, this chapter describes the use of POE by a major government organization, the United States Postal Service (USPS). The chapter outlines the influences that generated the POE program, and discusses how the program became established, early threats to the program, and how the initial results have been used. It further describes how the program responded to shifts in organizational goals and how it was used to reinforce and redefine new goals. The chapter concludes with a discussion of the extent to which the USPS's experiences with POE may be applicable to other organizations.

The Role of Facilities in Complex Organizations

Several writers have identified the roles which facilities play in organizations (Duffy, 1974; Farbstein, 1975; Steele, 1973). Facilities provide settings for the organization's activities, reflect organizational structures (such as status and functional differentiation), express organizational images, and can support or inhibit realization of organizational goals (such as productivity and employee satisfaction). Facilities also represent the largest single asset category of many organizations. The emergence of facilities management as an identified function and professional role within large organizations reflects both a recognition of their importance and a need for organizations to manage these assets actively and carefully (Margulis, 1987).

POE, as a set of systematic methods for evaluating the performance of facilities, can be a key tool for facilities managers to use in integrating information about facilities with other organizational concerns. For example, the design process has often been seen as an opportunity for organizations to prioritize goals and implement policy. Many models of design advocate evaluation as an important step toward "closing the loop" on information learned from design successes and failures for the benefit of subsequent projects (Zeisel, 1981). When this loop is successfully closed, evaluation makes an important contribution to the shaping and implementation of organizational policies.

Why Do Some Organizations Adopt POE While Others Do Not?

POE programs in large organizations have been described by a number of writers (Gray et al. for the government of New Zealand; Harvey for Health and Welfare Canada; and Shibley for the U. S. Army Corps of Engineers). Yet, there is no systematic assessment of the organizational factors which allow POE to flourish within a large organization. Given that POE can be an integral part of a facilities design and management program, certain questions arise about the use of POE in large organizations:

- Why do some organizations adopt and practice POE while others do not?
- What does it take to make a successful POE program within an organization? From what administrative level must support come?

To answer these questions, a number of key organizational features need to be analyzed and understood. An organizational inventory and assessment should examine the following topics:

- What are the key organizational goals and objectives that relate to facilities? This can indicate appropriate analytical issues for POEs. It leads to further questions, such as "What performance outcomes are important to the organization (that can and should be measured in POEs)?" and "What existing programs and efforts are related to POE (and could be 'tagged onto' by POE practitioners) or could be served by POE results?"
- How does the organization see itself in relation to its "environment" (i.e., its clients, markets, competition, regulators, etc.)? Is the organization facing pressures or changes which might make it amenable to new ways of thinking or responsive to new information? This can affect the perception of appropriate uses for POE.
- What is the overall organizational structure and, especially, where does facilities management fit in that structure? This can help identify potential supporters of POE (who may need to be educated or convinced about it). Parallel questions include:
  - What are the politics of the organization and of the part of it that might undertake POEs?
  - Who will support POE, why, and under what conditions? How much clout do they have?

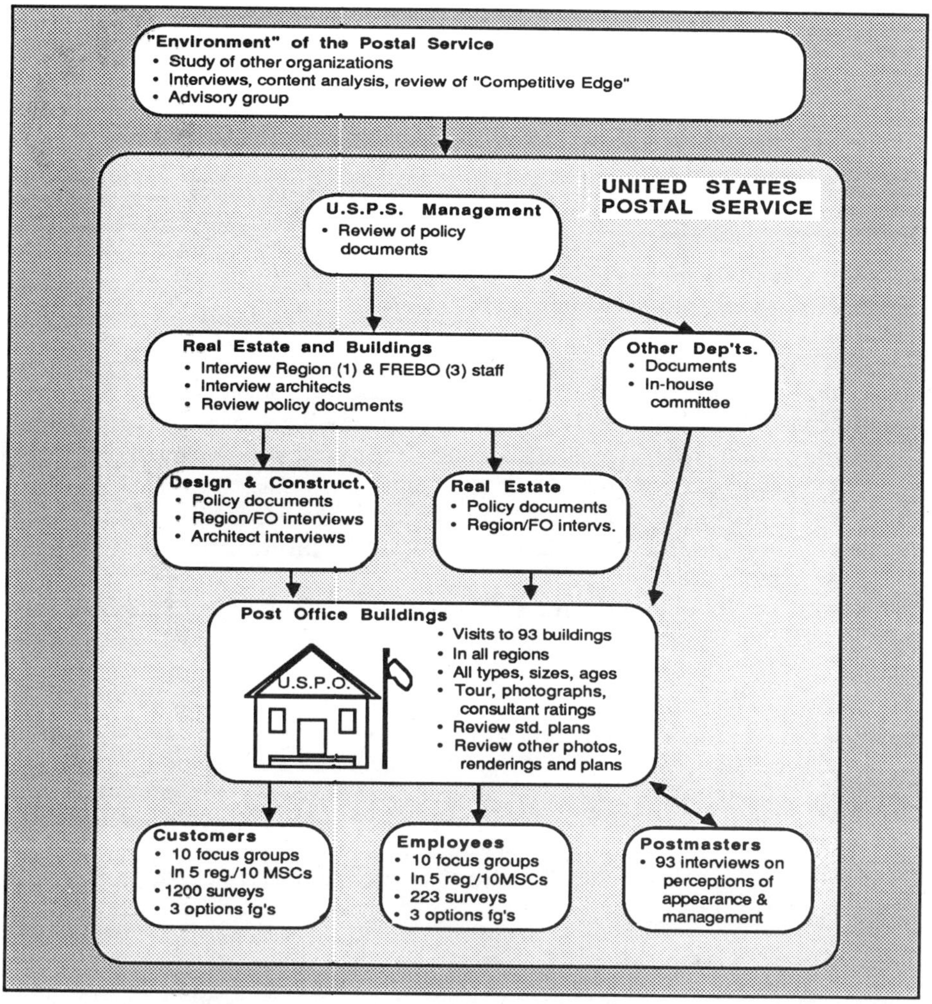

Figure 1. Image Study Methods.

Conversely, who might be threatened by, or choose to attack, POE (and under what circumstances)?

Insiders to the organization are most likely to be in a position to assess these variables. With answers to these questions, an appropriately placed manager can assess whether the organization is a suitable candidate for a POE program and, if so, develop a strategy to gain support for trying it out.

In the balance of this chapter, a post-occupancy evaluation program recently implemented by the United States Postal Service serves as a vehicle to illustrate and discuss these questions. It should be made clear, however, that many of the above issues are raised with "20-20 hindsight" and that the present example, while supporting the need for such an analysis, was not carried out in a linear fashion. Rather, the analysis was carried out as the strategy developed and many circumstances (and some luck) contributed to making it work. Still, much can be learned from examining the USPS program as it began, took root and, apparently, began to flourish.

THE USPS FACILITIES ORGANIZATION

Building Stock

The Postal Service is the largest non-military organization in the U.S., with approximately 700,000 employees and over 35,000 facilities. Most of the facilities are smaller leased buildings (many of which are built according to standard plans). Yet, because the larger buildings are owned, USPS owns more than half of the space it occupies. Buildings range from 25 square feet to over 1,250,000 square feet and are located in all parts of the country. The USPS's capital spending program for constructing and renovating buildings has recently been about $1 billion per year. Postal facilities accommodate a diverse, yet limited, range of functions which are housed in a rather repetitive set of building types. About 500 buildings are constructed each year, many of them relatively small, standard designs. The USPS is exploring further standardization of plans for medium and larger facilities.

The USPS is an obvious candidate for post-occupancy evaluation and related research, since it has the opportunity for great leverage of the results; the value of even a minor improvement or cost savings is magnified greatly when applied over such a large building stock.

In recent years – and in response to the need to quickly produce many new buildings within a limited budget – the primary goals for the building program have been to achieve economy and speed of construction, at a reasonable-to-good level of quality. Limited building and design evaluations had been conducted at the USPS, but solely in response to technical construction or code problems (roofs, floors, asbestos, accessibility). The general attitude could be paraphrased as: "design, build, and forget – unless it comes back to bite you."

Administrative Responsibility for Facilities

At the time the POE program began, the responsible entity was known as the Real Estate and Buildings Department. The Design Management Division at its headquarters office was in charge of the project. This group had a broad range of responsibilities, which included both central managerial functions (such as allocation of resources and development of standards) and direct management of major construction projects in all parts of the country. The Postal Service overall had a strong regional organization, with five Regional Postmasters General who had considerable autonomy. Field offices with significant amounts of design responsibility reported to the Regional Postmasters General (not to headquarters).

In a major reorganization, a considerable share of autonomy was delegated to a lower level, with some 75 administrative subcenters taking on more facilities development (and operational) authority. Project management was delegated to field Facilities Service Centers and Offices while the headquarters group was limited to central managerial functions. The name of the overall group was changed to the Facilities Department.

This reorganization, which at times seemed threatening to the POE program, in fact worked to its advantage. Power over field offices was consolidated somewhat (since

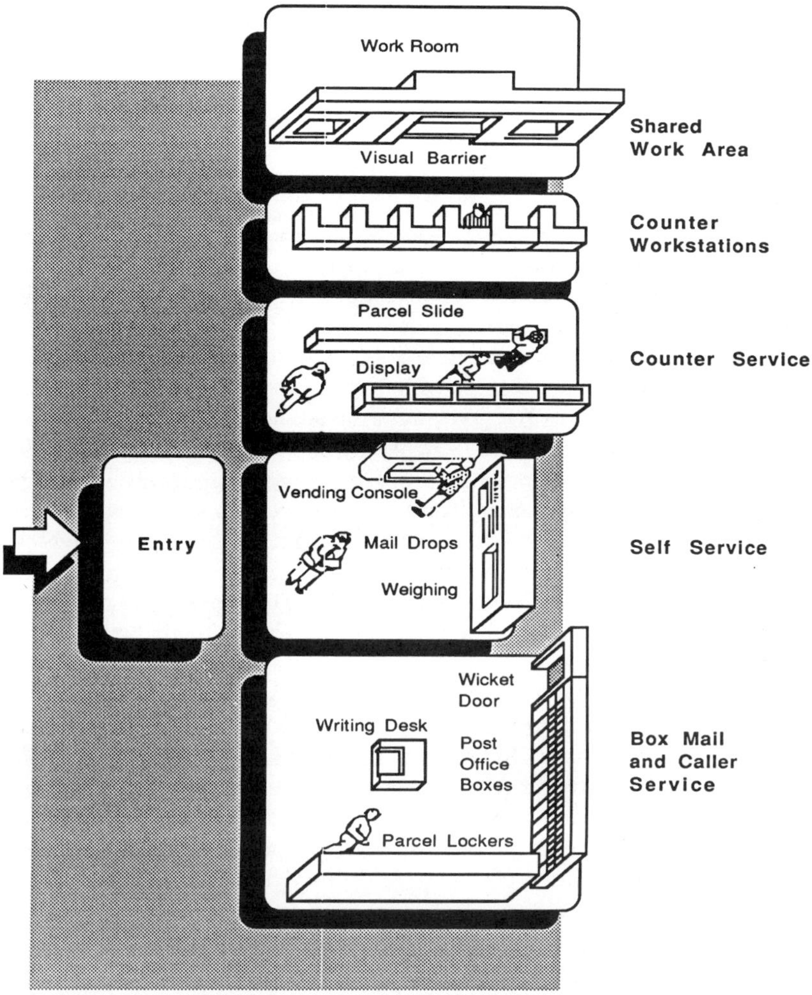

Figure 2. Overall Lobby Relationship Diagram (from Retail Design Guidelines)

they now reported to and were reviewed by the headquarters facilities executive, rather than the Regional Postmasters General) and a clearer role was defined for headquarters in terms of supporting field operations through the development of knowledge, standards and guidelines. Thus, POE became a natural function for the Facility Department's headquarters group.

THE USPS POE PROGRAM

Origins and Development of the POE Program

The USPS's POE program was initiated with a nationwide study of the image of post office buildings. Called for by a high level-management committee, the concern with image was consistent with shifts in overall direction in the USPS. Facing a newly competitive environment for delivery of parcels, overnight mail, and other services,

USPS could no longer take its monopoly on First Class mail for granted. Facility image was seen as one aspect of the way in which the Postal Service presents itself to its customers and employees – and a facet of design which, in the view of some, had been "benignly neglected" in recent years. Now, quality of service, customer and employee satisfaction, and economic performance were gaining emphasis. Therefore, the image study was carried out to define the image which postal facilities projected to customers and employees and how, if at all, the image could be improved.

The methods (illustrated below) and results of the image study are not described in this chapter (see references for more information), but several of its features (which are described below) contributed to the institution of a series of follow-up research and application projects which, in aggregate, constitute an organization-wide POE program.

Follow-ons from the image study (finished in May, 1986) include POEs of two sets of retail lobbies – one set intended to demonstrate new directions in lobby design, the other consisting of "baseline" or comparison lobbies (completed in November, 1986). Results from the lobby evaluations were incorporated into new *Retail Design Guidelines*. Completed in May, 1987, they are intended to provide design guidance for creating efficient, pleasant and businesslike environments for customers and employees, and an appropriate setting for marketing products and services. The guidelines, which include a large number of diagrams, photos and sketches to illustrate the text, are targeted toward a range of users including postal operations and design project managers as well as in-house and contract designers. They were the first completely "practical" product of the program.

Another completed POE project is a study of small, standard-plan post offices. This nationwide study was carried out by a team combining the POE consultants with a multi-disciplinary design firm. The research findings will contribute to new programmatic requirements and, ultimately, to a new family of designs (or perhaps a new approach to designing these facilities.)

Other POE efforts which have begun, or are planned and budgeted for implementation, include using POE to:

- refine the design of postal "business convenience centers," drive-through windows and lobby furniture;
- develop and test prototypes of the "store of the future," a new retail marketing concept; and
- serve as the basis for a program of continuously funded, on-going evaluations of lobbies, large mail handling facilities, and others.

A nationwide database for POE information is planned, as is a process for incorporating POE activities as part of the organization's normal facility development sequence.

How the POE Program Became Established

In retrospect, the institutionalization of the POE program from its beginnings with the image study was far from a certainty. Many aspects of the image study were problematical or presented threats to the development and acceptance of a POE program. The concept of building image is broad, relatively subjective, and not always easily grasped. Some individuals within the organization were not convinced that a study of building image was warranted. Others resisted, and may have felt threatened because potentially negative findings could have implied that they were not doing their jobs well enough. In addition, the notion of a high-quality design image seemed at odds with the prior emphases on economy and speed of construction. The study's involvement of users (postal customers and employees) was also threatening to, or viewed as irrelevant by, some professionals and managers used to a more insular mode of operations. Finally, there was a mistrust of research and of reports which some felt might not lead to useful solutions.

The effort to establish a formal POE program succeeded at the USPS by overcoming these and other obstacles. Among the reasons for its success are the following.

Comprehensiveness. The POE used a comprehensive and systematic approach. The team viewed facility "image" as resulting from many factors in addition to the building itself, including signs and graphics, furniture, plantings, and displayed items. Representatives of many levels and groups within the organization (from headquarters managers to field staff, line employees, and customers) contributed directly. The study also examined all phases of the construction process and identified policy and practice issues as well as design concerns. Thus, the final report included very broad recommendations, including further targeted POEs – the benefits of which had begun to become clear from the study itself (which demonstrated that tangible results could be obtained and addressed in concrete courses of action). Through this experience, "POE" became a term that people understood and adopted.

Advocacy. The POE program developed advocates and constituencies beyond the sponsoring group. The broad range of concerns mentioned above suggested the involvement of many departments in addition to the facilities department. Some, such as retail, marketing, consumer support, and communications, saw the study as a vehicle to pursue their own interests that related to building design. The research team solicited and responded to their input through an interdepartmental review board, established early in the POE program. Members of this group ensured that the research addressed issues that were important to their departments and responded to the study findings and recommendations. This gained substantial interdepartmental support for the project. Thus, the POE program galvanized energy from departments, many of which had not formally communicated about design issues or seen how design affected their particular missions. The interdepartmental group was able to function in a non-threatening advisory role which put them at low personal risk while offering a high potential payoff in terms of realizing their department's missions. We are informed that the presence of a "neutral" third party (the consultants) helped this to take place. The consultants' agenda, in fact, was to facilitate just this kind of process.

Increased design awareness. The studies raised design awareness throughout the organization and alerted people that many of their operational concerns had important design implications. The logical course of action to ensure that their interests were addressed was through systematic evaluations that included their input (e.g., through participation in field evaluations). Other departments began their own spin-off projects, such as graphics, landscaping, maintenance, etc.

Diffusion. As a result of involvement in the Facilities Department's image study, another department came to see POE's potential for contributing to their operational mandate and became a supporter of a second major POE effort. This department, responsible for retail functions, was interested in using POE to provide information on how to upgrade retail services and settings. The retail lobby studies were a good target for the first in-depth evaluations. By limiting the POE to retail areas, more effort could be concentrated on a smaller scale to demonstrate the effectiveness of the approach. Addressing this finite area was, in some ways, more realistic than trying to evaluate and overhaul the entire building program.

Improved service to clients. POE strengthened the ability of the facilities group to serve its client departments. It not only encouraged incorporation of user client's objectives into the planning cycle, but allowed the facilities' degree of success in meeting those objectives to be measured. This helped the clients to more effectively manage their resources, which they were eager to do. POE enabled (and perhaps forced) the client departments to be clear about what they wanted, allowing the Facilities Department to respond more effectively.

Strong project management. A strong project manager not only grasped the policy and design implications of the study, but also devoted considerable energy and savvy to selling the program throughout the organization. The project manager attended a conference of the Environmental Design Research Association (EDRA) where he was able to exchange ideas with facilities people from other large governmental organizations with POE programs. This sharpened and reinforced the focus of the overall program.

Nationwide input. Since results of the POEs were to be applied nationwide, it was important to assure that input from a great variety of employees, customers and field facilities staff would be integrated into the POE program. Thus the image study enlarged the sampling of users and buildings beyond the minimum required for statistical validity while other studies were careful to schedule nationwide visits. Facilities Department staff from around the country contributed input and assistance during research team visits to field offices, and through a national postal design conference. As a result of these efforts, the research team learned, for example, that design managers supported POE research, but were much more interested in clearly presented design guidance (based upon solid, research-based information) than they were in reading the actual research reports. These efforts to obtain nationwide input ensured that the findings were comprehensive and applicable to all types of postal buildings in all parts of the country, as well as helping to gain support for the research program at all levels within the organization.

Clear communication. Careful attention was paid to "selling" the program to management within the departments and all the way to the top of the organization. The presentation of study findings used effective communication tools, such as slide/tape and video-tape presentations of customer and employee focus group discussions as well as an introduction and overview of the *Retail Design Guidelines.* The former enabled Headquarters staff to see and hear employees and customers express their own concerns, which was far more immediate and convincing than the words of consultants or in-house professional staff.

Concrete recommendations. Results were accepted in part because recommendations from studies were tangible, giving managers specific steps to carry out. Material was relevant to management concerns and was presented as management reports (rather than as research studies); i.e., with clearly stated objectives, cost implications, and actionable items.

Measures of Success of the POE Program

Design research and POE have been accepted by USPS because they presented benefits to the organization which were clearly perceived as flowing from the process and results. These included the perception that knowledge gained from research could be applied to improve image, efficiency, employee relations, job satisfaction, and perhaps economic performance. Other organizational benefits, such as the ability to better manage facilities and improved communications and relations between departments, were described in the preceding section.

Among the measures of the acceptance and success of POE at the USPS are the following:

- *POE has been adopted* by the Facilities Department, not only as an accepted mode of inquiry, but, more importantly, as part of the normal procedure for assessing results and developing design guidance. The notion that building performance can be assessed on a broad range of indicators that reflect the diverse objectives of the Postal Service and the Facilities (and other) Department(s), has been accepted.
- This has led to a *new model of facility development*, which includes design innovation and testing. This is being applied both to rather clearly defined building types, such as small standard-design post offices and large mail processing facilities (which reflect "business as usual") as well as to the exploration and development of new operational modes and facility types, such as the "store of the future." For the latter, an interdisciplinary team of market researchers, architects and POE consultants are using a cyclical "test-image-test-refine" process to develop the next generation of retail facilities and operations, using field POEs and focus groups among other techniques.
- The approach to the development of *design standards* has also changed. Where standards used to be developed both regionally and centrally through a somewhat informal process of accretion, they are now being developed centrally using results from formal POEs and targeted special research. While standards will continue to evolve over time, they will become the major vehicle for field application of research results.
- POE processes and results have gained acceptance and support at the *highest levels of the organization*, including the Postmaster General and other senior managers. For example, these managers have promulgated both organization-wide and departmental image statements and established a on-going interdepartmental visual image review board (both were recommendations of the image study). In fact, *all* of the higher priority recommendations from the image study have been accepted and implemented, including the further use of POE.
- Other successes which were mentioned in previous sections include *support* from the facilities field organization, as well as from other "client" departments, for POE and its application to new standards.
- This has resulted in a degree of *continuity* of projects and funding. While all of the work reported on here has been on a project-by-project basis, the current budget includes funding for up to three years of POE work. USPS has been spending approximately $200,000 to $300,000 per year on POE and related services. Another measure of POE's robustness is that during the course of implementation the USPS and the Facilities Department underwent significant changes, including an overall reorganization, three different Postmasters General, and a new Assistant Postmaster General for facilities. In the new organization, POE helped to define one of the appropriate functions for the headquarters facilities group.

Future Potentials and Needs

Despite the overall success of the program, there are many ways in which POE can evolve at the the USPS. While POE is becoming part of standard practice in the facilities group, it is still largely tied to expense funding, making it vulnerable to periodic budget constraints. To integrate it more permanently, the Postal Service will consider tying POE to capital funds (as some other agencies do).

POE at the USPS started with image, but has since been focused on total building performance, including technical and economic performance measures. An acronym has been adopted to express the key aspects of facilities performance: "FACT" (functional, aesthetic, cost, technical). In the future, design will be called upon to produce a verifiable return-on-investment, and POEs will have to become more sophisticated in defining and measuring economic performance.

There has been and remains pressure and need for standardization of POE methods. There are many reasons for standardization, including obtaining consistent data for accumulation and future analysis in a data base, efficiency of training individuals who will be conducting POEs, and efficiency of carrying out the evaluations (rather than reinventing methods each time). There are, however, also problems and limitations that derive from standardization. Standard methods do not necessarily respond to the diversity of design issues and evaluation concerns. Some researchers feel that it is easier to standardize procedures than to standardize information-gathering formats (Gray et al., 1986). This issue has not yet been fully resolved at the USPS. Another need is for clearer definition of the varied levels of POE (Shibley, 1985; Zimring, Goldman, and Fuller, 1987). Since POEs cannot always be full blown and data-intensive, clearly defined levels of POE would improve the ability of managers and consultants to appropriately apply available resources to each project.

LESSONS FOR OTHER ORGANIZATIONS

A POE program holds potential payoffs for any organization with a large building stock or an (especially, repetitive) ongoing building program. Many of the lessons learned from, and factors which contributed to the success of, the USPS's POE program may be applicable to other organizations. These include:

- *Clarify goals and expectations* for the POE program. Ensure that the program is directed toward developing information on issues which are of concern to, and contribute to the goals of, the organization. Make sure that POE is relevant.
- *Carefully choose the POE project manager*. Assign someone who is energetic, understands the organization, and has good communication skills. Give that person the opportunity to spend enough time and resources to effectively manage and support the program.
- *Find a powerful champion for POE*, at least at first. This mentor's influence in the organization can win the resources and cooperation needed to initiate a new POE program. Once the program is successful and gains others' support, it will be less necessary to have a single champion.
- *Involve other departments*, not just the facilities group, and get them to "buy in" to the program. While a facilities unit can operate its own POE program, involvement and support from client departments strongly

justify the effort. Since facilities is a service function within the organization, establish POE as a means to better serve organizational clients.

- *Involve building users* (customers/clients and employees), at least by measuring their response to facilities. These large groups are the ultimate constituencies for the organization, and if POE is not measuring facilities' performance for them, it misses its potential richness.
- *Involve field facilities staff.* These are the people who really know how things get done "out there" and who will eventually be responsible for supporting and implementing many kinds of recommendations. If they have not been involved, the findings may not be germane to them and their support cannot be assured.
- *Use appropriate POE techniques.* While methodological questions are the primary consideration, include organizational determinants when choosing methods and survey samples. This can range from using techniques which give managers first-hand experience with evaluation, to selecting survey topics or sites which they feel are important.
- *Cast findings in practical, positive terms.* Demonstrate clearly what the research findings show managers and others to do. Create action plans and tangible products. Design guides, designs, prototypes, training tools, or other approaches, may be received more readily than research reports, depending on organizational culture. Avoid emphasizing failures or casting blame, which is the fear of many who resist POE. Rather, identify "missed opportunities" and point out successes along with suggestions for improvement.
- *Utilize effective communication* media and channels for disseminating (or even selling) POE results. Find the appropriate vehicles for dissemination of findings through such ways as design guidelines and standards. Make reports readable and germane, putting details of methods and data in appendices or separate volumes, since most people are probably too busy to read them. Use audio-visual techniques whenever possible.

CONCLUSION

This chapter has reported on the organizational development aspects of the USPS' POE program. Starting as an experimental approach, POE has moved toward increasing institutionalization at the USPS. It has been adopted as a multi-year funded program, has influenced organizational communication, and is currently being used to inform the processes of facilities design, construction, and management. While there are many aspects of the program which are unique to USPS, some of the lessons learned may be applicable to other organizations. This hypothesis is difficult to test at this time, since we lack a sufficiently descriptive, let alone predictive, model of the role of facilities and facilities research within large organizations. It is hoped that by describing the USPS experience, a small contribution will be made toward development of such a model.

NOTE

USPS reports and further information on this project can be obtained from Warren Walker, RA, Building Technology Branch – Room 4421, Facilities Department, U.S. Postal Service, 475 L'Enfant Plaza SW, Washington, DC 20260-6400, (202) 268-3887.

REFERENCES

Davis, G. (Ed.), 1986, *Building Performance: Function, Preservation and Rehabilitation*, ASTM STP 901, American Society for Testing Materials, Philadelphia, PA.

Duffy, F., October 1969, Role and Status in the Office, *Architectural Association Quarterly*, v.1, pp. 4-13.

Duffy, F., 1974, *Office Interiors and Organizations: A Comparative Study of the Relation Between Organizational Structure and the Use of Interior Space in Sixteen Office Organizations*, doctoral dissertation, Princeton University.

Farbstein, Jay, 1975, *Organization, Space and Activity: the Relationship of Task and Status to the Allocation and Use of Space in Certain Organizations*, doctoral dissertation, University of London.

Farbstein, Jay & Associates and Min Kantrowitz & Associates, May 1986, *Image Study Final Report*, United States Postal Service, Headquarters Facilities Department, Washington, DC.

Farbstein, Jay & Associates and Min Kantrowitz & Associates, October 1986, *Post-Occupancy Evaluations of New Lobby Designs* (New Directions in Lobby Design) United States Postal Service, Headquarters Facilities Department, Washington, DC.

Farbstein, Jay & Associates and Min Kantrowitz & Associates, May 1986, *Post-Occupancy Evaluations of Added Lobby Designs*, United States Postal Service, Headquarters Facilities Department, Washington, DC.

Farbstein, Jay & Associates and Min Kantrowitz & Associates, May 1986, *Retail Design Guidelines*, United States Postal Service, Headquarters Facilities Department, Washington, DC.

Gray, J. M., Daish, J. R., and Kernohan, D. Q., 1986, A Touring Interview Method of Building Evaluation; The Place of Evaluation in Building Rehabilitation, in: G. Davis (Ed.), *Building Performance: Function, Preservation and Rehabilitation*, ASTM STP 901, American Society for Testing Materials, Philadelphia, PA, pp. 46-68.

Margulis, S., 1987, Facilities Management: An Insider's View, workshop reported in J. Harvey and D. Henning (Eds.), *Public Environments*, proceedings of the Environmental Design Research Association Conference, Ottawa, Canada, p. 271.

Shibley, R., January 1985, Building Evaluation in the Main Stream, *Environment and Behavior*, 17:1, pp. 7-24.

Steele, F., 1973, *Physical Settings and Organization Development*, Addison Wesley: Reading, MA.

Zimring, C., Goldman, M., and Fuller, C., 1987, Getting it Done: Post-Occupancy Evaluation in the Public Sector, workshop reported in J. Harvey and D. Henning (Eds.), *Public Environments*, proceedings of the Environmental Design Research Association Conference, Ottawa, Canada.

Zeisel, J., 1981, *Inquiry By Design*, Monterey, CA, Brooks-Cole.

EPILOGUE:

ISSUES AND FUTURE DIRECTIONS

Jacqueline C. Vischer

Division of Capital Planning and Operation
State of Massachusetts
Boston, Massachusetts

I recently joined a large State agency to implement a program of Post-Occupancy Evaluation (POE). As the supplier of public buildings throughout the State, the agency spends several millions of dollars a year on design and construction services, retaining and providing work for architects, engineers, cost estimators, contractors and building trades. The agency is responsible for large and complex institutional buildings, such as mental institutions, courthouses, and jails. These buildings include many repeat elements such as institutional kitchens, secure areas, and visitors' parking. It also builds other public buildings which are small and repetitive, such as visitor centers, police stations, and child-care facilities.

A program of POE is appropriate to an organization of this nature in that building evaluation will provide feedback on building performance over time, will help generate optimal design for repeated elements, will improve programming and design for new buildings, and will provide the agency with some assessment of the quality of professional services received from its consultants. However, as we struggle to frame the POE program and design a structure for it, to provide it with resources, and to initiate specific building studies that will contribute to it, we must confront — in the immortal words of Bob Sommer — some "questions that will not go away."

In reading this book of collected papers, I realized that such questions are endemic to the field of POE, and that every author in this volume has implicitly or explicitly responded to them. It is possible to find some answers in these chapters, and to draw out of them the steps necessary to develop not only POE in a large public agency, but some thoughts and ideas about the future of POE. Some of them are listed below.

The "armchair" approach. The first question has to do with the investigation of buildings already built and occupied being a rather passive "armchair" activity: isn't it more exciting and creative to design and construct a new building than to study and analyze it after it has been built? Are we, as POE practitioners, taking an "armchair" approach to the grand challenge of actually building buildings, and therefore rather wasting the time of the *real* practitioners?

Information control. Another question concerns control over POE information and the determination of its applicability to new building design. Who in an organization decides which results of a POE are useful and which are not? Should it be the people using the building rather than the owner who built the building, or should it be the designers, who, after all, know what was intended, or should it be the researchers, whose detached position places them beyond praise or blame for how well the building works? And how should whoever controls the nature and flow of POE information define what is and is not "useful"?

Information versus knowledge. In third place is the problem of disseminating new information. A POE program's purpose is to generate new information and thereby new knowledge. Many action-oriented decision-makers feel they already have too much information to enable them to solve problems quickly and efficiently. If POE results in lengthy detailed research reports, decision-makers will not read them, and in spite of new *information*, the new knowledge will not be acquired or used.

Methodological integrity. Another important question is the importance of using a defensible methodological approach. People veer away from building research in an action-oriented building design and construction context preferring to leave research to academic institutions. One's peers in large building agencies are architects, engineers, planners and project managers. They do not use academic criteria to judge the quality of POE data collection and analysis. They do not know whether the data are good or not; they only judge results. How important is it in this kind of context to maintain a scholarly research approach?

The question that will not go away. And finally, but far from least important, does the knowledge about building performance yielded by POE really improve building quality? Is the information about existing buildings really applied to the creation of new ones? In fact, is it systematically applied to the improvement of existing ones? What many building researchers find is that building errors can often be traced to the building delivery process itself, and this and the industry context in which it occurs do not (yet) change as a result of POE activity.

The first question concerning POE as an armchair activity is repudiated by a number of highly action-oriented POE activities, such as the Pre-Occupancy Building Evaluation described by Molloy as a service to new tenants of New York office buildings. Other examples include using POE in evaluating earthquake damage as explained by Durkin, and Gustafsson's use of POE to provide more suitable animal environments. Each of these examples demonstrates how POE is an assertively product-oriented activity, an example of taking active control over the environment, of making POE work as a tool for effecting change. In applications like these, POE is as action-oriented as design and construction, and as many contributors to this volume point out, the era of POE as uniquely an academic endeavour is over.

Regarding the question of the usefulness of POE results, several of the chapters in this book recommend a thorough analysis of the organizational context in which POE is occurring in order to ensure best use of results. Joiner and Ellis point out that design quality is negotiable, and indeed implicit in their discussion is the belief that design quality ought to be negotiated. This is not a traditional view, inasmuch as members of the architectural profession are trained to believe that design quality is their responsibility and in their exclusive control. The negotiability of design quality involves by definition clients, managers, owners, occupants, and neighborhood residents. Planning, building, occupying and managing a building is a participatory process of which POE is a part.

Farbstein uses the example of the U.S. Post Office to illustrate how important the role of POE is in effecting organizational change. Pegrum and Bycroft on POE in Australia illustrate how the organizational context determines the relative priorities attached to POE results, and therefore how best use is made of them. Organizational priorities are not always the same as researchers' priorities, but if POE is to work as an integrated system, organizational priorities must be observed.

Several chapters raise the point that effective communication of POE results is crucial to successful POE. They agree that disseminating POE results is difficult and is often overlooked. The days when a large and important-looking research report could be placed on someone's desk as the final stage of the POE process are over. POE results are being transmitted through workshops and seminars with users, through design standards and guidelines to the professions, through computerized databases and databanks to researchers and students, and through technical reports, published papers and books.

White's paper pays careful attention to the political sensitivity of POE activities, with advice on the management and dissemination of information. He lists some of the players in a typical POE project who should be involved and informed on an ongoing basis if a POE is to pay off. The failure to observe the sensitivities of the POE process result in neglect of POE results, as Zimring, Francescato et al., and others point out.

The question of how to turn POE information into knowledge is addressed by Rabinowitz, Preiser et al., and Zeisel, who emphasize that POE can be product-oriented or research-oriented, and is usually somewhere in-between. Rabinowitz describes POE as "a successful commercial activity", and predicts that the "next generation" of POE will be split between the commercial sphere and pure research. Becker states this succinctly in his title "POE: Research Paradigm or Diagnostic Tool?". He emphasizes the distinctions between POE as a scholarly undertaking, and POE as a legitimate activity for the emerging profession of facilities management. The resolution lies in one type of POE for the doers (diagnostic POE) and another for the academics (research POE) — a resolution echoed in several other chapters. Zeisel calls for a shared definition of POE that will enable different types and varieties of POE performed in different settings to be comparable and therefore communicable. The definition of knowledge implied in this discussion is not so much results-based as process-based: it is at the level of the POE *process* that knowledge will be disseminated.

Several authors provide historical overviews of POE, which show how POE developed out of the substantial amount of government-sponsored building evaluation, using public funds almost on a research grant basis to evaluate large multi-user public buildings such as public housing, prisons, hospitals and airports. The product-oriented POE, on the other hand, is a more recent development, catering to commercial expectations, or, as Farbstein characterizes it, "a demand-driven mode". For this POE, clients pay a fee for a product, for example an indoor air quality evaluation of office space, and expect to use the product to design environmental improvements, improve worker productivity, etc.

Whereas public agencies are more likely to accept the potential usefulness of POE and tolerate a somewhat traditional research approach (the "supply-side" mode), commercial clients have to be sold on POE and have to believe they are buying a cost-effective product. Ventre points out that evaluation "is a logical component of the process of managing resources", and that some form of POE is a natural endeavour for agencies and organizations who have to make decisions about the allocation of their resources.

Many of the chapters in this book indicate a pre-occupation with the methodology of POE which perhaps reflects an anxiety borne of the tension between the research-oriented and the product-oriented POE. Caught between two stools, POE strains towards academically defensible research *and* useful marketable product. Some of the authors advance POE approaches that they believe can do both, such as my (Vischer's) Building-In-Use Assessment System, Bechtel's overview of methodological innovations in POE, and Parshall's Problem-Seeking Method.

Loftness et al are not alone in their emphasis on the interactive nature of building systems performance and human behavior in a built environment. In their call for a POE approach that recognizes the interactive effects of buildings on people they assert that the most far-reaching of the methodological innovations is the development and use of the comparative database.

A degree of computerization in most modern organizations means that data collected from a number of different buildings, and/or data on a range of building conditions, can be integrated, stored and analyzed for a variety of different purposes. Whereas the comprehensive POE is directed towards detailed study of all aspects of a building's performance, for example, the ecological approach advanced by Machado, the comparative POE extrapolates one or two key topics, such as the office lighting data analyzed by Marans, and studies them across a number of different buildings. If a POE database is designed appropriately, data analysis should not be an either/or proposition. The long-standing distinction between comprehensive and comparative POE's will dissolve as the full power of the database approach becomes evident.

The final major question is implied if not actually expressed by all the contributors, each of whom feels it incumbent upon him or her to explain why POE is worth doing. This would not be the case if we were writing about cancer research, or even about studies of the properties of different types of concrete. In my view, we would be a healthier group of POE practitioners if we stopped justifying our work. We should be anticipating the pivotal role of POE in the emergence of the discipline of building science, a field of study and knowledge which incorporates multi-disciplinary analyses of building performance as well as a psychology and a sociology of the human use of space.

Although the papers are highly positive in their statements about POE, some criticisms are expressed. These authors are concerned at its lack of strong theoretical models, at the elementary nature of its practical applications, and at the slowness of its integration into organizational contexts. The call is for more theoretical rigor and the proliferation of models. It is for wider practice and more far-reaching applications; it is for process-oriented POE in existing organizations that will make use of it; improved and innovative POE *methods* will solve problems of theory, practice and process.

Zimring's discussion of "Implicit Theory" and his preliminary typology of POE illustrates the oft-cited saying that there is nothing so practical as a good theory. One of the more fruitful areas of methodological innovation is the development of accessible, available research tools and techniques that draw on the best scientific traditions and can be used by people not trained in research. For example, architects and interior designers, facilities managers and engineers can be trained in a certain amount of building evaluation, enabling them not only to do some POE of their own, but also to make competent judgments about the kinds of services they may buy from experts.

Of course the answer to the question of tangible improvements to building quality is that the value of knowledge accrued cannot be measured in terms of its immediate usefulness. As Preiser points out, if the study is good, new information will broaden knowledge. This is a desirable end in itself. To see immediate changes in building design is only one possible result of new knowledge. To many POE practitioners, change in the industry itself and how it operates are a bigger key to environmental change than disseminating POE results to the design professions; the pressure for social change develops in part from new knowledge.

The requirements of the State agency in which POE is starting up, and the concerns, questions and conclusions of the contributors to this book are similar. I have summarized these into eight conclusions and future predictions for POE.

1. *Increasing Diversity*: More building studies will be carried out in different ways, using different methods, working with different data, and performed by people from a range of training and disciplines. As these diverse POE activities proliferate, POE itself will need a new name. Some likely new names drawn from this book include building evaluation, building-in-use assessment, building science, and building technology transfer.

2. *Less Rigorous Approach*: The pressure on POE to conform to existing paradigms of social science research is decreasing. The shift towards a broader definition of POE as part of building science allows POE researchers to make use of a greater variety of methods, approaches and concepts. This is fruitful, because with the increased diversity comes a greater variety of new knowledge, but it should not be at the cost of a POE research paradigm of its own. Recognizing the need for a paradigm creates more pressure to disseminate new knowledge throughout the building research community. As more POE information is shared, so the POE paradigm has a greater chance of developing.

3. *Effect Change*: Using building change to effect social and organizational change has long been a popular topic for discussion, and as POE becomes more commonplace, we will see more examples of this in practice. Many enlightened organizations already use large office moves or building renovations to spark some self-analysis and organizational overhaul. Integrating POE into the organizational context provides systematic opportunities for organizational change: what Peters and Waterman call the "hands-on, value-driven" approach.

4. *POE as a Business*: In this Epilogue, I have distinguished between research-oriented and product-oriented POE. Whereas the past has favored research-oriented POE, the future will favor commercial POE with tangible products. The product will often be in the form of new information as well as actual building improvements and solutions to building problems. There is an important counterpart to POE as a business, and this is POE as applied research. The refinement of knowledge and methods inherent in the scholarly approach will stabilize and root "building science" in the empirical tradition, will provide the practitioners with better tools and more knowledge to sell, and will command the respectability and respect that has eluded POE for so long and may still be whisked away.

5. *Institutional POE*: The POE as a case study of a single building is decreasingly acceptable unless it is oriented narrowly to solving a building problem. POE will develop increasingly into an institutionalized process, emerging out of organizational need, and tailored to each organizational system. Being context-responsive, POE will by definition be diverse. Few of the authors in this volume, for example, include building costs and cost monitoring in their definition of POE, yet the potential importance of this topic to the evolution of institutional POE is hard to overestimate, and is likely to be next on the POE agenda.

6. *U.S. and non-U.S. POE*: POE is this country has developed somewhat differently from POE elsewhere. European researchers have favored theory and concept development. POE in Canada and Australia is organization-driven and has emerged out of a cultural sense of responsibility for the welfare of the group and improving the quality of life for the collective. POE in the U.S. is made more difficult by the traditional emphasis on the survival of the individual and the preoccupation with individual good and individual advancement. POE is an activity that places the welfare of the group ahead of that of the individual, and it is therefore culturally inimical to the U.S. context. It has been harder for practitioners in this country to be heard, but in the near future the unavoidable good sense of POE activity will triumph even over cultural biases.

7. *Future Types of POE*: We have tended to discuss POE as evaluation of individual buildings, or of groups of individual buildings. However, other types of POE are possible, for example, Serra's POE of a nation-wide urban design program, Preiser's POE of activation of a new hospital building, and Jockusch's use of POE in architectural competitions. Urban design POE that focuses on neighborhoods, small communities, or mixed use developments is likely to become more important in the future. As well as diversity of method and diversity of results, future POE will be characterized by diversity of study topic.

8. *Design Information Systems*: Most design decisions are still made in the traditional way even though the design process itself is becoming computerized through CAD/CAM systems. By increasing the amount of building information available, feedback on design decisions, product performance information, and other types of building performance data will be accessed by designers involved in large projects and by their clients. One of the goals of the State government agency which is about to start a POE program is to implement a design information system that integrates information on cost, materials performance, human behavioral response, building systems, and other data. The design information system will direct, assist and inform the design decision-maker of the future, not, as architects tend to fear, by rationalizing the design process at the expense of design creativity, but by ensuring that the creative element of the design process takes place in an informed and rational context so that it has the best chance of success.

Reviewing these ideas and nurturing this Epilogue into being has provided many directions for the development of our POE program. The positive commentary, constructive thinking, and impressive display of achievements set forth in these papers have done much to dispel the ghosts of difficult and unanswerable questions that have dogged POE throughout its brief life; questions about its usefulness, its relevance, its scientific validity, its disciplinary affiliations, and its stumbling steps along the road to authority and acceptance. Having come so far we must not lose our momentum: the survival of POE lies in its redefinition in broader, more far-reaching terms. The tools of building analysis, an understanding of building performance, and systematic knowledge about how built space does and will affect human behavior are the cornerstones of POE for the future and the contributors to this book have had a major role in building them.

INDEX

A

B

C

D

E

N

O

P

V

W